Analysis and Design of Networked Control Systems under Attack

Analysis and Design of Networked Control Systems under Attack

Yuan Yuan
Hongjiu Yang
Lei Guo
Fuchun Sun

CRC Press is an imprint of the
Taylor & Francis Group, an **informa** business

CRC Press
Taylor & Francis Group
6000 Broken Sound Parkway NW, Suite 300
Boca Raton, FL 33487-2742

First issued in paperback 2020

Version Date: 20180825

ISBN 13: 978-0-367-57099-6 (pbk)
ISBN 13: 978-1-138-61275-4 (hbk)

Library of Congress Cataloging-in-Publication Data

Names: Yuan, Yuan (Systems engineer), author. | Yang, Hongjiu, author. | Guo, Lei, author. | Sun, Fuchun, 1964- author.
Title: Analysis and design of networked control systems under attacks / by Yuan Yuan, Hongjiu Yang, Lei Guo and Fuchun Sun.
Description: First edition. | Boca Raton, FL : CRC Press/Taylor & Francis Group, [2019] | Includes bibliographical references and index.
Identifiers: LCCN 2018023408| ISBN 9781138612754 (hardback : acid-free paper) | ISBN 9780429443503 (e-book)
Subjects: LCSH: Supervisory control systems--Security measures. | Automatic control--Security measures.
Classification: LCC TJ222 .Y83 2019 | DDC 629.8/9--dc23
LC record available at https://lccn.loc.gov/2018023408

Visit the Taylor & Francis Web site at
http://www.taylorandfrancis.com

and the CRC Press Web site at
http://www.crcpress.com

Network security affects and changes the world and life

Game theory deals with network attack effectively

For researchers devoted to optimal controller

Contents

Preface

With the rapid development of network and control technologies, the combination of communication networks and control systems has become an inevitable trend now. Recent years have witnessed the widespread applications of Networked Control Systems (NCSs) in critical infrastructures such as power systems, chemical industries, manufacturing, water supply, and natural gas industries, etc. Signals of control systems are transmitted via public networks, which increases flexibility, interoperability and resource sharing of a control system. However, the introduced networks also bring some challenges on NCSs. The control system will not be an isolated system anymore if control and measurement signals are transmitted over public networks, which largely increases the attack probability. Since control systems can be regarded as connections of information world and physical world, any successful attacks on NCSs will lead to significant loss of properties or even human lives.

At present, the design and analysis of NCSs under cyber attacks are a cross field in which there exist a large number of original, serious problems and challenges to exploit. Insertion of communication networks in a feedback-loop system breaks the integrity of the control system and results in a lot of challenging issues, such as time delays, packets dropout, packets disorder, communication limitation, and network attacks. Controller design should be analyzed for NCSs under the consideration of aforementioned issues. Many original problems still need to be addressed in NCSs. Therefore, analysis and design of NCSs under attack is of great importance.

Chapter 1 provides a motivation on the research and its history and an overview on recent development of NCSs under attack.

Then, this book consists of four parts:

Part I: The attacks in networked control systems are introduced. In Chapter 2, two types of optimal control strategies are developed in delta domain by using game theoretic tools. In Chapter 2, multiple-tasking and central-tasking optimal control strategies subject to DoS attacks are developed in delta domain, respectively. Meanwhile, strategy design is provided for finite and infinite time horizon cost-to-go functions, respectively. In Chapter 3, $\epsilon-$optimal control for NCSs with disturbances is presented. In Chapter 4, a defence strategy based on networked predictive control is proposed for a system with actuator saturation.

Part II: Some resilient control strategies are provided for wireless NCSs. In Chapter 5, a two-player zero-sum Markov game scheme is established for wireless NCSs and an H_∞ optimal controller is designed by solving the minmax problem in delta domain. In Chapter 6, a Bayesian Stackelberg game is used to design an H_∞ minmax resilient controller.

Part III: Resilient control strategies are applied in power systems. In Chapter 7, an upper bound of epsilon level under a novel attacks model is provided explicitly to quantify the impact of attacks on NCSs. In Chapter 8, interaction of intelligent denial-of-service attackers and Intrusion Detection Systems (IDSs) is modeled as a static infinite Stackelberg game. On the other hand, some algorithms are given to find a joint optimal defense strategy. In Chapter 9, a multitasking optimal control problem is solved for the NCSs with packets dropout and time delays.

Part IV: Coupled design of cyber-physical systems under attacks is investigated for further in-depth consideration. In Chapter 10, a game-in-game structure for the coupled design of resilient control systems is proposed to develop defense strategies. In Chapter 11, an iterative adaptive dynamic programming algorithm is derived for the NCSs with actuator saturation. In Chapter 12, cross layer design is proposed for NCSs based on the IDSs configuration policy at the cyber layer and controller for a physical layer. In Chapter 13, a coupled design methodology is proposed to achieve a goal of joint optimality.

We would like to thank some of our colleagues who, through collaboration on the topics of this book, motivated and helped us in many ways. First of all, we would like to extend our sincere gratitude to Min Shi, for her instructive advice and useful suggestions on our book. We are deeply grateful to Peng Zhang for help in completion of this book. Also, we are greatly indebted to Huanhuan Yuan, for her valuable instructions and suggestions on our book as well as her careful writing of the manuscript. Meanwhile, high tribute shall be paid to Hao Xu and Ying Li, who have instructed and helped us a lot. And last but not least, our special thanks go to our families, without their sacrifice, encouragement and support, this book would not have been completed.

Haidian District, Beijing, China, *Yuan Yuan*
Haigang District, Qinhuangdao, China, *Hongjiu Yang*
Haidian District, Beijing, China, *Lei Guo*
Haidian District, Beijing, China, *Fuchun Sun*
July 2017

Symbols and Acronyms

$\mathtt{T}_s$	sampling period
R^n	n-dimensional real Euclidean space
$R^{n\times m}$	space of $n \times m$ real matrices
I	identity matrix
A	system matrix
A^{-1}	inverse of matrix A
A^T	transpose of matrix A
$A \geq 0$	symmetric positive semi-definite
$A > 0$	symmetric positive definite
$A \leq 0$	symmetric negative semi-definite
$A < 0$	symmetric negative definite
min	minimum
max	maximum
DoS	denial-of-service
NCS	networked control system
WNCSs	wireless networked control systems
CPSs	cyber-physical systems
SINR	signal-to-interference-plus-noise ratio
IDSs	intrusion detection systems
NPC	networked predictive control
NE	Nash Equilibrium
LIMs	linear matrix inequalities
$l_2[0,\infty)$	the space of square integrable vectors
$\{M_i\}_{i=1}^r$	a series of matrices $M_1,\ M_2,\ \cdots,\ M_r$
$\det(A)$	determinant of matrix A
$\mathrm{rank}(A)$	rank of matrix A
$0_{n\times m}$	zero matrix of dimension $n \times m$
$\lambda(A)$	eigenvalue of matrix A
$\lambda_{\min}(A)$	minimum eigenvalue of matrix A
$\lambda_{\max}(A)$	maximum cigenvalue of matrix A
$\mathrm{sign}(x)$	sign of x

$\lvert x \rvert$	absolute value (or modulus) of x
$\lVert x \rVert$	Euclidean norm
$\lVert P \rVert$	induced norm $\sup_{\lVert x \rVert=1} \lVert Px \rVert$
$\forall$	for all
$\in$	belong to
$\rightarrow$	tend to, or mapping to (case sensitive)
$\otimes$	matrix Kronecker product
$\sum$	sum
$\mathbf{E}\{\cdot\}$	mathematical expectation operator
$\mathcal{G}_1$	Cyber layer security game
$\mathcal{G}_2$	Physical layer game
$\mathbf{P}_{\mathbf{a}_i}$	Player i of $\mathcal{G}_1$, $i \in \{1, 2\}$
$\mathbf{P}_{\mathbf{b}_j}$	Player j of $\mathcal{G}_2$, $j \in \{1, 2\}$
J_a^1	Cost function for Player $\mathbf{P}_{\mathbf{a}_1}$ in $\mathcal{G}_1$
J_a^2	Cost function for Player $\mathbf{P}_{\mathbf{a}_2}$ in $\mathcal{G}_1$
J_b	Cost function in $\mathcal{G}_2$
$\mathbf{u}_a^i$	Strategy vector for $\mathbf{P}_{\mathbf{a}_i}$ in $\mathcal{G}_1$
u_k	Strategy vector for $\mathbf{P}_{\mathbf{b}_1}$ at k in $\mathcal{G}_2$
w_k	Strategy vector for $\mathbf{P}_{\mathbf{b}_2}$ at k in $\mathcal{G}_2$

Chapter 1
Introduction

1.1 Background

In recent years, networks have received considerable attention with the rapid development of network technologies. It has become an inevitable trend for combining networks with control systems. At present, NCSs have been widely applied in strategic and significant infrastructure fields such as electrical power systems, chemical industry, manufacturing industry, natural gas systems, etc [54, 183]. Equipped with networks, control systems have many advantages in mobility and flexibility. However, the introduced networks also bring some new challenging problems on control systems. Although it reduces costs to transmit control commands or measurement signals via public networks, inherent closeness of control systems is inevitably broken. Traditional control systems adopted dedicated signal transmission protocols. At present, standard transmission protocols and commercial operation systems are used for control systems, which will seriously increase attacked abilities. Since control systems have high requirements for real time and availability, a lot of control systems ignore or even deliberately decrease security protection from security protection perspective. From external environment perspective, attacked means and techniques gain increasing developments. Attacks aiming at industry control systems are emerging in an endless stream now [194, 141].

Very recently, there exist the following examples under attacks as:

- In 2010, the first nuclear power station in Iran was attacked by Stux-net, which was a malicious computer worm targeting industrial computer systems. The nuclear program of Iran has been delayed seriously by the btux-net attack [72].
- In April 2016, a nuclear power plant in Germany was attacked by "Conficke" and "W32.Ramnit" viruses, which were discovered at the nuclear power plant's Block B IT networks that handled the fuel handling system.

- In November 2016, San Francisco's Municipal Railway was hacked, which seriously resulted in the unavailability of the railway fare system.
- In December 2016, Ukraine Electric Grid was attacked simultaneously at three regional power firms, which led to an electricity blackout for 225,000 Ukrainian power customers. Before attacks, adversaries prepared for six months of reconnaissance; then they broke into the utility's networks via a phishing attack.

Among the aforementioned control system security events, Stuxnet viruses specifically target industrial control systems by infecting Programmable Logic Controllers (PLCs). According to statistics, at least 60% personal computers have been affected with Stuxnet viruses. Moreover, Stuxnet has generated a lot of homeotic viruses such as Duqu viruses, Flame viruses, and so on [134]. It is shown from Figure 1.1 that the numbers of security events is shown for industrial control systems, which is reported by the industrial control systems cyber emergency response team.

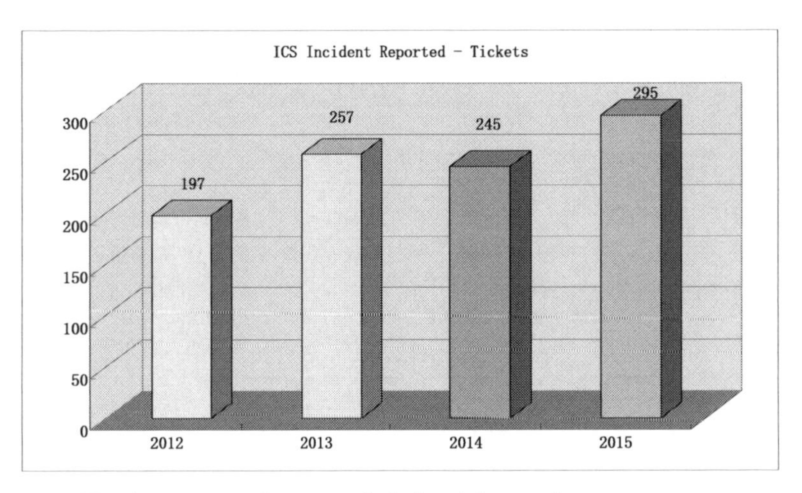

Figure 1.1 Numbers on security events in industrial control systems.

From Figure 1.1, it is seen that security events in industrial control systems are increasing year after year. Due to the fact that industrial control systems play key roles in national infrastructure, the poor security of control systems seriously threatens republic lives. Therefore, many countries have adopted essential steps to enhance the security of industrial control systems. In America, the Department of Energy has established the national Supervisory Control And Data Acquisition (SCADA) test bed program and a 10-year outline for the protection of industrial control systems [30]. Oak Ridge National Laboratory, Edward National Laboratory, and some other universities together have investigated the security of control systems. In 2012, Japan has also established the center of industrial control systems for the purpose of enhancing the network security for key infrastructures. In 2013, the European union agency for network and information security has published the white paper on industrial control system network security. IEEE Transactions on Cybernetics, IEEE Transactions on Automatic Control, and IEEE Transactions on Industrial Informatics have held special issues on industrial control system security. At present, industrial control system securities have received significant attentions in the world.

In the following, we present some typical examples of control systems that are vulnerable to network attacks.

An Unmanned Aerial Vehicle (UAV) communication and control system is shown in Figure 1.2. It consists of UAV, navigation satellite, mobile ground

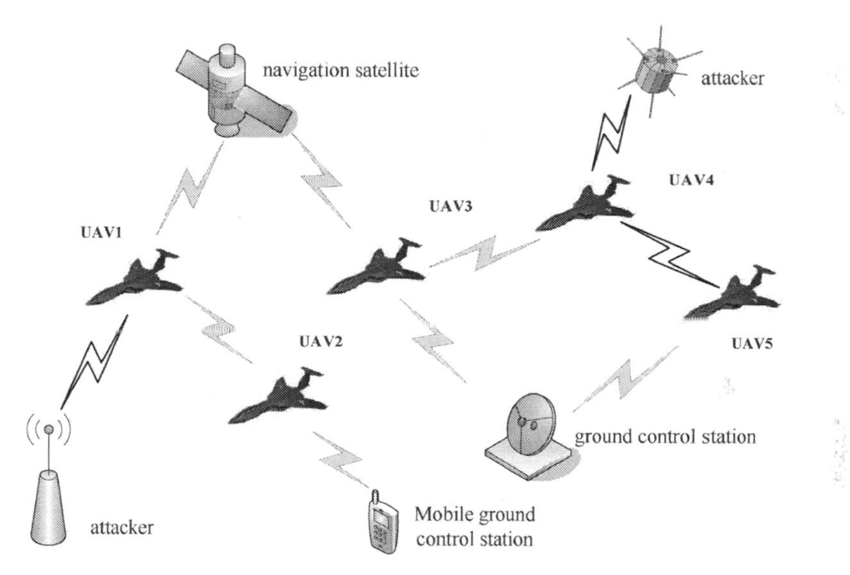

Figure 1.2 Network attack UAV control system.

control station, ground control station, and so on. Among the aforementioned portions, there exist communication links. Network adversaries are able to attack the communication links to affect UAV control. For example, in the "RQ-170 Sentinel Event" of the United States, an unmanned aerial vehicle was captured because the navigation communication networks were attacked. In addition, since the UAVs can communicate with each other, if an unmanned aerial vehicle is attacked, then the other UAVs will also get affected.

Figure 1.3 is a SCADA system architecture diagram that generally includes a data acquisition and a control terminal equipment, which are also named the slave computer and the host computer, respectively. The slave computer usually consists of Remote Terminal Cells (RTCs) and PLCs. On the other hand, the typical host computer system comprises a workstation, data ser-ver,

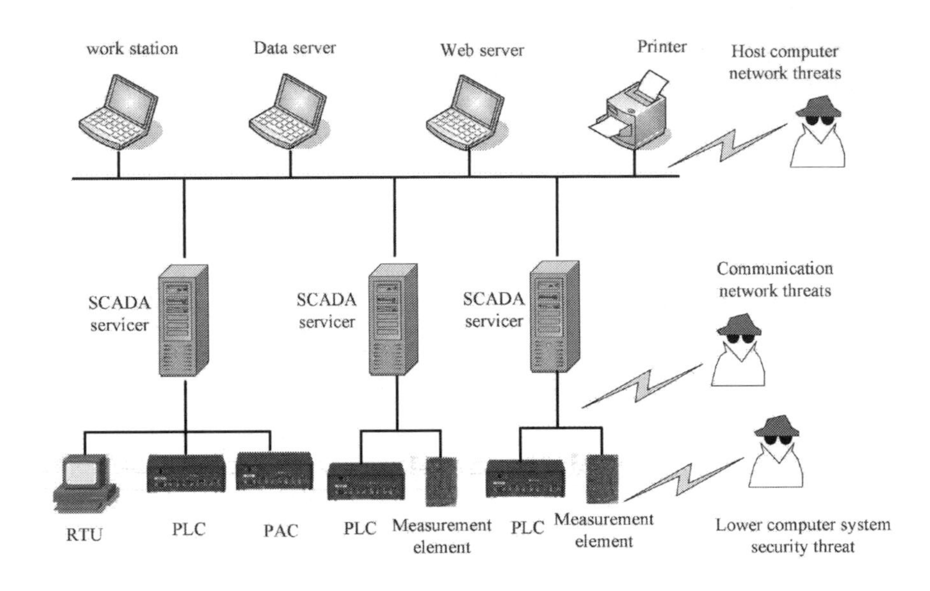

Figure 1.3 SCADA system architecture diagram under attack.

web server, SCADA server, and so on. Due to the wide deployments of SCADA systems, it is easy for adversaries to get access to the SCADA systems. They are capable of directly attacking actuators or sensors in the slave computer, or attacking the networks between the slave computer and the host computer, or even invading the interior of host computer. For example, the adversary modifies the value displayed on man-machine interfaces, which makes the operator unaware of attacks.

Summarizing the above discussion, it is of great urgency and necessity to develop the research on the security of NCSs.

1.2 Form of Attacks in NCSs

This section presents potential attack forms in NCSs, and some typical attack forms are illustrated emphatically in the following.

- Attacks against physical objects [107]: Attacks against physical objects are a kind of attack form which is directed against physical structures such as controllers, actuators, sensors, or plants. The attack model of attacks against physical objects is shown as follows.

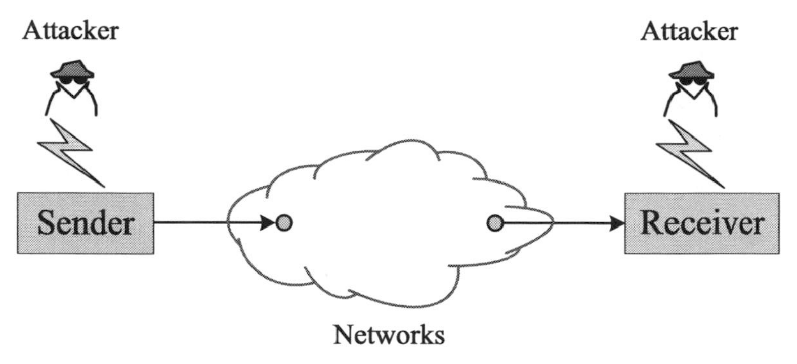

Figure 1.4 Attacks against physical objects.

- Integrity attacks [98]: For integrity attacks, attackers intentionally modify control commands or measurement data to compromise the NCSs. The NCSs are affected by wrong operations for the reason that wrong external information is obtained. Moveover, the integrity attack can be further subdivided into deception attack, cover attack, replay attack, and data injection attack [188]. Among the attacks, deception attacks compromise the NCSs via fault detection and isolation systems mainly. In a fault detection and isolation system, filtering algorithms are usually used to calculate an estimated value of sensor measurement. If the difference between the measured value and estimated value is larger than a given threshold value, then the fault detection and isolation system will trigger alarm. In fact, deception attacks are to interfere with control or measurement processes of the NCSs without triggering an alarm. The attack model of integrity attacks is shown in Figure 1.5.

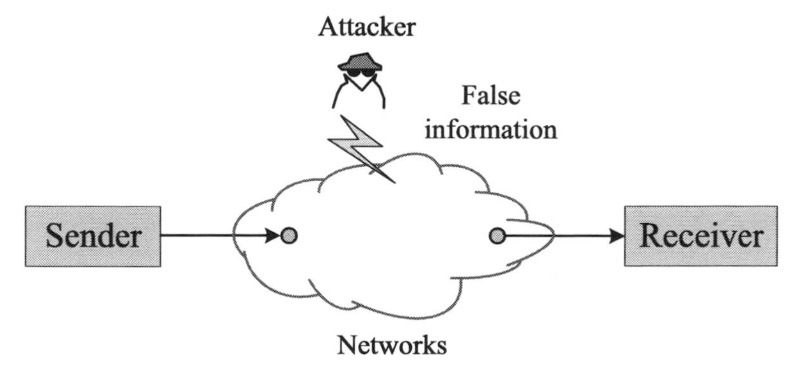

Figure 1.5 Deception attack.

- Availability attacks [21, 170, 61]: Availability attacks are also called denial-of-service (DoS) /jamming attacks which aim at preventing the control command or sensor measurement from being sent to intended users by interfering communication channels. When DoS attacks interfere with transmission channels of NCSs, the additional time delays and packets dropout are caused. Note that the current NCSs have high requirements for real-time properties. Any additional time delays or packets dropout will have a serious impact on the performance of NCSs, even lead to instability of the NCSs. The attack model of DoS attacks is shown in Figure 1.6.

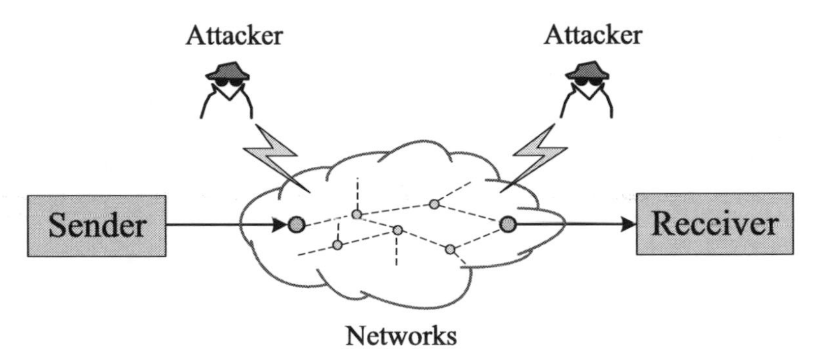

Figure 1.6 DoS attack.

1.3 Problem Studied in This Book

1.3.1 Attacks in Networked Control Systems

In 1998, G. C. Walsh put forward a concept of NCSs for the first time at the university of Maryland. Since problems of modeling for the NCSs have been studied deeply, such as time delays, packets dropout, data confusion and other issues. In particular, the inherently limited bandwidth of communication channels has led to a number of network-induced phenomena, which is worth exploiting. Note that the phenomena on packets dropout and communication delays have attracted much attention for the reason that they are considered to be two of the main causes of the performance degradation or even instability of the NCSs [156, 53].

Compared with traditional point-to-point systems, NCSs have many advantages, such as fewer expenses, higher flexibility, and better resource sharing; please refer to [78, 142, 142, 101, 20], and the references therein. Nevertheless, an ever-increasing popularity of communication networks also brings new challenges. The exposure to public networks renders control systems as targets of potential cyber attacks. As connection of information and reality world, control systems targeted by cyber attacks can lead to serious incidents, which have been verified during the past decade [105, 62]. By targeting different components of control systems, attackers can launch various types of attacks. Most of these control-system-oriented attacks can be categorized as deception attacks, and DoS attacks that compromise data integrity and data availability, respectively. The deception attack is launched by directly modifying the control or measurement signal and it is further categorized as cover attacks [122], data injection attacks [73], stealthy attacks [31] and replay attacks [98]. While DoS attacks or jamming attacks are launched by corrupting the communication channels of NCSs. DoS attacks usually lead to congestions in communication networks, causing time delays and packets dropout. It is worth mentioning that DoS attacks, which compromise the data availability, are vital for the reason of that all control systems operate in real time. For example, control systems using deadline corrective control may be driven to instability under DoS attacks [171]. Unlike deception attacks, DoS attacks that require little prior knowledge on control systems are also easy to apply. Hence, the DoS attacks have been listed as the most financially expensive security incidents [90]. Thus, securing NCSs under DoS attacks raise major concerns. In [7], a class of DoS attack models have been considered to find an optimal causal feedback controller by minimizing a given objective function subject to safety and power constraints. When NCSs with multitasking and central-tasking structures suffer DoS attacks, optimal control strategies have been presented by game theory in delta domain [173]. In [84], a game theoretic approach has been utilized to analyze a Nash equilibrium problem between sensors and attackers. Considering a Markov modulated DoS attack strategy, attackers stochastically jam control packets in NCSs with a hidden Markov model [18]. When an energy-constrained attacker jams a network channel, DoS attack schedules are provided to degrade system performances in an optimal attack pattern [177]. Though various attack schemes have been researched, optimal DoS attack schemes that are dangerous to NCSs have not been studied in depth yet. Moreover, it is very interesting to analyze optimal attack schemes for their serious harms on NCSs. Actually, there have been a number of literatures addressing the problem on resilient control under DoS attacks [171, 170, 7, 79].

As far as we know, most of these literatures can be categorized as attack tolerant resilient control methods and attack compensation resilient control methods. For the first category, the resilient control strategies can tolerate a certain level of negative effects caused by DoS attacks. To be specific, resilient control strategies are developed such that NCSs remain within the safety zone in spite of DoS attack induced time delays or packets dropout. For example, a semi-definite programming method has been used to minimize the objective function subject to power and safety constraints in [7]. Stability conditions of an event trigger system under DoS attacks have been exploited in [79]. A model predictive resilient control method has been proposed in [188], where predictive values are used if DoS attack occurred. For the second category, resilient control methods are employed to compensate for the control performance degradation caused by DoS attacks. In [171, 170, 193], IDSs have been deployed in the cyber layer which can defend against DoS attack and improve performances of underlying control systems. Data-sending strategies to contradict the negative influence of DoS attacks has been developed in [83]. From the aforementioned literatures, it is concluded that game theory employed intensively in resilient control is a powerful tool in characterizing cooperation and contradiction among agents. Furthermore, some works investigate Networked Predictive Control (NPC) schemes to cope with DoS attacks on NCSs [35]. NPC schemes have been to used to compensate random delays and consecutive packets dropout [104, 165, 35]. Based on a switched system approach, stability analysis on NPC systems has been established via an average dwell time technique in [182]. The NPC scheme has been utilized well on NCSs under deception attacks [103]. Unfortunately, optimal control by using NPC approaches has not adequately investigated on the security issues of NCSs yet.

In practice, all real-time NCSs operate in the presence of disturbances caused by a number of factors [45], including the fluctuation of communication environment [167], channel fading [148], quantization effects [133], load variation [38], friction [186], and measurement noises [168]. Therefore, the study on NCSs with external disturbances is of great importance from both theoretical and engineering points of view [46, 139, 164]. So far, a number of advanced control approaches have been developed to deal with the optimal control problem on NCSs. It is worth mentioning that the disturbances acting on the underlying dynamics will impact on the optimum of cost functions. Nevertheless, it has been implicitly assumed that no disturbance exists or all the disturbances are fully estimated and compensated in most literature concerning the optimal control of NCSs. It is shown that the influences from the disturbances on the optimum is largely neglected [126, 151, 136]).

Summarizing the above results, we arrive at the conclusion that several challenges still remain despite all the reported literatures on securing NCSs. One of such challenges is to develop optimal control strategies subject to DoS attacks in delta-domain and provide optimal defense and attack strategies of designed NCSs. The second challenging problem is how to quantify the influences of disturbances and packets dropout from the concept of ϵ-level, which are equally important for the NCSs. Another challenge is to find a defense strategy based on NPC to cope with DoS attacks under optimal schemes.

1.3.2 Resilient Control of WNCSs

In recent years, Wireless Networked Control Systems (WNCSs) have experienced a great development on fields of theory and practice. In WNCSs, the sensor and actuator communicate with the controller through wireless networks. Compared with traditional NCSs, WNCSs have considerable advantages, such as reduced wires, much flexibility, and low installation and maintenance cost. However, the wireless networks are more vulnerable than wired networks, which may be caused by weather changing, multi-path propagation, doppler shift, networked attacks, and so on. Considering the dynamic wireless networks that result in poor communication performance are vital in the design of WNCSs [28]. Additionally, utilizing the inherent nature of "openness" of WNCSs, malicious attackers can destroy communication communities and control systems [155]. Thus, a number of literatures focusing on the security of WNCSs have been found as [84, 177, 154, 81], and the re-ferences therein. Some advanced results have been presented on the security problem of NCSs in recent years. In [127], attack scenarios have been modeled and analyzed according to a three-dimensional resources framework. In [102], two-channel false data injection attacks against output tracking problem of NCSs have been researched. To detect integrity attacks, the probability of detection has been optimized by conceding system performance [98].

Specially for WNCSs, security issues for remote state estimation communicating by wireless channels have been studied in [84, 81], and Markov game theoretic approaches have been used to obtain the optimal attack and defense strategies with energy constrained sensor and attacker. Then multiple power levels have been available for sensor and attacker in remote state estimation system, and the mixed Nash equilibrium strategies have been obtained under the framework of Signal-to-Interference-plus-Noise Ratio (SINR)-based game [84]. In [173], the closed-loop system performance degradation caused by DoS attacks has been compensated by inverse game pricing method. The optimal attack and defense strategies have been obtained by modeling the attacker and defender as a Stackelberg game [173]. It makes practical sense to investigate SINR-based attack scheme for enhancing resilience of the closed-loop WNCSs. To analyze jamming attacks on cyber-layers of WNCSs, game theory which acts as a powerful tool has been employed to model interactions between legitimate users and malicious jammers [22]. A stochastic game framework for anti-jamming defense design is proposed with time-varying spectrum environment in a cognitive radio network [135]. In [117], a Bayesian jamming game between a legitimate transmitter and a smart jammer is discussed when there exists incomplete information for every network user. In [147], a power control strategy of a legitimate user against a smart jammer under power constraints is handled as a Stackelberg game with observation errors. The jammer, which acts as a follower, chooses a jamming power according to an observed ongoing transmission, while the user as a leader determines its transmitting power based on an estimated jamming power. The Stackelberg game is a well-developed and appropriate method to cope with hierarchical interactions among players in the anti-jamming field [155]. Furthermore, an anti-jamming Bayesian Stackelberg game with incomplete information is proposed in [66].

Virtually, all the WNCSs operate in the presence of disturbances that are caused by many factors such as load variation [38], friction [186], and measurement noise [168]. Therefore, considering the influence of disturbances on WNCSs is of great importance [45]. The H_∞ minimax control theory [16], which addresses the worst controller design for plants makes systems achieve the optimal performance under disturbances. The H_∞ minimax control has advantages compared with the traditional observer-based disturbance control method when it is difficult to model the disturbances. A number of advanced results for NCSs or WNCSs have been presented, e.g., [171], [170], [77] and the references therein. Furthermore, a resilient control problem is investigated when WNCSs suffer malicious DoS or jamming attacks in cyber-layers with H_∞ minimax control theory.

Summarizing the above discussions, although the security of WNCSs has been widely investigated in many literatures, several urgent issues still exist. The first one is to analyze the interaction between defender and attacker via a hierarchical game approach and design an H_∞ minimax controller in delta-domain to guarantee the optimal system performance for high frequency sampled WNCS under disturbance. The second one is to comprehensively analyze interconnections between the cyber layer and physical layer and to design cross layer control strategies such that the studied WNCS can remain stable in spite of the DoS attacks. The last one is how to establish a Bayesian Stackelberg game framework between a malicious jammer and a legitimate user due to the incomplete information, and how to design an H_∞ minimax controller to guarantee the optimal WNCS performance under disturbances.

1.3.3 Application of Resilient Control to Power System

In recent years, NCSs have received an increasing research interest due to their wide applicabilities to smart grids [74], intelligent transportation [177], industrial control systems, navigation systems [69], teleoperation or remote systems [132], and so on. For the NCSs, sensors, actuators, control processing units, and communication devices are connected via networks [51], which renders several challenging problems such as time delays [179], [110], packets dropout [75], packets disorder [88], quantization [153, 108], cyber attacks [171], and so on.

Among these network-induced factors, two significant challenges are time delays and packets dropout, which lead to degradation of the control performance or even destabilize the whole controlled system [111]. For instance, deterministic time delays in NCSs have been considered in [182, 179, 24, 44, 42], and the time delays in the stochastic setting have been addressed in [34, 23, 185]. Specifically, the time delayed control system, which can only accommodate a subset of actuators at any time, has been exploit in [44]. In [23], an H_∞ control problem has been addressed for NCSs with stochastic time delays subjected to Markovian distributions. On the other hand, the NCSs with packets dropout have been exploited in [92, 86, 169, 119, 43, 44]. For instance, in [119], an optimal Linear Quadratic (LQ) gaussian control problem has been addressed with signal estimation subjected to packets dropout. In [43], a stochastic optimal control problem of nonlinear NCSs with packets dropout and long time delays has been studied. In simultaneous presence of time delays, packets dropout and measurement quantization effects, a coupled design of networked controller has been addressed in [44].

Since control systems can be regarded as a connection between the information world and physical world, any successful attacks on NCSs may lead to significant loss of properties or even human lives. Actually, it has already been reported in [81] and [103] that systems in a number of critical infrastructures have been compromised by a series of attacks. Hence, many researchers have exploited the security of NCSs from both control and communication communities; please refer to [127, 118, 84, 177, 47] and the references therein. To avoid attacks, some IDSs are deployed in a cyber layer to raise alarms once an anomaly behavior is detected such that it can be removed automatically. Thus, DoS attackers have to go through the IDSs firstly before they compromise a control system.

Owing to the rapid development of sensing techniques, sampling intervals of modern industrial control systems are normally quite small, and a sampled-data problem becomes very critical in system design. The delta operator approach has been well recognized in addressing sampling issues for NCSs [160]. Numerical-stiffness problems resulting from the fast sampling protocol can be circumvented by using the delta operator approach [162]. Furthermore, some related results for both discrete-and continuous-time systems can be unified in the delta operator systems. As such, it is of vital importance to develop delta-domain results for discrete-time systems with a high sampling rate. Due to its theoretical significance and practical importance, the delta operator approach has been extensively exploited in NCSs (see, e.g., [55, 76, 161, 160, 163]). Specifically, some inspiring results have been reported for some control and filtering problems of NCSs [162], a control problem on Markovian jump systems, and some robust control problems with actuator saturation [163]. It has been shown in [161] that the delta-domain results can not only deal with the inherent numerical stiffness caused by the fast sampling protocol in the discrete-time NCSs, but can also adapt to dynamic network environment. In comparison with reported literatures, the delta-domain results for dynamic games have been scattered, especially within the framework of Riccati recursions [162].

Summarizing the above results, we arrive at the conclusion that several challenges still remain despite all the reported literatures on securing NCSs. One such challenge is to estimate the attack-induced performance degradation such that the loss does not exceed the limitation of designed NCSs. This is of vital importance for the reason that one can verify whether a system remains within safety regions with applied securities or control strategies by assessing the security level before adversarial incidents occur. The second challenging problem is how to quantify the influences of disturbances, long delays, and packets dropout on NE, which is equally important for the NCSs. Another challenge is to find a tradeoff between system performances and security enforcement levels under the coupled design of the IDSs and controllers for the reason that higher security level IDSs lead to control performance degradation. Electric Power Systems (EPSs) are a kind of typical NCSs in national basic industry. In this part, we will mainly investigate the aforementioned issues on the EPSs.

1.3.4 Coupled Design of CPS under Attacks

In recent decades, the coupled design of Cyber-Physical Systems (CPSs) under attacks has been paid wide attention, since a number of critical infrastructures have been compromised by DoS attacks, as reported in [62, 70] and [37]. Design of such systems requires a system perspective towards cyber-physical systems against threats and malicious behavior. Note that state awareness of ICSs under attacks has been discussed in [95]. A passivity combined with an adaptive sampling approach to design a control architecture is proposed in [36], and the method shows certain robustness to network uncertainties. As mentioned in [194], the Siemens SCADA systems have been compromised by Stuxnet, which is a computer worm. It has also been reported in [94] that a hacker intruded and shut down a traffic air control system tower at Worcester Regional Airport USA.

Virtually, the attackers become much smarter at present. The traditional separate design of cyber layers and physical layers appears weaker than before. CPSs are a combination of physical systems with cyber systems, where the cyber systems receive real-time data from the physical systems. Then a processing and final response is made to the physical systems in real-time. CPSs integrate networked computational resources into physical processes in order to add new capabilities into an original system. Due to the development of networks, the concept of resilient control emphasizing controller design in adversarial cyber environment has been proposed in [115]. The aim of resilient control is to maintain an accepted level of operational normalcy in response to both external disturbances in the physical layer and malicious attacks in the cyber layer. Thus, Resilient Control Systems (RCSs), which have fully coupled cyber module and control module, require a holistic view and cooperation between Information Technology expert and control expert. RCSs have been studied under replay attacks in [189], and a class of competitive resource allocation problems are characterized as convex games. RCSs have been modeled as a two-level receding-horizon dynamic Stackelberg game, in which there is coupled decision-making process between control system operator and jammer [189].

In order to reduce the loss of DoS attack, many approaches have been proposed for CPSs. A game-theoretical method, which is a powerful tool to model the interactions among agents, has been also used to model the cyber attack and defense [4]. Attack-resilient control built within a framework of game theory, has been tackled in recent years. Hybrid models are proposed for RCSs, in which stochastic switching is governed by a Markov security game [191] and [175]. It is worth mentioning that the application of game theory to cyber security issue have background in the configuration of IDSs [4, 175, 190]. IDSs are used to raise alarms once an anomaly behavior such as packets dropout or overlong time delays are detected, so that malicious attacks can be removed automatically. In the attack model, game theory has been used to describe interaction between the IDSs and DoS attacks, and to get the best delivery package rate. By Nash Equilibrium (NE) strategies, the IDSs obtain an appropriate tradeoff between system performances and security enforcement levels [190]. Some critical issues on cyber security in IDSs require a holistic and cross layer design approach for controller design of integrated cyber physical systems. In [39], an adaptive neural control architecture is used for NCSs within a resilient control framework. Parameters of an attacked plant are changed to match a reference model. However, few efforts have been made to consider integrated design of defense mechanisms in the cyber layers and controller design in the physical layers.

Summarizing current situation, coupled designs of the cyber layers and physical layers meet the following challenges. Firstly, some methodologies and principles are needed for integrated design because the cyber systems of the IDSs are not isolated from the physical systems for defense against malicious adversaries in practical situations. Secondly, how to establish a game-in-game structure for the coupled design of RCS with the aim at obtaining the tradeoff between an outcome of inner game and a solution of out game. Thirdly, practical control systems are subject to actuator saturations that bring challenging problems for the coupled design of CPSs. In this part, we will investigate the coupled design of CPSs details.

Part I
The Attacks in Networked Control Systems

Chapter 2
A Unified Game Approach for NCSs under DoS Attacks

2.1 Introduction

Recent years have witnessed a wide application of network technologies to control systems. Integrating with communication core, NCSs have increased mobility and interoperability, and reduced maintenance and installation costs [62]. The exposure to public networks renders control systems the targets of potential cyber attacks. Since there is a connection between the information world and reality, control systems are targeted by cyber attacks resulting in serious incidents, which has been verified during the past decades [105, 62]. By targeting at different components of control systems, attackers can launch various types of attacks. Actually, there have been a number of literatures addressing the problem of resilient control under DoS attacks [171, 170, 7, 79]. From the aforementioned literatures, it is concluded that game theory has been employed intensively in resilient control. On the other hand, it has been well recognized that delta operators can overcome numerical mistakes on discrete-time systems with fast sampling frequency. Then they unify some previous related results of continuous and discrete systems into the delta operator systems framework. For delta operator systems, they are much easier to observe and analyze the control effect over dynamic networks for the reason of that sampling period is explicitly expressed. To the best of the authors' knowledge, game theoretic resilient control has not been investigated in the delta domain and still remains challenging.

In this chapter, the multiple-tasking and central-tasking optimal control strategies subject to DoS attacks are developed in the delta domain, respectively. The algorithms to obtain optimal defense and attack strategies are also provided. Specifically, the main contributions of this chapter are summarized as follows: First, the conditions for MTOC and CTOC strategies are derived in terms of backward recursions in delta domain. The numerical advantages of using delta operators are verified theoretically. Second, an inverse game approach on the defense side is proposed to compensate for the attack induced performance loss. Last, to provide a worst case scenario for the defender, the optimal strategies of the DoS attacker are provided that can drive the control system out of the safety zone by using lowest attacking intensity.

In Section 2.2, a brief introduction of the delta operator is provided. The control objectives of MTOC and CTOC are provided, and the design objective of optimal defense and attack strategies are given. In Section 2.3, the conditions and analytical form of MTOC and CTOC strategies are derived. The algorithms to develop optimal defense and attack strategies are provided in Section 2.4. In Section 2.5, the proposed methodologies are applied to HVAC system and the validity is verified by simulation results. Conclusions are drawn in Section 2.6.

2.2 Problem Formulation

The main idea of this chapter is to develop control strategies for MTOC and CTOC under DoS attacks. To compensate for the attack induced packets dropout, the defense strategies are also to be designed. On the opposite, the optimal attack strategies are developed that drive the NCS out of a safety zone using the least resources. Keeping the aforementioned idea in mind, we provide the design objectives for defense and attack strategies, and control strategies of MTOC and CTOC.

2.2.1 The Model of NCS Subject to DoS Attack

The block diagram of a complex NCS with multiple channels is shown in Figure 2.1, where forward and backward communication networks are both subject to DoS attacks. DoS attacks may cause congestions in the network that further leads to packets dropout of the control or sensor signal. We assume here that the buffer located with the actuator can check the time stamps of the collected signal. When a control packet arrives at the actuator, it will be used to update the buffer if its timestamp is newer than that of the control signal in the buffer. If no newer control packet arrives, the actuator will apply 'zero-control' strategy. For analysis purpose, we introduce variable α_k^{FN}, $\alpha_k^{BN} \in \{0, 1\}$ to indicate whether the packet is lost in forward or backward networks.

The dynamics of the delta domain NCS under DoS attack is given as follows

$$\delta x_k = A_\delta x_k + \sum_{i=1}^{S} \alpha_k^i B_\delta^i u_k^i, \tag{2.1}$$

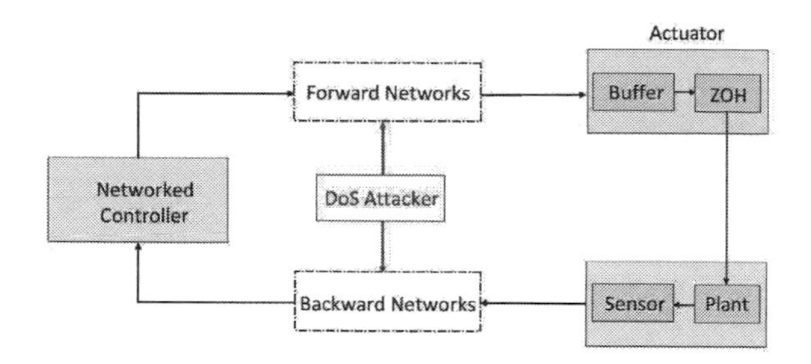

Figure 2.1 The NCS under DoS attack.

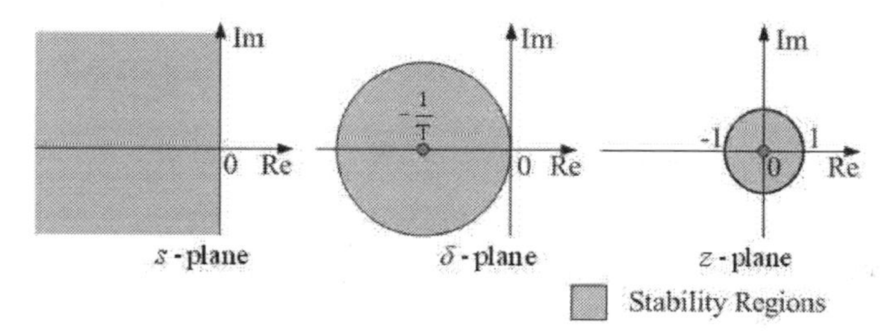

Figure 2.2 Stability regions for the continuous Laplace plane, the discrete delta and shift planes.

where $x_k = x_{k\mathrm{T}_s}$, T_s is the sampling interval, A_δ and B_δ^i are matrices in the delta domain with appropriate dimensions. The indicator variable $\alpha_k^i = \alpha_k^{FN}\alpha_k^{BN}$, $i \in \mathbf{S} := \{1, 2, \cdots, S\}$ expresses the round-trip packets dropout encountered by the k th sensor packet and is also used to indicate whether 'zero-control' input strategy is applied.

Remark 2.1. Note that delta operator is defined as [49]

$$\delta x_k = \begin{cases} dx_t/dt, & \mathrm{T}_s = 0, \\ \frac{x_{k+1} - x_k}{\mathrm{T}_s}, & \mathrm{T}_s \neq 0. \end{cases} \tag{2.2}$$

System (2.1) is considered obtained from continuous model

$$\dot{x}_t = A_s x_t + \mathbf{I}_t \sum_{i=1}^{S} B_s^i u_t^i, \tag{2.3}$$

where A_s and B_s^i are matrices in the continuous domain with appropriate dimensions. $\mathbf{I}_t$ is the indicator function. We have $\mathbf{I}_t = I^{n\times n}$ if the buffer is updated at time t and $\mathbf{I}_t = \mathbf{0}^{n\times n}$ if not. The matrices in (2.1) are obtained by

$$A_\delta = \frac{e^{A_s \mathrm{T}_s} - I}{\mathrm{T}_s},$$

We assume that α_k^i is a random variable that is distributed according to Bernoulli distribution. Suppose that α_k^i is i.i.d and, for $i \neq j$, $i, j \in \mathbf{S}$, α_k^i is independent of α_k^j. Let us denote

$$\mathbb{P}\{\alpha_k^i = 0\} = \alpha^i, \ \ \mathbb{P}\{\alpha_k^i = 1\} = 1 - \alpha^i = \bar{\alpha}^i, \forall i \in \mathbf{S}, k \in \mathbf{K} := \{1, 2, \cdots, K\}.$$

Note that α^i here is viewed as Intensity of Attack (IoA). A more intense DoS attack will lead to larger IoA α^i. Let us define the strategy set of the DoS attacker to be $\mathcal{G} := \{\alpha^i\}_{i=1}^S$.

Remark 2.2. Resilient control under DoS attacks should be able to address the problem of packets dropout that is also common in the traditional NCS [171, 7, 8, 178]. However, it is worth mentioning that the packets dropout rate caused by inherent communication failure is much smaller comparing with the one caused by malicious DoS attack [75]. This puts forward a higher requirement for resilient control, which should be able to tolerate serious congestions in the communication channel, not just occasionally occurring information loss.

2.2.2 MTOC and CTOC Design

The architecture of MTOC and CTOC is shown in Figure 2.3. It can be seen

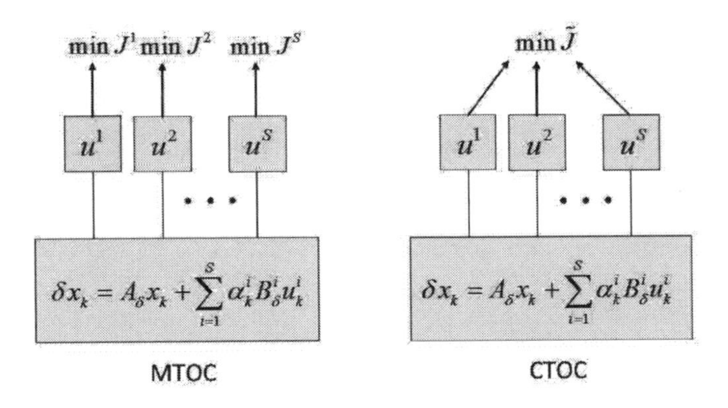

Figure 2.3 The architecture of MTOC and CTOC.

from Figure 2.3 that the controllers of CTOC are coordinated to reach a common objective, while those of the MTOC do not cooperate with each other and only minimize the individual cost function. Thus, MTOC can be used to model some complex and distributed systems, where each decision maker is selfish and only needs to pursue its own interest [126]. The CTOC structure, on the other hand, is able to address problems where central coordination and cooperation between decision makers are possible. Note that the controller design of MTOC falls within the framework of non-cooperative dynamic game in delta domain with each controller (Player) minimizing the cost-to-go function

$$J^i = \mathbb{E}\left\{x_K^{\mathrm{T}} Q_\delta^K x_K + \mathrm{T}_s \sum_{k=0}^{K-1} \Psi_0\right\}, \quad i \in \mathbf{S}, \tag{2.4}$$

where $\Psi_0 = x_k^{\mathrm{T}} Q_\delta^i x_k + \alpha_k^i {u_k^i}^T R_\delta^i u_k^i$. We assume that $Q_\delta^K \geq 0$, $Q_\delta^i \geq 0$ and $R_\delta^i > 0$ for all $i \in \mathbf{S}$. For CTOC, there is a central coordination that all players cooperate to minimize a common cost-to-go function which is also named as sum social welfare. The sum social welfare is defined as $\tilde{J} = \sum_{i=1}^S \eta^i J^i$, where $\eta^i > 0$ is the weighting factor on Player i's cost-to-go function and satisfies the normalization condition $\sum_{i=1}^S \eta^i = 1$. For both MTOC and CTOC, Transmission Control Protocol (TCP) is applied where each packet is acknowledged, and the information set is defined as $\mathcal{I}_0 = \{x_0\}$, $\mathcal{I}_k = \{x_1, \cdots, x_k, \alpha_0, \cdots, \alpha_{k-1}\}$. For each controller i ($i \in \mathbf{S}$), Γ^i and Λ^i are introduced as the set of control strategies with MTOC and CTOC structures, respectively. The admissible strategies μ_k^i for MTOC and $\tilde{\mu}_k^i$ for CTOC are seen as functional to map information set $\mathcal{I}_k$ to u_k^i, that is, $u_k^i = \mu_k^i(\mathcal{I}_k)$ or $u_k^i = \tilde{\mu}_k^i(\mathcal{I}_k)$. Let us further define

$$\mu^i := \{\mu_0^i, \mu_1^i, \cdots, \mu_{K-1}^i\}, \quad \tilde{\mu}^i := \{\tilde{\mu}_0^i, \tilde{\mu}_1^i, \cdots, \tilde{\mu}_{K-1}^i\}.$$

Towards this end, two optimization problems should be addressed. For Problem 1, all the Players are selfish and noncooperative. Let μ^{-i} denote the collection of strategies of all players except Player i, i.e., $\mu^{-i} = \{\mu^1, \cdots, \mu^{i-1}, \mu^{i+1}, \cdots, \mu^S\}$. Player i is faced with minimizing its own associated cost function by solving the following dynamic optimization problem.

Problem 2.3. Find control strategies μ^{i*}, $i \in \mathbf{S}$ such that the following optimization problem is solved for all $i \in \mathbf{S}$.

$$(\mathrm{OC(i)}) \min_{\mu^i} \quad J^i(\mu^i, \mu^{-i*}) := \mathbb{E}\{x_K^{\mathrm{T}} Q_\delta^K x_K + \mathrm{T}_s \sum_{k=0}^{K-1} \Psi_0\}, \tag{2.5}$$

$$s.t. \quad \delta x_k = A_\delta x_k + \sum_{i=1}^S \alpha_k^i B_\delta^i u_k^i.$$

If optimization is carried out for each i, Nash Equilibrium (NE) will be obtained which is $J^{i*}(\mu^{i*}, \mu^{-i*}) \leq J^i(\mu^i, \mu^{-i*})$. The corresponding total cost achieved is given by $J^* = \sum_{i=1}^S \eta^i J^{i*}$.

In the same vein, the associate optimization problem of CTOC is shown as follows.

Problem 2.4. Find control strategies $\tilde{\mu}^* := \{\tilde{\mu}^{i*}, \tilde{\mu}^{-i*}\}$ such that the following optimization problem is solved

$$\text{(COC)} \min_{\tilde{\mu}} \tilde{J}(\tilde{\mu}^i, \tilde{\mu}^{-i}) := \sum\nolimits_{i=1}^{S} \eta^i \mathbb{E}\left\{ x_K^{\mathrm{T}} Q_\delta^K x_K + \mathrm{T}_s \sum\nolimits_{k=0}^{K-1} \Psi_0 \right\}, \quad (2.6)$$
$$s.t. \quad \delta x_k = A_\delta x_k + \sum\nolimits_{i=1}^{S} \alpha_k^i B_\delta^i u_k^i.$$

This minimization is essentially an optimal control problem. The optimal value $\tilde{J}^*(\tilde{\mu}^{i*}, \tilde{\mu}^{-i*}) \leq \tilde{J}(\tilde{\mu}^i, \tilde{\mu}^{-i})$ will be obtained if the optimization problem is carried out.

Before ending this section, the following instrumental lemma, is introduced, to which will be used subsequently.

Lemma 2.5. *[158] The property of delta operator, for any time function x_k and y_k, there exists*

$$\delta(x_k y_k) = y_k \delta x_k + x_k \delta y_k + T_s \delta x_k \delta y_k.$$

Remark 2.6. The objective functions of OC(i) and COC shown above have a finite time horizon. The case of the infinite time horizon will also be considered subsequently where OC(i) and COC are carried out with $K \to \infty$.

2.2.3 Impact Metrics

Before exploiting the resilient compensation strategy, we perform a vulnerability analysis for MTOC and CTOC and provide metrics to characterize the impact of non-cooperative behavior and DoS attack, respectively. First, Price of Anarchy (PoA) [178], which is a terminology from economics, is introduced to quantify the performance degradation caused by noncooperative behaviors of Players.

Definition 2.7. (PoA) Consider MTOC and CTOC without DoS attack. With the same K, S, $\{\eta^i\}_{i=1}^S$, $\{Q_\delta^i\}_{i=1}^S$, $\{R_\delta^i\}_{i=1}^S$, Q_δ^K and T_s, PoA is defined as

$$\text{PoA} = \max_{\mu_{\mathcal{G}_0}^* \in \Gamma_{\mathcal{G}_0}^*} J_{\mathcal{G}_0}^* / \tilde{J}_{\mathcal{G}_0}^*,$$

where $\mu := \{\mu^i\}_{i=1}^S$, $\mathcal{G}_0 := \{\alpha^i = 0\}_{i=1}^S$, and the subscript $\mathcal{G}_0$ denotes the dependence on attacking strategy.

Note that $\max J^*$ represents the worst performance in the set of Nash equlibrium Γ^*. PoA is lower bounded by 1 since the lack of central coordination leads to performance degradation. Similarly, we can introduce Price of DoS Attack (PoDA) to quantify the negative impact of DoS attack on NCS with MTOC structure.

Definition 2.8. (PoDA) Consider NCS with MTOC structure and two different attacking strategies $\mathcal{G}_1$ and $\mathcal{G}_2$. With the same K, S, $\{\eta^i\}_{i=1}^S$, $\{Q_\delta^i\}_{i=1}^S$, $\{R_\delta^i\}_{i=1}^S$, Q_δ^K and $\mathtt{T}_s$, PoDA is defined as

$$\text{PoDA} = \max_{\mu^*_{\mathcal{G}_1} \in \Gamma^*_{\mathcal{G}_1}} J^*_{\mathcal{G}_1} / \max_{\mu^*_{\mathcal{G}_2} \in \Gamma^*_{\mathcal{G}_2}} J^*_{\mathcal{G}_2}.$$

If PoDA > 1, it is evident that DoS attacking strategy $\mathcal{G}_1$ causes more damage to NCS with MTOC structure. Since the DoS attack induced packets dropout leads to control performance degradation, we have

$$\text{PoDA} = \max_{\mu^*_{\mathcal{G}_1} \in \Gamma^*_{\mathcal{G}_1}} J^*_{\mathcal{G}_1} / \max_{\mu^*_{\mathcal{G}_0} \in \Gamma^*_{\mathcal{G}_0}} J^*_{\mathcal{G}_0} > 1,$$

where $\mathcal{G}_0 := \{\alpha^i = 0\}_{i=1}^S$ and $\mathcal{G}_1 := \{0 < \alpha^i < 1\}_{i=1}^S$.

From the vulnerability analysis above, we conclude that, for MTOC, the degradation of control performance is a mutual effect of DoS attack (PoDA) and non-cooperative behavior of the Players (PoA). Thus, the overall performance degradation of MTOC is denoted as OPD $:= \max_{\mu^*_{\mathcal{G}_1} \in \Gamma^*_{\mathcal{G}_1}} J^*_{\mathcal{G}_1} / \tilde{J}^*_{\mathcal{G}_0} > 1$.

Remark 2.9. If conditions of the unique NE of MTOC structure are satisfied, then the index PoA, PoDA, and OPD reduce to PoA $= J^*_{\mathcal{G}_0} / \tilde{J}^*_{\mathcal{G}_0}$, PoDA $= J^*_{\mathcal{G}_1} / J^*_{\mathcal{G}_2}$ and OPD $:= J^*_{\mathcal{G}_1} / \tilde{J}^*_{\mathcal{G}_0}$, respectively.

2.2.4 Defense and Attack Strategy Design

In this chapter, we assume that the defender has the freedom to participate in choosing the weighting matrices Q_δ^i, $i \in \mathbf{S}$ to compensate for OPD of NCS with MTOC structure [29]. Thus, the set of defense strategies is defined as $\mathcal{F} := \{Q_\delta^i\}_{i=1}^S$ and the optimal defense strategies are obtained by solving the following problem.

Problem 2.10. The optimal defense strategy $\mathcal{F}^*$ is the solution of

$$\min_{\mathcal{F}} \quad DIST := |\text{OPD} - 1|.$$

The corresponding overall performance degradation yields OPD*.

For DoS attackers, a scenario is considered here that the the attacker is fully aware of the defender's compensation strategy $\mathcal{F}^*$, that is, the attacker and defender possess asymmetric information. To save the attacking cost, the DoS attacker intends to drive the underlying NCS out of a safety zone using attacking intensity $\mathcal{G} := \{\alpha^i\}_{i=1}^S$ as low as possible. The optimal attack strategies $\mathcal{G}^*$ are obtained by solving the following problem.

Problem 2.11. The optimal attacking strategy $\mathcal{G}^*$ is the solution of the following optimization problem

$$\begin{aligned} &\min \quad \alpha^i \\ s.t. \quad &|\text{OPD}^* - 1| > \mathcal{S}_o, \quad \alpha^1 = \alpha^2 = \cdots = \alpha^S, \end{aligned} \tag{2.7}$$

where $\mathcal{S}_o$ is a scalar representing the safety zone [7, 89]. The same IoA constraint is reasonable when communication environments are assumed to be the same.

2.3 MTOC and CTOC Control Strategies

In this section, the conditions and analytical form of optimal control strategies for MTOC and CTOC are provided, respectively. Let us first consider the finite time horizon case and then extend the results to infinite time horizon case.

We give the preliminary notations for theorems in this section.

$$\Theta_0 = (\mathtt{T}_s A_\delta + I) - \mathtt{T}_s \sum_{j \neq i}^{S} \bar{\alpha}^j B_\delta^j L_k^j,$$

$$\Theta_1 = A_\delta - \sum_{i=1}^{S} \bar{\alpha}^i B_\delta{}^i L_k^i,$$

$$\Theta_2 = (\mathtt{T}_s A_\delta + I) x_k + \mathtt{T}_s \sum_{i=1}^{S} \alpha_k^i B_\delta^i u_k^i,$$

$$\Theta_3 = (\mathtt{T}_s A_\delta + I) x_k + \mathtt{T}_s \sum_{j \neq i}^{S} \bar{\alpha}^j B_\delta^j u_k^j,$$

$$\Omega_0 = \sum_{i=1}^{S} \eta^i Q_\delta^i, \quad \Omega_1 = (\mathtt{T}_s A_\delta + I),$$

$$\begin{aligned} \Omega_2 = \text{diag} \Big(\Big[&\bar{\alpha}^1 (1 - \bar{\alpha}^1) B_\delta^{1^T} \tilde{P}_{k+1} B_\delta^1, \bar{\alpha}^2 (1 - \bar{\alpha}^2) \\ &\times B_\delta^{2^T} \tilde{P}_{k+1} B_\delta^2, \cdots, \bar{\alpha}^S (1 - \bar{\alpha}^S) B_\delta^{S^T} \tilde{P}_{k+1} B_\delta^S \Big] \Big) . \end{aligned}$$

2.3.1 Finite Time Horizon Case

In this subsection, we consider the strategy design for MTOC and CTOC with the finite time horizon cost-to-go function. The following theorem is shown first to provide the solution to Problem 2.3 with a finite time horizon.

Theorem 2.12. *For finite time horizon MTOC, with a given attack strategy $\mathcal{G}$, the following conclusions are presented*

1. *There exists a unique NE if*

$$R_\delta^i + T_s {B_\delta^i}^T P_{k+1}^i B_\delta^i > 0, \tag{2.8}$$

and the matrix Φ_k is invertible, where $\Phi_k(i,i) = R_\delta^i + T_s {B_\delta^i}^T P_{k+1}^i B_\delta^i$, $\Phi_k(i,j) = T_s \bar{\alpha}^j {B_\delta^i}^T P_{k+1}^i B_\delta^j$.

2. *Under condition 1, the optimal control strategies of MTOC is given by $u_k^i = \mu_k^{i*}(\mathcal{I}_k) = -L_k^i x_k$ for all $i \in \boldsymbol{S}$, where*

$$L_k^i = (R_\delta^i + T_s {B_\delta^i}^T P_{k+1}^i B_\delta^i)^{-1} B_\delta^{iT} P_{k+1}^i \Theta_0. \tag{2.9}$$

3. *The backward iterations are carried out with $P_K^i = Q_\delta^K$ and*

$$\begin{aligned} -\delta P_k^i &= Q_\delta^i + \bar{\alpha}^i {L_k^i}^T R_\delta^i L_k^i \\ &\quad + T_s \Theta_1^T P_{k+1}^i \Theta_1 + \Theta_1^T P_{k+1}^i + P_{k+1}^i \Theta_1 \\ &\quad + T_s \sum_{j=1}^{S} \left(\left(\bar{\alpha}^j - {\bar{\alpha}^j}^2 \right) {L_k^j}^T {B_\delta^j}^T P_{k+1}^i B_\delta^j L_k^j \right), \end{aligned} \tag{2.10}$$

$$P_k^i = P_{k+1}^i - T_s \delta P_k^i. \tag{2.11}$$

4. *Under condition 1, the NE values under the MTOC is $J^{i^*} = x_0^T P_0^i x_0$, $i \in \boldsymbol{S}$, where x_0 is the initial value.*

Proof. An induction method is employed here. The claim is clearly true for $k = K$ with parameters $P_K^i = Q_\delta^K$. Let us suppose that the claim is now true for $k+1$

$$V^i(x_{k+1}) = \mathbb{E}\{x_{k+1}^T P_{k+1}^i x_{k+1}\}, \tag{2.12}$$

with $P_{k+1}^i > 0$. According to the definition of delta operator and Lemma 2.5, the above equation can be rewritten in the delta domain as

$$\begin{aligned} V^i(x_{k+1}) &= \mathtt{T}_s \delta(x_k^T P_{k+1}^i x_k) + x_k^T P_{k+1}^i x_k \\ &= \mathtt{T}_s \delta x_k^T P_{k+1}^i x_k + \mathtt{T}_s x_k^T P_{k+1}^i \delta x_k \\ &\quad + \mathtt{T}_s^2 \delta x_k P_{k+1}^i \delta x_k + x_k^T P_{k+1}^i x_k. \end{aligned}$$

By using the dynamic programming, the cost at time k is obtained by

$$\begin{aligned}
V^i(x_k) &= \min_{u_k^i} \mathbb{E}\left\{\mathtt{T}_s x_k^T Q_\delta^i x_k + \mathtt{T}_s \alpha_k^i {u_k^i}^T R_\delta^i u_k^i + V^i(x_{k+1})\right\} \\
&= \min_{u_k^i} \mathbb{E}\left\{\mathtt{T}_s x_k^T Q_\delta^i x_k + \mathtt{T}_s \alpha_k^i {u_k^i}^T R_\delta^i u_k^i + \mathtt{T}_s \delta x_k^T P_{k+1}^i x_k\right. \\
&\quad \left. + \mathtt{T}_s x_k^T P_{k+1}^i \delta x_k + \mathtt{T}_s^2 \delta x_k P_{k+1}^i \delta x_k + x_k^T P_{k+1}^i x_k \right\} \\
&= \min_{u_k^i} \mathbb{E}\left\{\mathtt{T}_s x_k^T Q_\delta^i x_k + \mathtt{T}_s \alpha_k^i {u_k^i}^T R_\delta^i u_k^i + \Theta_2^T P_{k+1}^i \Theta_2\right\}. \qquad (2.13)
\end{aligned}$$

The cost-to-go function $V^i(x_k)$ is strictly convex of u_k^i, since the second derivative of (2.13) yields $R_\delta^i + \mathtt{T}_s {B_\delta^i}^T P_{k+1}^i B_\delta^i > 0$. The minimizer is obtained by solving $\partial V^i(x_k)/\partial u_k^i = 0$

$$\begin{aligned}
\partial V^i(x_k)/\partial u_k^i &= R_\delta^i u_k^i + {u_k^i}^T R_\delta^i + {B_\delta^i}^T P_{k+1}^i \Theta_3 \\
&\quad + \Theta_3^T P_{k+1}^i B_\delta^i + \mathtt{T}_s {B_\delta^i}^T P_{k+1}^i B_\delta^i u_k^i \\
&\quad + \mathtt{T}_s {u_k^i}^T {B_\delta^i}^T P_{k+1}^i B_\delta^i = 0 \qquad (2.14)
\end{aligned}$$

The optimal control strategies are obtained by ${u_k^i}^* = -L_k^i x_k$, with

$$L_k^i = (R_\delta^i + \mathtt{T}_s {B_\delta^i}^T P_{k+1}^i B_\delta^i)^{-1} B_\delta^{iT} P_{k+1}^i \Theta_0. \qquad (2.15)$$

The equality sets up for all players, so there exists the unique NE if the invertible matrix Φ_k satisfies

$$\Phi_k \bar{L}_k = \Pi_k \qquad (2.16)$$

with $\Phi_k(i,i) = R_\delta^i + \mathtt{T}_s {B_\delta^i}^T P_{k+1}^i B_\delta^i$, $\Phi_k(i,j) = \mathtt{T}_s \bar{\alpha}^j {B_\delta^i}^T P_{k+1}^i B_\delta^j$, $\Pi_k(i,i) = {B_\delta^i}^T P_{k+1}^i (\mathtt{T}_s A_\delta + I)$, $\Pi_k(i,j) = 0$, $\bar{L}_k = \left[L_k^{1T}, L_k^{2T}, \cdots L_k^{ST}\right]^T$.

Substituting ${u_k^i}^* = -L_k^i x_k$ into (2.13), we have

$$\begin{aligned}
-\delta P_k^i &= Q_\delta^i + \bar{\alpha}^i {L_k^i}^T R_\delta^i L_k^i + \mathtt{T}_s \Theta_1^T P_{k+1}^i \Theta_1 + \Theta_1^T P_{k+1}^i + P_{k+1}^i \Theta_1 \\
&\quad + \mathtt{T}_s \sum_{j=1}^{S} \left(\left(\bar{\alpha}^j - {\bar{\alpha}^j}^2\right) {L_k^j}^T {B_\delta^j}^T P_{k+1}^i B_\delta^j L_k^j\right), \qquad (2.17)
\end{aligned}$$

and $P_k^i = P_{k+1}^i - \mathtt{T}_s \delta P_k^i$. The backward induction is carried out and the NE value will be obtained if $k = 0$. This completes the proof.

Towards this end, the following theorem is given to provide solutions to Problem 2.4 with a finite time horizon.

Theorem 2.13. *First, let us define*

$$\begin{aligned}
\tilde{R}_\delta &:= diag\left(\left[\bar{\alpha}^1\eta^1 R_\delta^1, \bar{\alpha}^2\eta^2 R_\delta^2, \cdots, \bar{\alpha}^S\eta^S R_\delta^S\right]\right)\\
\tilde{B}_\delta &:= \left[\bar{\alpha}^1 B_\delta^1, \bar{\alpha}^2 B_\delta^2, \cdots, \bar{\alpha}^S B_\delta^S\right]\\
\Xi &:= \tilde{R}_\delta + diag\left(\left[T_s\bar{\alpha}^1(1-\bar{\alpha}^1){B_\delta^1}^T\tilde{P}_{k+1}B_\delta^1,\right.\right.\\
&\quad T_s\bar{\alpha}^2(1-\bar{\alpha}^2){B_\delta^2}^T\tilde{P}_{k+1}B_\delta^2, \cdots, T_s\bar{\alpha}^S(1-\bar{\alpha}^S)\\
&\quad \left.\left.\times {B_\delta^S}^T\tilde{P}_{k+1}B_\delta^S\right]\right) + T_s\tilde{B}_k^T\tilde{P}_{k+1}\tilde{B}_k.
\end{aligned}$$

For finite time horizon CTOC, given attack strategy $\mathcal{G}$, the following conclusions are presented

1. *If we have invertible matrix $\Xi > 0$, there exists the unique optimal solution.*
2. *Denote $\tilde{u}_k = [u_k^1, u_k^2, \cdots, u_k^S]^T$. Under condition 1, the optimal control strategy of CTOC $\tilde{\mu}^*$ is given by $\tilde{u}_k = \tilde{\mu}^*(\mathcal{I}_k) = -\tilde{L}_k x_k$, where $\tilde{L}_k = \Xi^{-1}\tilde{B}_\delta^T\tilde{P}_{k+1}\Omega_1$.*
3. *The backward recursions are carried out with $\tilde{P}_K = \sum\limits_{i=1}^{S}\eta^i Q_\delta^K$ and*

$$\begin{aligned}
-\delta\tilde{P}_k &= \Omega_0 + T_s A_\delta^T\tilde{P}_{k+1}A_\delta + \tilde{P}_{k+1}A_\delta\\
&\quad + A_\delta^T\tilde{P}_{k+1} - \Omega_1^T\tilde{P}_{k+1}\tilde{B}_\delta\Xi^{-1}\tilde{B}_\delta^T\tilde{P}_{k+1}\Omega_1, \qquad (2.18)\\
\tilde{P}_k &= \tilde{P}_{k+1} - T_s\delta\tilde{P}_k. \qquad (2.19)
\end{aligned}$$

4. *Under condition 1, the optimal value with CTOC is $\tilde{J}^* = x_0^T\tilde{P}_0 x_0$, where x_0 is the initial value.*

Proof. An induction method is employed in the proof. The claim in Theorem 2.13 is clearly true for $k = K$ with $\tilde{P}_K = \sum_{i=1}^{S}\eta^i Q_\delta^K$. Suppose that the claim is true for $k+1$.

$$\tilde{V}(x_{k+1}) = \mathbb{E}\{x_{k+1}^T\tilde{P}_{k+1}x_{k+1}\}. \qquad (2.20)$$

With the definition of delta operator and Lemma 2.17, the above equation can be rewritten in the delta domain as

$$\begin{aligned}
\tilde{V}(x_{k+1}) &= \mathrm{T}_s\delta(x_k^T\tilde{P}_{k+1}x_k) + x_k^T\tilde{P}_{k+1}x_k\\
&= \mathrm{T}_s\delta x_k^T\tilde{P}_{k+1}x_k + \mathrm{T}_s x_k^T\tilde{P}_{k+1}\delta x_k\\
&\quad + \mathrm{T}_s^2\delta x_k\tilde{P}_{k+1}\delta x_k + x_k^T\tilde{P}_{k+1}x_k.
\end{aligned}$$

According to the principle of dynamic programming, the cost at time k is then

$$\begin{aligned}\tilde{V}(x_k) &= \min_{\tilde{u}_k} \mathbb{E}\left\{\mathtt{T}_s x_k^T \Omega_0 x_k + \mathtt{T}_s \tilde{u}_k^T \tilde{R}_\delta \tilde{u}_k + \tilde{V}(x_{k+1})\right\} \\ &= \min_{\tilde{u}_k} \mathbb{E}\left\{\mathtt{T}_s x_k^T \Omega_0 x_k + \mathtt{T}_s \tilde{u}_k^T \tilde{R}_\delta \tilde{u}_k + \mathtt{T}_s \delta x_k^T \tilde{P}_{k+1} x_k \right. \\ &\quad \left. + \mathtt{T}_s x_k^T \tilde{P}_{k+1} \delta x_k + \mathtt{T}_s^2 \delta x_k \tilde{P}_{k+1} \delta x_k + x_k^T \tilde{P}_{k+1} x_k\right\} \\ &= \min_{\tilde{u}_k} \mathbb{E}\left\{\mathtt{T}_s x_k^T \Omega_0 x_k + \mathtt{T}_s \tilde{u}_k^T \tilde{R}_\delta \tilde{u}_k + \Theta_2^T \tilde{P}_{k+1} \Theta_2\right\}. \end{aligned} \tag{2.21}$$

For further derivation, we present the equivalent form of $\tilde{V}(x_k)$ as follows

$$\begin{aligned}\tilde{V}(x_k) = \min_{\tilde{u}_k} &\left\{\mathtt{T}_s x_k^T \Omega_0 x_k + \mathtt{T}_s \tilde{u}_k^T \tilde{R}_\delta \tilde{u}_k + x_k^T \Omega_1^T \tilde{P}_{k+1} \Omega_1 x_k + \mathtt{T}_s x_k^T \Omega_1^T \tilde{P}_{k+1} \right. \\ &\left. \times \tilde{B}_\delta \tilde{u}_k + \mathtt{T}_s \tilde{u}_k^T \tilde{B}_\delta^T \tilde{P}_{k+1} \Omega_1 x_k + \mathtt{T}_s^2 \tilde{u}_k^T \left(\tilde{B}_\delta^T \tilde{P}_{k+1} \tilde{B}_\delta + \Omega_2\right) \tilde{u}_k\right\}. \end{aligned} \tag{2.22}$$

The cost-to-go function $\tilde{V}(x_k)$ is strictly convex with respect to $\tilde{u}_k$, since the second derivative of the above equation yields $\Xi_k > 0$. The minimizer is obtained by $\partial\tilde{V}(x_k)/\partial\tilde{u}_k = 0$,

$$\begin{aligned}\partial\tilde{V}(x_k)/\partial\tilde{u}_k = &\tilde{R}_\delta \tilde{u}_k + \tilde{u}_k^T \tilde{R}_\delta + \tilde{B}_\delta^T \tilde{P}_{k+1} \Omega_1 x_k + x_k^T \Omega_1^T \tilde{P}_{k+1} \tilde{B}_\delta \\ &+ \mathtt{T}_s \tilde{B}_\delta^T \tilde{P}_{k+1} \tilde{B}_\delta \tilde{u}_k + \mathtt{T}_s \tilde{u}_k^T \tilde{B}_\delta^T \tilde{P}_{k+1} \tilde{B}_\delta \\ &+ \mathtt{T}_s \Omega_2^T \tilde{u}_k + \mathtt{T}_s \tilde{u}_k^T \Omega_2 = 0. \end{aligned}$$

Then the optimal control strategy is obtained as $\tilde{u}_k = -\tilde{L}_k x_k$, with

$$\tilde{L}_k = \Xi^{-1} \tilde{B}_\delta^T \tilde{P}_{k+1} \Omega_1,$$

from which we know the matrix Ξ must be invertible to guarantee the existence of a unique optimal solution. Substituting $\tilde{u}_k = -\tilde{L}_k x_k$ back into (2.21), we have

$$\begin{aligned}-\delta\tilde{P}_k = &\Omega_0 + \mathtt{T}_s A_\delta^T \tilde{P}_{k+1} A_\delta + \tilde{P}_{k+1} A_\delta \\ &+ A_\delta^T \tilde{P}_{k+1} - \Omega_1^T \tilde{P}_{k+1} \tilde{B}_\delta \Xi^{-1} \tilde{B}_\delta^T \tilde{P}_{k+1} \Omega_1, \end{aligned} \tag{2.23}$$

and $\tilde{P}_k = \tilde{P}_{k+1} - \mathtt{T}_s \delta \tilde{P}_k$. The backward induction is carried out and the optimal value will be obtained if $k = 0$. The proof is completed.

Remark 2.14. The results in Theorem 2.12 and 2.13 can be directly extended to the time-variant case where state matrices are $A_{\delta k}$ and $B_{\delta k}^i$, $i \in \mathbf{S}$, $k \in \mathbf{K}$.

Remark 2.15. Note that Theorem 2.12 and 2.13 develop control strategies using game theory and dynamic programming. It is different from most of the existing literatures on the control of delta operator system, which falls within the framework of Linear Matrix Inequalities (LMIs) [49, 158, 157]. The delta domain results in this chapter is different from the aforementioned literatures mainly for two reasons: (1) By using dynamic programming, the optimal solution can be obtained instead of suboptimal results obtained by using LMI. (2) As mentioned in Remark 2.14, our results can be used to develop time-variant optimal control strategies for time-variant system. The LMI method normally obtains time-invariant control strategies.

Remark 2.16. The delta domain results in both Theorem 2.12 and 2.13 are seen as a unified form of the results in both discrete and continuous domain. We take the results in Theorem 2.12 for example here. The transformation of the delta domain results in Theorem 2.12 to the discrete domain is very simple by just assigning the sampling interval $\mathtt{T}_s = 1$ in the cost-to-go function and dynamic programming recursion. The equivalence of the results in Theorem 2.12 with $\mathtt{T}_s \to 0$ to continuous time results are verified by the following arguments:

Without loss of generality, we assume that $\alpha^i = 0,\ i \in \mathbf{S}$. When $\mathtt{T}_s \to 0$, we have $A_\delta \to A_s$, $B^i_\delta \to B^i_s$. The feedback strategies (2.9) are

$$L^i_t = R^{-1}_\delta B^{iT}_\delta P^i_t. \tag{2.24}$$

From Remark 2.2, we have

$$\lim_{\mathtt{T}_s \to 0} \delta P^i_k = \dot{P}^i_t. \tag{2.25}$$

By using (2.24) and (2.25), when $\mathtt{T}_s \to 0$ the backward iteration (2.10) can be rewritten as

$$\begin{aligned}
0 = Q^i_\delta + P^i_t B^i_\delta {R^i_\delta}^{-1} B^{iT}_\delta P^i_t + \dot{P}^i_t + P^i_t A_\delta - P^i_t \left\{ \sum_{i=1}^{S} B^i_\delta {R^i_\delta}^{-1} B^{iT}_\delta P^i_t \right\} \\
+ A^T_\delta P^i_t - \left\{ \sum_{i=1}^{S} B^i_\delta {R^i_\delta}^{-1} B^{iT}_\delta P^i_t \right\}^T P^i_t.
\end{aligned} \tag{2.26}$$

Note that equations (2.24) and (2.26) are consistent with feedback strategies and Riccati recursions in continuous domain [16]. Thus, it is evident Theorem 2.12 and 2.13 unify the results in both discrete and continuous domain.

2.3.2 Infinite Time Horizon Case

For the infinite time horizon case, the cost-to-go function for MTOC and CTOC yield (2.5) and (2.6) with $K \to \infty$, respectively. With the finite time horizon results shown in Theorem 2.12 and 2.13, the infinite-horizon case can be solved by taking the horizon length $K \to \infty$. However, this requires that as $K \to \infty$, the matrix P^i_0, $i \in \mathbf{S}$ (resp. $\tilde{P}_0$) solved by the backward iteration of (2.10) (resp. (2.18)) converges to a matrix P^i_∞ (resp. $\tilde{P}_\infty$). Before providing the convergent condition of P^i_k, $i \in \mathbf{S}$ and $\tilde{P}_k$, the following instrumental lemma is provided first.

Lemma 2.17. *[64] Let $(A, Q^{1/2})$ be observable and $\bar{\alpha}$ be the packet delivery rate. Suppose that B is invertible. Then, Φ_k satisfying the following recursion is convergent, i.e. $\lim_{k\to\infty} \Phi_k = \Phi_\infty$, for every $\Phi_0 > 0$, if and only if $\bar{\alpha} > 1 - 1/\rho_m^2$, where $\rho_m = \max_l |\lambda_l(A)|$.*

$$\Phi_{k+1} = A^T \Phi_k A + Q - \bar{\alpha} A^T \Phi_k B (R + B^T \Phi_k B)^{-1} B^T \Phi_k A. \tag{2.27}$$

Now we are in a position to provide the convergent condition for P_k^i and provide the proof for stability.

Theorem 2.18. *We denote*

$$\lim_{k\to\infty} L_k^i = L_\infty^i$$
$$\bar{F}^i := T_s A_\delta + I - T_s \sum_{j\neq i}^{S} B_\delta^j L_\infty^j.$$

Let $(\bar{F}^i, (T_s Q_\delta^i)^{1/2})$ be observable and B_δ^i be invertible. Given $\{L_\infty^j\}_{j=1, j\neq i}^S$, the following closed-loop system is stable in the mean-square (m.s.) sense if we have $\bar{\alpha}^i > 1 - 1/\rho_m^2$, where $\rho_m = \max_l \left|\lambda_l(\bar{F}^i)\right|$.

$$\delta x_k = (A_\delta - \sum_{j\neq i}^{S} B_\delta^j L_\infty^j - \alpha_k^i B_\delta^i L_\infty^i) x_k \tag{2.28}$$

Proof. The backward recursions (2.10) is rewritten as

$$\begin{aligned} P_k^i &= \bar{F}^{iT} P_{k+1}^i \bar{F}^i + \mathrm{T}_s Q_\delta^i \\ &\quad - \bar{\alpha}^i \mathrm{T}_s \bar{F}^{iT} P_{k+1}^i B_\delta^i (R_\delta^i + \mathrm{T}_s B_\delta^{iT} P_{k+1}^i B_\delta^i)^{-1} B_\delta^{iT} P_{k+1}^i \bar{F}^i. \end{aligned} \tag{2.29}$$

Let us denote $\hat{B}_\delta^i = \sqrt{\mathrm{T}_s} B_\delta^i$ and reverse the time index. Then we have

$$P_{\tau+1}^i = \bar{F}^{iT} P_\tau^i \bar{F}^i + \mathrm{T}_s Q_\delta^i - \bar{\alpha}^i \bar{F}^{iT} P_\tau^i \hat{B}_\delta^i (R_\delta^i + \hat{B}_\delta^{iT} P_\tau^i \hat{B}_\delta^i)^{-1} \hat{B}_\delta^{iT} P_\tau^i \bar{F}^i.$$

From Lemma 2.17, it can be deduced that $\lim_{\tau\to\infty} P_\tau^i = P_\infty^i$ if $\bar{\alpha} > 1 - 1/\rho_m^2$, where $\rho_m = \max_l |\lambda_l(A)|$. Thus, we can rewrite (2.10) as

$$\begin{aligned} P_\infty^i &= \mathrm{T}_s Q_\delta^i + \mathrm{T}_s \bar{\alpha}^i L_\infty^{iT} B_\delta^{iT} R_\delta^i B_\delta^i L_\infty^i \\ &\quad + \bar{\alpha}^i (\bar{F}^i - \mathrm{T}_s B_\delta^i L_\infty^i)^T P_\infty^i (\bar{F}^i - \mathrm{T}_s B_\delta^i L_\infty^i) + \alpha^i \bar{F}^{iT} P_\infty^i \bar{F}^i \end{aligned} \tag{2.30}$$

We will show system (2.28) is stable in the m.s. sense. By using equation (2.2) and (2.30), we have

$$\begin{aligned}
&\mathbb{E}\{x_{k+1}^T P_\infty^i x_{k+1} - x_k^T P_\infty^i x_k\} \\
&= \mathbb{E}\{x_k^T(\alpha^i \bar{F}^{iT} P_\infty^i \bar{F}^i - P_\infty^i)x_k \\
&\quad + \bar{\alpha}^i x_k^T(\bar{F}^i - T_s B_\delta^i L_\infty^i)^T P_\infty^i (\bar{F}^i - T_s B_\delta^i L_\infty^i)x_k\} \\
&= -\mathbb{E}\{x_k^{\mathrm{T}}(T_s Q_\delta^i + \mathrm{T}_s \bar{\alpha}^i L_\infty^{iT} R_\delta^i L_\infty^i)x_k\}.
\end{aligned} \tag{2.31}$$

Thereby, calculating equation (2.31) for all k, we have

$$\begin{aligned}
&\mathbb{E}\{x_{k+1}^T P_\infty^i x_{k+1}\} \\
&= \mathbb{E}\{x_0^T P_\infty^i x_0\} - \sum_{l=0}^{k} \mathbb{E}\{x_l^T(T_s Q_\delta^i + \mathrm{T}_s \bar{\alpha}^i L_\infty^{iT} R_\delta^i L_\infty^i)x_l\}.
\end{aligned} \tag{2.32}$$

It is concluded that $\lim_{k\to\infty} \mathbb{E}\{x_k^{\mathrm{T}}(\mathrm{T}_s Q_\delta^i + \mathrm{T}_s \bar{\alpha}^i L_\infty^{iT} R_\delta^i L_\infty^i)x_k\} = 0$. Since $R_\delta^i > 0$ and $Q_\delta^i > 0$, we have $\mathbb{E}\{\|x_k\|^2\} \to 0$ as $k \to \infty$. Therefore, system (2.28) is stable. This completes the proof.

Theorem 2.19. *Let $(\Omega_1, (T_s\Omega_0)^{1/2})$ be observable. Suppose we have $\tilde{\rho}_m = \max_l |\lambda_l(\Omega_1)|$ and $\hat{B}_{\delta 1}$ is invertible. If $\min_i(\{\bar{\alpha}^i\}_{i=1}^S) > 1 - 1/\tilde{\rho}_m^2$, and denote $\lim_{k\to\infty} \tilde{L}_k = \tilde{L}_\infty$, then the following closed-loop system is stable in the m.s. sense.*

$$\delta x_k = (A_\delta - \boldsymbol{a}_k \ddot{B}_\delta \tilde{L}_\infty)x_k, \tag{2.33}$$

where

$$\begin{aligned}
\boldsymbol{a}_k &= diag\{\alpha_k^1, \cdots, \alpha_k^S\}, \hat{B}_\delta = [B_\delta^1,\ B_\delta^2, \cdots, B_\delta^S], \\
\hat{B}_{\delta 1} &= [\sqrt{T_s} B_\delta^1, \sqrt{T_s} B_\delta^2, \cdots, \sqrt{T_s} B_\delta^S].
\end{aligned}$$

Proof. Let us denote

$$\bar{\boldsymbol{a}} := diag\{\bar{\alpha}^1, \cdots, \bar{\alpha}^S\}, \boldsymbol{a} := diag\{\alpha^1, \cdots, \alpha^S\}, \boldsymbol{R}_\delta := diag\{\eta^1 R_\delta^1, \cdots, \eta^S R_\delta^S\}.$$

The matrix Ξ^{-1} is rewritten as

$$\Xi^{-1} = \{\boldsymbol{R}_\delta + T_s \hat{B}_\delta^T \tilde{P}_{k+1} \hat{B}_\delta\}^{-1} \bar{\boldsymbol{a}}^{-1}.$$

Thus, the feedback gain $\tilde{L}_k$ yields

$$\tilde{L}_k = \{\boldsymbol{R}_\delta + T_s \hat{B}_\delta^T \tilde{P}_{k+1} \hat{B}_\delta\}^{-1} \hat{B}_\delta^T \tilde{P}_{k+1} \Omega_1. \tag{2.34}$$

By substituting $\hat{B}_{\delta 1}$ and reversing the time index, the Riccati recursion (2.18) is rewritten as

$$\tilde{P}_{\tau+1} = T_s\Omega_0 + \Omega_1^T \tilde{P}_\tau \Omega_1 - \bar{\boldsymbol{a}} \Omega_1^T \tilde{P}_\tau \hat{B}_{\delta 1} \{\boldsymbol{R}_\delta + \hat{B}_{\delta 1}^T \tilde{P}_\tau \hat{B}_{\delta 1}\}^{-1} \hat{B}_{\delta 1}^T \tilde{P}_\tau \Omega_1.$$

From Lemma 2.17, the Riccati equation converges, i.e., $\lim_{\mathcal{T}\to\infty} \tilde{P}_{\mathcal{T}} = \tilde{P}_\infty$. *Thereby, by using the dynamic programming equation we have the following equality*

$$\begin{aligned}\tilde{P}_\infty &= T_s\Omega_0 + T_s\bar{\boldsymbol{a}}\tilde{L}_\infty^T \boldsymbol{R}_\delta \tilde{L}_\infty + \boldsymbol{a}\Omega_1^T \tilde{P}_\infty \Omega_1 \\ &\quad + \bar{\boldsymbol{a}}(\Omega_1 - T_s\hat{B}_\delta\tilde{L}_\infty)^T \tilde{P}_\infty(\Omega_1 - T_s\hat{B}_\delta\tilde{L}_\infty). \end{aligned} \tag{2.35}$$

We will show system (2.33) is stable in the m.s. sense. With equation (2.2) and (2.35), we have

$$\begin{aligned}&\mathbb{E}\{x_{k+1}^T\tilde{P}_\infty x_{k+1} - x_k^T\tilde{P}_\infty x_k\} \\ &= \mathbb{E}\{x_k^T(\boldsymbol{a}\Omega_1^T\tilde{P}_\infty\Omega_1 - \tilde{P}_\infty)x_k + \bar{\boldsymbol{a}}x_k^T(\Omega_1 - T_s\hat{B}_\delta\tilde{L}_\infty)^T\tilde{P}_\infty(\Omega_1 - T_s\hat{B}_\delta\tilde{L}_\infty)x_k\} \\ &= -\mathbb{E}\{x_k^T(T_s\Omega_0 + T_s\bar{\boldsymbol{a}}\tilde{L}_\infty^T\boldsymbol{R}_\delta\tilde{L}_\infty)x_k\}. \end{aligned} \tag{2.36}$$

Hence,

$$\mathbb{E}\{x_{k+1}^T\tilde{P}_\infty x_{k+1}\} \tag{2.37}$$

$$= \mathbb{E}\{x_0^T\tilde{P}_\infty x_0\} - \sum_{l=1}^{k}\mathbb{E}\{x_l^T(T_s\Omega_0 + T_s\bar{\boldsymbol{a}}\tilde{L}_\infty^T\boldsymbol{R}_\delta\tilde{L}_\infty)x_l\}. \tag{2.38}$$

It is concluded that

$$\lim_{k\to\infty}\mathbb{E}\{x_k^T(T_s\Omega_0 + T_s\bar{\boldsymbol{a}}\tilde{L}_\infty^T\boldsymbol{R}_\delta\tilde{L}_\infty)x_k\} = 0.$$

Since $\Omega_0 > 0$ *and* $\boldsymbol{R}_\delta > 0$, *we have* $\mathbb{E}\{\|x_k\|^2\} \to 0$ *as* $k \to \infty$. *System (2.33) of CTOC is stable. The proof is completed.*

Remark 2.20. From Theorem 2.18 and 2.19, we can see that, the attacker must choose an IoA bigger than $1/\rho^2$ or $1/\tilde{\rho}^2$ to destabilize the corresponding closed-loop system. In the following, a more general case will be considered, where the attacker aims to drive the control system out of the safety zone (not necessarily the stability region).

2.4 Defense and Attack Strategies

In this section, an intelligent but resource-limited attacker is considered. The attacker is intelligent by being able to capture the defence strategies adopted by the defender. Thus, the attacker and defender possess asymmetric information and the interactions can be described by a Stackelberg game [171], where the attacker acts as the leader and the defender acts as the follower. The defender aims to minimize the performance degradation caused by the attacker. Being fully aware of the defender's strategies, the attacker aims to drive the control system out of the safety zone while minimizing the attacking intensity. The details of the Stackelberg game and the derivation of the corresponding Stackelberg solutions are shown sequentially.

2.4.1 Development of Defense Strategies

According to the previous section, NCS with CTOC structure under no DoS attack is viewed as a nominal case. Thus, the desirable cost-to-go function yields $\tilde{J}^*_{\mathcal{G}_0}$ and the corresponding desirable feedback gains are $\{\tilde{L}^i_k\}_{i=1}^S$. The algorithm to achieve both desirable cost-to-go function and feedback gains is provided in the following theorem.

Theorem 2.21. *For given $\mathcal{G} = \{\alpha^i\}_{i=1}^S$ and desired strategies $\{\tilde{L}^i_k\}_{i=1}^S$, if $\{Q^i_\delta\}_{i=1}^S$ and $\{P^i_0, P^i_1, \cdots, P^i_{K-1}\}_{i=1}^S$ exists such that the following convex problem is feasible for all $i \in \boldsymbol{S}$, then $\{u^{i*}_k = \tilde{L}^i_k x_k\}$ is the NE of OC(i), thereby achieving the defender's goal.*

$$\begin{cases} Q^i_\delta > 0 \\ (R^i_\delta + T_s B^{iT}_\delta P^i_{k+1} B^i_\delta)\tilde{L}^i_k - B^{iT}_\delta P^i_{k+1} \Upsilon_1 = 0 \\ T_s Q^i_\delta + T_s \bar{\alpha}^i \tilde{L}^{iT}_k R^i_\delta \tilde{L}^i_k + \Upsilon_2^T P^i_{k+1} \Upsilon_2 - P^i_k \\ + T_s^2 \sum_{j=1}^S ((\bar{\alpha}^j - \bar{\alpha}^{j2}) \tilde{L}^{jT}_k B^{jT}_\delta P^i_{k+1} B^j_\delta \tilde{L}^j_k) = 0 \end{cases} \tag{2.39}$$

where

$$\Upsilon_1 = (T_s A_\delta + I) - T_s \sum_{j \neq i}^S \bar{\alpha}^j B^j_\delta \tilde{L}^j_k,$$

$$\Upsilon_2 = (T_s A_\delta + I) - T_s \sum_{i=1}^S \bar{\alpha}^i B^i_\delta \tilde{L}^i_k.$$

Proof. The result is easily obtained by using Theorem 2.12 and the proof process is omitted here.

Problem 2.10 is reformulated as the following convex optimization problem

$$\begin{aligned} &\min \quad |\text{OPD} - 1| \\ &s.t. \; \left(\{Q^i_\delta\}_{i=1}^S, \;\; \{P^i_0, P^i_1, \cdots, P^i_{K-1}\}_{i=1}^S\right) \in \mathcal{C} \\ &\mathcal{C} := \left\{ \left(\{Q^i_\delta\}_{i=1}^S, \;\; \{P^i_0, P^i_1, \cdots, P^i_{K-1}\}_{i=1}^S\right) : (2.39) \text{ hold} \right\}. \end{aligned} \tag{2.40}$$

The convex optimization problem is solved by Matlab software SDPT3 [130], which is an effective Matlab software package for semidefinite programming.

Remark 2.22. The defence algorithm in this subsection is actually the pricing mechanism design [5, 114] in game theory. By finding appropriate pricing parameters, the final outcome of the game can be driven to a desired target. The weighting matrices $Q^i_\delta, i \in \mathbf{S}$ are the so-called pricing parameters here, and the cost function of each player can be compensated if $Q^i_\delta, i \in \mathbf{S}$ is appropriately tuned. On the other hand, the proposed pricing mechanism in this chapter can be regarded as an extension to the inverse LQR method [19, 27], where the desired control strategies are known in advance and the parameter Q or R is to be determined.

2.4.2 Development of Attack Strategies

It should be noticed that the traditional work on NCS normally regards the packets dropout as a constraint [17] and seldom studies how to enhance the packets dropout rate to degrade the system performance. However, in this subsection, a strategic and resource-limited attacker is considered, and furthermore, the corresponding attacking strategy is provided. This actually provides a worst case for the defender and is helpful in the defense mechanism design.

The IoA α^i, $i \in \mathbf{S}$ is a scalar located in a finite interval $[0,1]$. Algorithm 1 is a standard dichotomy algorithm that can provide the optimal attacking strategy $\mathcal{G}^*$ such that the underlying NCS is driven out of the safety zone with smallest α^i, $i \in \mathbf{S}$.

Algorithm 1 The algorithm for optimal attacking strategy $\mathcal{G}^*$

Require: Set $a = 1$; $b = 0$; $DIST = 0$, $0 < \varepsilon \ll 1$.
1: Denote $\alpha = \bar{\alpha}^1 = \bar{\alpha}^2 = \cdots = \bar{\alpha}^S$, and $\alpha = (a+b)/2$.
2: **while** $|b - a| > \varepsilon$ **do**
3: Calculate the optimal defense strategy $(\{Q^{i*}_\delta\}_{i=1}^S)$ by solving 2.40.
4: Calculate $|\text{OPD}^* - 1|$ using Q^{i*}_δ, and compare it with $\mathcal{S}_0$.
5: **if** $|\text{OPD}^* - 1| < \mathcal{S}_0$ **then**
6: $a = \alpha$, $b = b$
7: **else if** $|\text{OPD}^* - 1| > \mathcal{S}_0$ **then**
8: $a = a$, $b = \alpha$
9: **end if**
10: Set $\alpha = (a+b)/2$.
11: **end while**

Remark 2.23. The condition for existence of solutions of Algorithm 1 is that the open-loop system $\delta x_k = A_\delta x_k$ should be unstable. The reasons are that the closed-loop system will be driven to instability again if all the control commands are lost, i.e. $\alpha^1 = \alpha^2 = \cdots = \alpha^S = 1$. Then, we will have OPD$^* \to \infty$ and the constraint (2.11) must be feasible.

Remark 2.24. It can be deduced that the time complexity of the proposed dichotomy algorithm is $O(\log_2 n)$, where $n = 1/\varepsilon$, while the time complexity of the exhaustive method is $O(n)$. Thus, the proposed attacking algorithm saves more time for the attackers to find out the attacking strategy online. For example, if the safety zone $\mathcal{S}_o$ for the control system changes, the attacker can quickly adapt to such change and find out the corresponding attacking strategy.

2.5 Numerical Simulation

In this section, the proposed methodology is applied to the thermodynamic model of a building and simulation is provided to verify the validity.

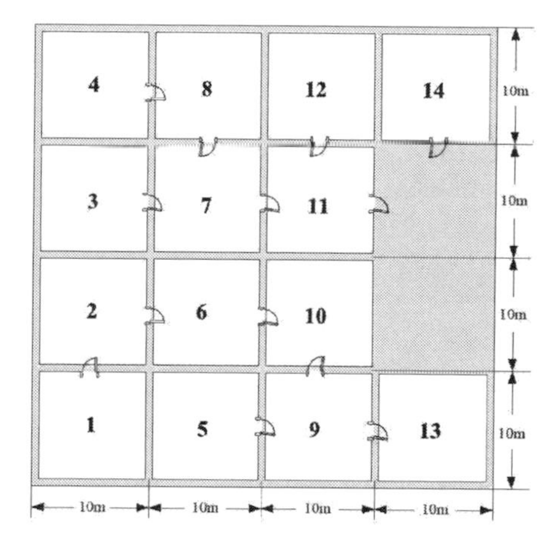

Figure 2.4 Floor plant for building example.

2.5.1 Building Model Description

If we consider transient conduction and convection as well the flow of air injected into each zone by the HVAC system as shown in Figure 2.4, the temperature in zone i, denoted as $\mathcal{T}_i$, evolves according to the following dynamics

$$\rho v_i C_p \frac{d\mathcal{T}_i}{dt} = \sum_{j \neq i}^{S} h_{i,j} a_{i,j} (\mathcal{T}_j - \mathcal{T}_i) + T_0,$$

with $T_0 = h_{i,o} a_{i,o} (\mathcal{T}_\infty - \mathcal{T}_i) + \dot{m}_i C_p (\mathcal{T}_i^{sup} - \mathcal{T}_i)$, and where ρ is the density of air, C_p is the specific heat of air and $\mathcal{T}_\infty$ is the outside air temperature. Parameter v_i is the volume of air in the ith zone, $a_{i,j}$ is the area of the wall between zone j and i, $a_{i,o}$ is the total area of the exterior walls and roof of zone i, $h_{i,j}$ and $h_{i,o}$ are the heat transfer coefficients of the wall between zone j and i and the heat coefficient of the exterior walls (which is determined by the material properties), respectively, $\dot{m}_i$ is the mass flow rate of air into zone i, and $\mathcal{T}_i^{sup}$ is the supply air temperature for zone i. The parameter values are shown in Table 2.1 [29], where $h_{i,j} = h_{i,o}$ and $a_{i,j} = a_{i,o}$. $\dot{V}$ is the volumetric flow and the value corresponds to 500 CFM.

Table 2.1 Parameters for Numerical Simulations

Parameter	Value
$\dot{V}$	$13.8\ \mathrm{m^3 \cdot s^{-1}}$
ρ	$1.2\ \mathrm{kg \cdot m^{-3}}$
C_p	$1000\ \mathrm{W \cdot s \cdot kg^{-1} \cdot K^{-1}}$
T_∞	$18\,^\circ\mathrm{C}$
$\dot{m}$	$1.035\mathrm{kg \cdot s^{-1}}$
v_i	$1000\ \mathrm{m^3}$
$a_{i,o}$	$100\ \mathrm{m^2}$
$h_{i,o}$	$8.9\mathrm{W \cdot m^{-2} \cdot K^{-1}}$

Note that $\mathcal{T}_{di}$ is the set temperature of zone i. Then, temperature error of zone i is denoted as $\mathcal{T}_{ei} = \mathcal{T}_i - \mathcal{T}_{di}$. Define $x = [x^1, x^2, \cdots, x^S]^T \triangleq [\mathcal{T}_{e1}, \mathcal{T}_{e2}, \cdots, \mathcal{T}_{eS}]^T$ to be the vector of zone temperatures and $\tilde{u} = [u^1, u^2, \cdots, u^S]^T \triangleq [\mathcal{T}_1^{sup}, \mathcal{T}_2^{sup}, \cdots, \mathcal{T}_S^{sup}]^T$ to be the control input, i.e., vector of supply air temperatures. Then the system dynamics are

$$\dot{x}_t = A_s x_t + \sum_{i=1}^{S} B_s^i u_t^i + d_s,$$

where

$$d_{si} = \frac{h_{i,o} a_{i,o}}{\rho v_i C_p} (\mathcal{T}_\infty - \mathcal{T}_{di}) - \frac{\dot{m}}{\rho v_i} \mathcal{T}_{di} + \sum_{j=1, j \neq i}^{S} \mathrm{T}_j,$$

with $\mathtt{T}_j = \frac{h_{i,j}a_{i,j}}{\rho v_i C_p}(\mathcal{T}_{dj} - \mathcal{T}_{di})$. Matrices A_s and B_s are given as follows

$$A_s(i,j) = \begin{cases} -\left(\sum_{j\in\mathcal{N}_i} \frac{h_{i,j}a_{i,j}}{\rho v_i C_p} + \frac{\dot{m}_i}{\rho v_i} + \frac{h_{i,o}a_{i,o}}{\rho v_i C_p}\right), & i = j, \\ \frac{h_{i,j}a_{i,j}}{\rho v_i C_p}, & j \in \mathcal{N}_i \text{ and } j \neq i, \\ 0, & \text{otherwise}, \end{cases}$$

where $\mathcal{N}_i$ denotes the neighborhood of Player i. $B_s = [B_s^1, B_s^2, \cdots, B_s^S] \in \mathbb{R}^{S\times S}$ is an invertible diagonal matrix with the ith entry being $\frac{\dot{m}_i}{\rho v_i}$. Suppose $\hat{u}$ is composed of a feedback term u and a compensation term that is $\hat{u} = u - B_s^{-1}d_s$. Then we have $\dot{x} = A_s x + B_s(u - B_s^{-1}d_s) + d_s = A_s x + B_s u$, where $u = [u^{1T}, u^{2T}, \cdots, u^{ST}]^T$.

Consider the packets dropout and assign $\mathtt{T}_s = 0.05$s, the delta operator system is described as follows

$$\delta x_k = A_\delta x_k + \sum_{i=1}^{S} \alpha_k^i B_\delta^i u_k^i.$$

2.5.2 Strategy Design

Let us denote IoA $\alpha^i = 0.2$ and $R_\delta^i = I$ for all $i \in \mathbf{S}$. We assign Q_δ^i, $i \in \mathbf{S}$ and Q_δ^K of the desired cost-to-go function of CTOC and the MTOC before compensation as identity matrix. The resultant curves of MTOC with and without compensation are shown in Figure 2.5 and 2.6, respectively. It is evident that the control performance improves after the defense strategy is applied. Figure 2.7 shows the convergent curve of parameter $\bar{\alpha}^i$ using Algorithm 1.

Then, we stand at the side of the attacker and assign $\mathcal{S}_0 = 0.38$, $R_\delta^i = I$, and $\varepsilon = 1\times 10^{-6}$. Let us assign Q_δ^i, $i \in \mathbf{S}$ and Q_δ^K of the desired cost-to-go function of CTOC and the MTOC before compensation as identity matrix. By using Algorithm 1, we arrive at the results in Table 2.2, where we conclude that the optimal IoA for the attacker yields $\alpha^i = 0.4639$ if we have $\alpha^1 = \alpha^2 = \cdots = \alpha^S$.

Table 2.2 Optimal Values for Network Security

Parameter	Value
Cost of CTOC $\tilde{J}^*$	74.9575
Cost with Defense Strategy of MTOC J^*	103.4415
The Overall Performance Degradation OPD*	1.3800
The Optimal Attack Intense α^{i*}	0.4639

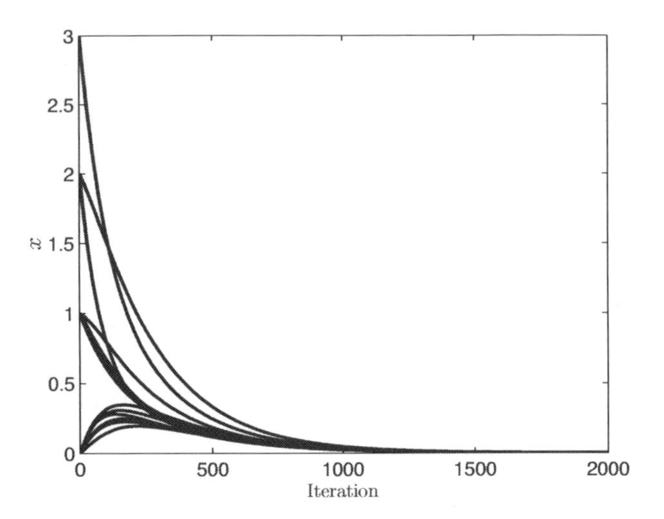

Figure 2.5 The resultant curves of MTOC without compensation.

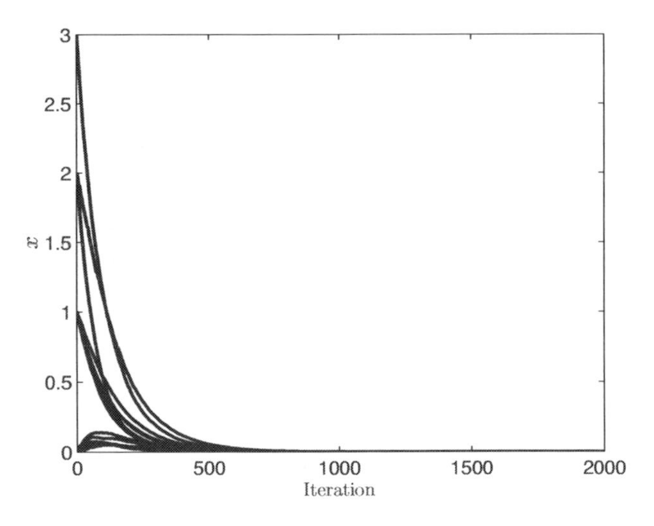

Figure 2.6 The resultant curves of MTOC with compensation.

2.5.3 Robust Study

In this subsection, the robustness of the proposed defense algorithm is verified. It should be noticed that the proposed defense algorithm requires that the defender should estimate the IoA of the attacker, which is not an easy task in practice. Therefore, it is necessary to test the robustness of the defense algorithm by checking whether it still works when the estimation of IoA is

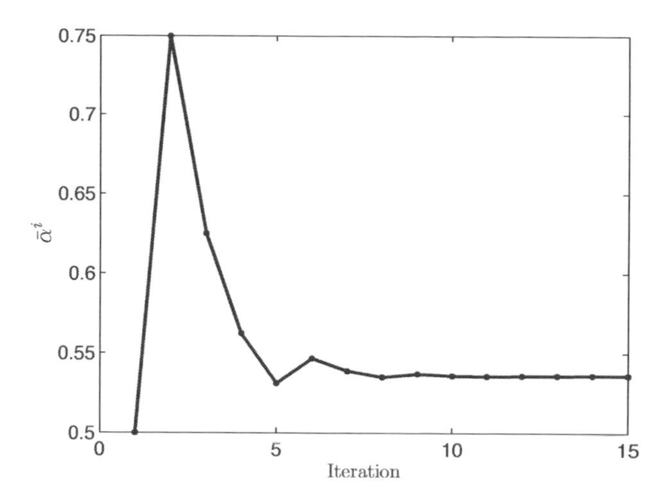

Figure 2.7 The convergent curve of parameter $\bar{\alpha}^i$ using Algorithm 1.

Table 2.3 Comparisons for MTOC System with Different Estimation Errors

IoA Dev	0	10%	20%	40%
J^*	117.1279	118.9997	120.9949	125.4236
J'^*	102.0808	102.0810	102.0832	102.0910

not accurate. As in Table 2.3, where 'IoA Dev' is denoted as the estimation error of IoA, 'J^*' is denoted as the original cost of MTOC system and 'J'^*' is denoted as the cost of MTOC system with defense strategies. We assume that the actual value of IoA is $\alpha^i = 0.2$, and the estimation error is described by a certain percentage of the actual value. From Table 2.3, we can see that the proposed defense mechanism still works even with the existence of the estimation error.

2.5.4 Comparative Study

In this chapter, the following four scenarios are also considered for comparison.

1. Optimal: The optimal attack and defense strategies are used as proposed in this chapter;
2. Random: Both the attacker and defender set their strategies randomly, regardless of the existence of the other;

3. Unaware: The attacker doesn't know the existence of defender and chooses IoA randomly, but the defender chooses optimal strategy Q^{i*}, $i \in \mathbf{S}$ according to α^i the attacker chooses.
4. Misjudge: The attacker believes that a defender exists, but the defense strategy is just picked randomly.

The comparison results are shown in Figure 2.8, where we can see that the cases of 'Optimal' and 'Unaware' are better than 'Random' and 'Misjudge'. Thus, the proposed defense strategy can improve the system performance no matter what strategies the attacker adopts. On the other hand, it should be noticed that the proposed attacking strategy is able to degrade the system performance more, especially when there are no defense mechanisms.

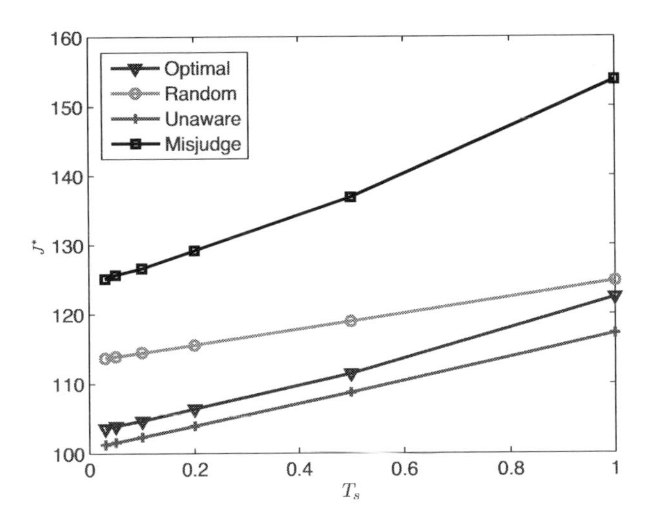

Figure 2.8 Performance comparison of different scenarios.

2.5.5 Experiment Verification

To illustrate the effectiveness and applicability, we use the proposed method of this chapter in a practical experiment. The platform is selected as the ball and beam system shown in Figure 2.9. From the physical structure of the system as in Figure 2.10, let $x = [\gamma \; \dot{\gamma} \; \theta \; \dot{\theta}]^T$, where γ and θ are the ball position and the beam angle, respectively. Then, the ball and beam system is expressed as

$$\dot{x}_t = A_s x_t + B_s u_t,$$

where

$$A_s = \begin{bmatrix} 0 & 1 & 0 & 0 \\ 0 & 0 & -7.007 & 0 \\ 0 & 0 & 0 & 1 \\ 0 & 0 & 0 & 0 \end{bmatrix}, \quad B_s = \begin{bmatrix} 0 \\ 0 \\ 0 \\ 1 \end{bmatrix}.$$

The initial state is chosen as $x_0 = [-0.2\ 0\ 0\ 0]$ and the sampling period is $\mathtt{T}_s = 0.02s$. According to Theorem 2.13, the control input is addressed as $u_k^* = [-9.2713\ \ -8.4746\ 27.1400\ 7.2887]$ when there is no packets dropout, i.e., $\alpha = 0$ and $u_k^* = [-9.3113\ \ -8.5235\ 27.3359\ 7.3519]$ when $\alpha = 0.01$. The goal of the design is to balance the ball in the center of the beam. From these experiment results, it is clear that the designed controllers can ensure the stability of the ball and beam system with and without packets dropout. What's more, worse system performance is resulted with packets dropout. Experiment result without packets dropout and packets dropout $\alpha = 0.01$ are as shown in Figure 2.11 and 2.12. It is worth mentioning that there are small steady-state errors in the experiment results, which is inevitable because of some nonlinearity such as dead zone and fabrication in the experimental process.

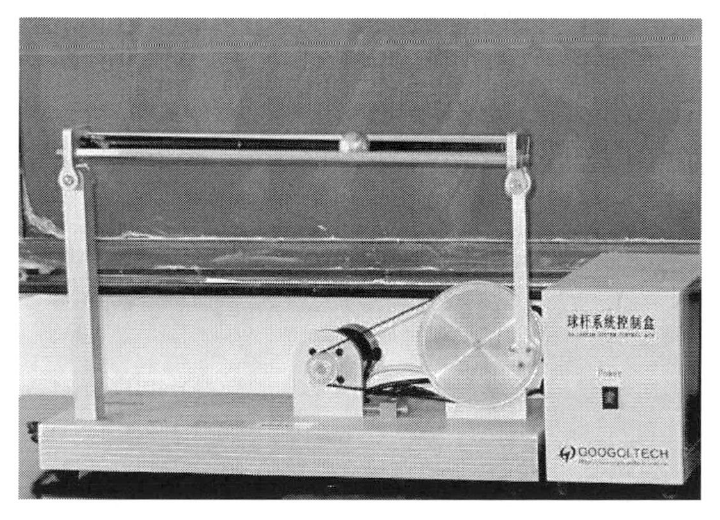

Figure 2.9 Ball and beam platform.

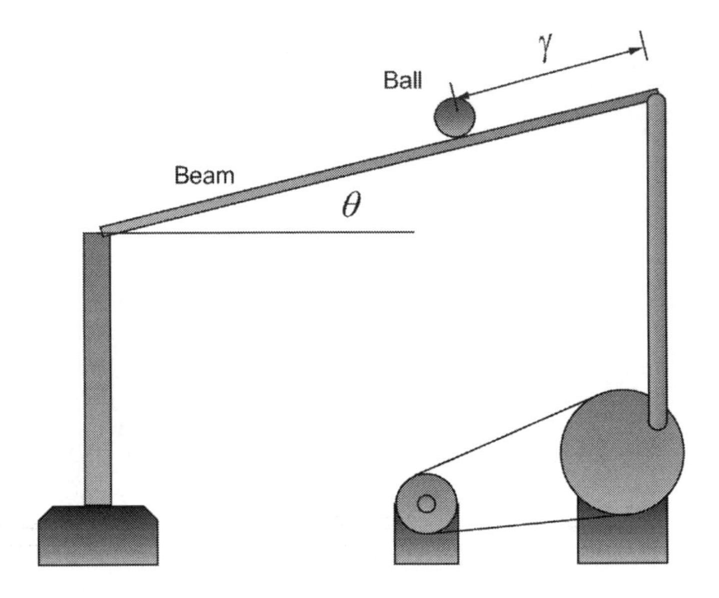

Figure 2.10 Physical structure of ball and beam system.

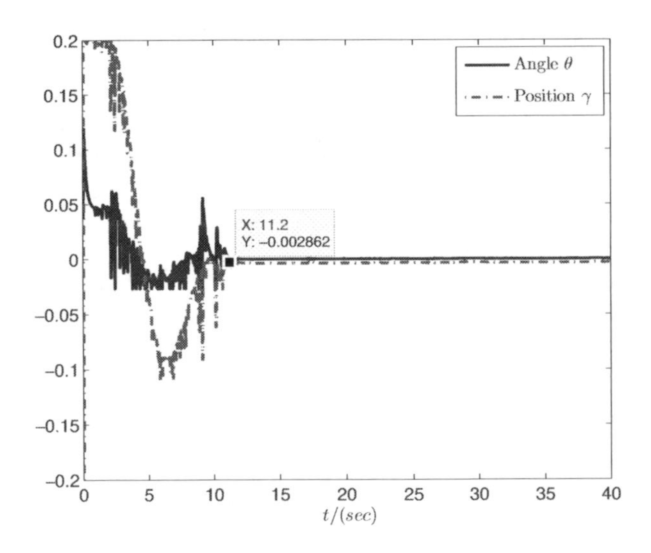

Figure 2.11 Experiment result without packets dropout.

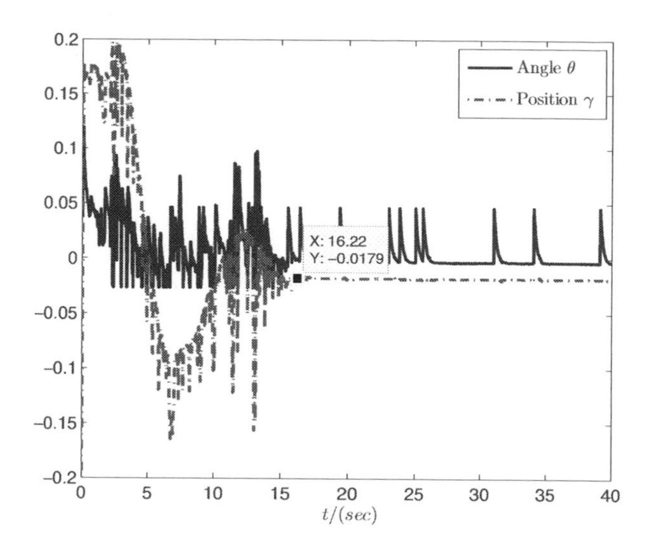

Figure 2.12 Experiment result with packets dropout $\alpha = 0.01$.

2.6 Conclusion

In this chapter, two types of optimal control strategies have been developed in the delta domain by using game theoretic tools. With the sampling interval $\mathtt{T}_s$ explicitly expressed in the backward recursions, the related results in both discrete and continuous domain have been unified by using the delta operator. The algorithms to develop optimal defense and attack strategies have been provided. The validity and advantage of the proposed method have been verified by numerical simulations.

The resilient control approach in this chapter is still an off-line design method. In our future work, some adaptive mechanisms [150] will be employed that can deal with uncertainties and randomness caused by cyber attacks on line.

Chapter 3
Optimal Control for NCSs with Disturbances

3.1 Introduction

NCSs have recently stirred much interest in the control community. Compared with traditional point-to-point systems, NCSs have advantages of fewer expenses, higher flexibility, better resource sharing, and so on [78, 142, 142, 101, 20]. Nevertheless, the inherently limited bandwidth of communication channels has led to a number of network-induced phenomena that is worth exploiting. The phenomena of time delays and packets dropout has attracted special attention because they are considered to be two of the main causes of performance degradation or even instability of controlled systems [156, 53]. On the other hand, sampling rates of modern industrial control systems are usually high. It has been well recognized that a delta operator is an effective tool in dealing with fast sampling systems. In practice, all real-time NCSs operate in the presence of disturbances [45] caused by a number of factors, including the fluctuation of communication environment [167], channel fading [148], quantization effects [133], load variation [38], friction [186], and noise measurement [168]. Therefore, studies of NCSs with external disturbances are of great importance from both theoretical and engineering points of view [46, 139, 164]. To the best of authors' knowledge, an optimal control problem subject to disturbances and network-induced constraints has not been adequately investigated, not to mention the system is described in the delta domain setting. The purpose of this chapter is to shorten such a gap.

In this chapter, an optimal control problem is investigated for a class of NCSs with disturbances and probabilistic packets dropout in the delta domain. The underlying system model of NCS is quite comprehensive and covers both the disturbances and network-induced packets dropout, thereby reflecting the reality closely. The problem addressed represents the first of a few attempts to deal with the optimal control problem of delta domain NCSs with disturbances.

The problem formulation is presented in Section 3.2. In Section 3.3, the delta domain optimal control strategy is developed in both finite-horizon and infinite-horizon cases. An upper bound for the epsilon level of the ϵ-optimum is provided explicitly, and the corresponding algorithms are given to compute such upper bound in Section 3.4. In Section 3.5, both simulations and experimental results are given to demonstrate the effectiveness and validity of the proposed approach. Conclusions are drawn in Section 3.6.

3.2 Problem Formulations

Consider the following delta domain NCS with disturbances and packets dropout phenomenon:

$$\begin{aligned}&(\Sigma_e): \\ &\delta x(t_k) = Ax(t_k) + \alpha(t_k)B(u(t_k) + d_m(t_k)) + Hd_u(t_k),\end{aligned} \tag{3.1}$$

where A, B, and H are matrices in the delta domain with appropriate dimensions, $x(t_k) \in \mathbb{R}^n$ and $u(t_k) \in \mathbb{R}^m$ are the state vector and control input, respectively, and $d_m(t_k)$ and $d_u(t_k)$ is the matched and unmatched disturbance, respectively. Both the matched and unmatched disturbances are norm-bounded, which satisfy $\|d_m(t_k)\|^2 \leq \zeta$ and $\|d_u(t_k)\|^2 \leq \xi$, $\forall k \in \mathbf{K} := \{1, 2, \cdots, K\}$, respectively. The packets dropout is described by $\alpha(t_k)$, and we assume that $\alpha(t_k)$ is a Bernoulli distributed white sequence with

$$\mathbb{P}\{\alpha(t_k) = 1\} = \bar{\alpha}, \ \ \mathbb{P}\{\alpha(t_k) = 0\} = 1 - \bar{\alpha} = \alpha, \ \forall k \in \mathbf{K}.$$

Remark 3.1. It follows from [162] that the definition of delta operator yields

$$\delta x(t_k) = \begin{cases} \frac{\mathrm{d}x(t)}{\mathrm{d}t}, & \mathtt{T}_s = 0, \\ \frac{x(t_{k+1}) - x(t_k)}{\mathtt{T}_s}, & \mathtt{T}_s \neq 0, \end{cases} \tag{3.2}$$

where $\mathtt{T}_s$ is the sampling period, t is the continuous-time index, and k is the time step with $t_k = k\mathtt{T}_s$. The NCS model (3.1) can be obtained from the following continuous-time model

$$\dot{x}(t) = A_s x(t) + \alpha(t)B_s(u(t) + d_m(t)) + H_s d_u(t), \tag{3.3}$$

where A_s, B_s, and H_s are matrices with appropriate dimensions in the continuous-domain. According to [162], the matrices in (3.1) can be obtained by $A = \frac{e^{A_s \mathtt{T}_s} - I}{\mathtt{T}_s}$, $B = \frac{1}{\mathtt{T}_s}\int_0^{\mathtt{T}_s} e^{A_s(\mathtt{T}_s - \tau)} B_s d\tau$, and $H = \frac{1}{\mathtt{T}_s}\int_0^{\mathtt{T}_s} e^{A_s(\mathtt{T}_s - \tau)} H_s d\tau$.

The system (Σ_e) is said to be the nominal/ideal system (Σ_n) if $d_u(t_k) = d_m(t_k) \equiv 0$ and $\alpha(t_k) \equiv 0$. The corresponding LQ cost function of system (Σ_n) yields

$$J(u) = x^T(t_K)Q_\delta^K x(t_K) + \mathrm{T}_s \sum_{k=0}^{K-1} \mathcal{W}(t_k), \tag{3.4}$$

where $\mathcal{W}(t_k) = x^T(t_k)Q_\delta x(t_k) + u^T(t_k)R_\delta u(t_k)$, $Q_\delta^K \geq 0$, $Q_\delta \geq 0$ and $R_\delta > 0$. In system (Σ_n), the classic optimum can be employed as the design objective, which is $J^* := J^*(u^*) \leq J(u)$ [12]. However, it should be noticed that, for system (Σ_e), the disturbances $d_m(t_k)$, $d_u(t_k)$ and the packets dropout phenomenon exist simultaneously, which can lead to a deviation from the traditional optimum. In order to tackle the optimal control problem for system (Σ_e), we are ready to introduce the concept of ϵ-optimum as follows:

$$\hat{J}(u^*) \leq J(u^*) + \epsilon, \tag{3.5}$$

where $\hat{J}(u)$ having the same structure as $J(u)$ is the cost function of system (Σ_e). As can be seen in (3.5), a deviation is caused due to the effect of disturbances and packets dropout, and the scalar ϵ is employed to describe how far the optimum deviates.

The objective of this chapter is to design the optimal control strategy u^* such that the cost function (3.4) is minimized for system (Σ_n) in both finite- and infinite-horizon. Furthermore, an upper bound for the scalar ϵ will be given explicitly. Before ending this section, we provide the following instrumental lemmas which will be used subsequently.

Lemma 3.2. *[162] The property of delta operator, for any time function $x(t_k)$ and $y(t_k)$, we have*

$$\delta(x(t_k)y(t_k)) = y(t_k)\delta x(t_k) + x(t_k)\delta y(t_k) + T_s\delta x(t_k)\delta y(t_k),$$

where T_s is the sampling period.

Lemma 3.3. *[65] Let x, y be any n_1-dimensional real vectors, and let Π be an $n_1 \times n_1$ symmetric positive semi-definite matrix. Then, we have $x^T y + y^T x \leq x^T \Pi x + y^T \Pi^{-1} y$.*

Lemma 3.4. *[26] If $g(\cdot)$ is a concave and finite function with random variable X, then the following inequality sets up*

$$\mathbb{E}\{g(X)\} \leq g(\mathbb{E}\{X\}).$$

3.3 Optimal Controller Design in the Delta Domain

In this section, the sufficient conditions and explicit expressions of the control strategy are provided such that system (Σ_n) can achieve the optimum over the finite- and infinite-horizon, respectively.

3.3.1 Finite-Time Horizon Case

Here, the finite-time horizon optimal control problem is considered. The following theorem provides the existence and uniqueness of the optimum, and the explicit expressions of the optimal control strategy.

Theorem 3.5. *Consider the system (Σ_n) with the associated cost function (3.4). There exists a unique feedback NE solution if the following conditions are simultaneously satisfied:*

1. *The matrix $\Theta(t_k)$ is positive definite, where $\Theta(t_k) := R_\delta + T_s B^T Z(t_{k+1})B$.*
2. *$Z(t_k)$ obeys the following backward recursion*

$$\begin{aligned}-\delta Z(t_k) = &Q_\delta + P^T(t_k)R_\delta P(t_k)\\&+T_s(A - BP(t_k))^T Z(t_{k+1})(A - BP(t_k))\\&+Z(t_{k+1})(A - BP(t_k)) + (A - BP(t_k))^T Z(t_{k+1}), \quad (3.6)\\Z(t_k) = &Z(t_{k+1}) - T_s\delta Z(t_k),\ Z(t_K) = Q_\delta^K, \quad (3.7)\end{aligned}$$

where $P(t_k) = (R_\delta + T_s B^T Z(t_{k+1})B)^{-1}B^T Z(t_{k+1})(T_s A + I)$. In this case, the controller takes the form of $u^*(t_k) = -P(t_k)x(t_k)$. Furthermore, the optimal value is $J^* = x^T(t_0)Z(t_0)x(t_0)$, where $x(t_0)$ is the initial value.

Proof: Let us choose the delta-domain cost function at time instant t_{k+1} as $V(x(t_{k+1})) = x^T(t_{k+1})Z(t_{k+1})x(t_{k+1})$ with $Z(t_{k+1}) > 0$. It follows from Lemma 3.2 that

$$\begin{aligned}&V(x(t_{k+1}))\\&= T_s\delta(x^T(t_k)Z(t_k)x(t_k)) + x^T(t_k)Z(t_k)x(t_k)\\&= T_s\delta x^T(t_k)Z(t_{k+1})x(t_k) + T_s x^T(t_k)Z(t_{k+1})\delta x(t_k)\\&\quad +T_s^2\delta x^T(t_k)Z(t_{k+1})\delta x(t_k) + x^T(t_k)Z(t_{k+1})x(t_k). \quad (3.8)\end{aligned}$$

In virtue of the principle of dynamic programming [12], the cost function at time t_k is obtained as

$$
\begin{aligned}
&V(x(t_k)) \\
&= \min_{u(t_k)}\{\mathtt{T}_s x^T(t_k)Q_\delta x(t_k) + \mathtt{T}_s u^T(t_k)R_\delta u(t_k) \\
&\quad + x^T(t_k)Z(t_{k+1})x(t_k) + \mathtt{T}_s x^T(t_k)Z(t_{k+1})(Ax(t_k) + Bu(t_k)) \\
&\quad + \mathtt{T}_s^2(Ax(t_k) + Bu(t_k))^T Z(t_{k+1})(Ax(t_k) + Bu(t_k))\} \\
&\quad + \mathtt{T}_s(Ax(t_k) + Bu(t_k))^T Z(t_{k+1})x(t_k). \qquad (3.9)
\end{aligned}
$$

The cost-to-go function is strictly convex with respect to $u(t_k)$ for the reason that the second-derivative of equation (3.9) yields $R_\delta + \mathtt{T}_s B^T Z(t_{k+1})B > 0$. Furthermore, by $\partial V(x(t_k))/\partial u(t_k) = 0$, one has the optimal controller $u^*(t_k) = -P(t_k)x(t_k)$. Substituting the optimal strategy $u^*(t_k) = -P(t_k)x(t_k)$ back into (3.9), we have (3.6). The proof is completed.

Remark 3.6. The results of delta domain in Theorem 3.5 are regarded as a unified form of the results in both discrete- and continuous-time systems. The delta-domain results in Theorem 3.5 are easily converted to its discrete domain analogue by substituting $A_z = \mathtt{T}_s A + I$, $B_z = \mathtt{T}_s B$, $Z(t_k) = Z(t_{k+1}) - \mathtt{T}_s\delta Z(t_k)$, and $P(k) = P(t_k)$ in the dynamic programming recursion (3.6). On the other hand, the equivalence of the results in Theorem (3.5) to those of the continuous-time systems can be verified by the following arguments: With the definition of delta operator (3.2), one has that

$$
\begin{aligned}
&\lim_{\mathtt{T}_s \to 0} \delta Z(t_k) = \dot{Z}(t), \\
&P(t) = R_\delta^{-1} B_s^T Z(t). \qquad (3.10)
\end{aligned}
$$

Letting the sampling period $\mathtt{T}_s$ approach, it implies from (3.10) that

$$
\begin{aligned}
&Q_\delta + Z(t)A_s + A_s^{\mathrm{T}} Z(t) \\
&-Z(t)B_s R_\delta^{-1} B_s^T Z(t) + \dot{Z}(t) = 0. \qquad (3.11)
\end{aligned}
$$

Note that (3.10) and (3.11) are consistent with the results of the continuous-time systems [12], which verifies the fact that Theorem 3.5 could unify the results in the continuous- and discrete-time systems.

3.3.2 Infinite-Time Horizon Case

In this part, we will continue to explore the sufficient conditions for minimizing the cost function (3.4) over the infinite-horizon. The infinite-horizon version of Theorem 3.7 is presented as follows.

Theorem 3.7. *Suppose that* $(T_s A + I, (T_s Q_\delta)^{1/2})$ *is observable. Using the time reverse notation, we have* $\tilde{Z}(t_k) = Z(t_{K-k})$ *and* $\tilde{P}(t_k) = P(t_{K-k})$. *Let* $\lim_{k\to\infty} \tilde{P}(t_k) = P_\infty$ *be given, then* $\tilde{Z}(t_k)$ *will converge, i.e.,* $\lim_{k\to\infty} \tilde{Z}(t_k) = Z_\infty$, *and the following closed-loop system will be stable.*

$$\delta x(t_k) = (A - BP_\infty)x(t_k), \tag{3.12}$$

where

$$\begin{aligned} &P_\infty = (R_\delta + T_s B^T Z_\infty B)^{-1} B^T Z_\infty (T_s A + I), \\ &Q_\delta + P_\infty^T R_\delta P_\infty + T_s (A - BP_\infty)^T Z_\infty (A - BP_\infty) \\ &+ Z_\infty (A - BP_\infty) + (A - BP_\infty)^T Z_\infty = 0 \end{aligned} \tag{3.13}$$

Proof: The backward recursion (3.6) is rewritten as

$$\begin{aligned} \tilde{Z}(t_{k+1}) = &\mathtt{T}_s Q_\delta + (\mathtt{T}_s A + I)^T \tilde{Z}(t_k)(\mathtt{T}_s A + I) \\ &-(\mathtt{T}_s A + I)^T \tilde{Z}(t_k)\hat{B}(R_\delta + \hat{B}^T \tilde{Z}(t_k)\hat{B})^{-1} \\ &\times \hat{B}^T \tilde{Z}(t_k)(\mathtt{T}_s A + I) \end{aligned} \tag{3.14}$$

where $\hat{B} = \sqrt{\mathtt{T}_s} B$. By using Lemma 3.3 in [64], $\tilde{Z}(t_k)$ converges to Z_∞ with $t_k \to \infty$. The infinite-time Riccati equation yields

$$\begin{aligned} Z_\infty = &\mathtt{T}_s Q_\delta + \mathtt{T}_s P_\infty^T R_\delta P_\infty \\ &+(\mathtt{T}_s A + I - \mathtt{T}_s BP_\infty)^T Z_\infty (\mathtt{T}_s A + I - \mathtt{T}_s BP_\infty) \end{aligned} \tag{3.15}$$

Noting 3.12, one has

$$\begin{aligned} &x^T(t_{k+1}) Z_\infty x(t_{k+1}) - x^T(t_k) Z_\infty x(t_k) \\ =&x^T(t_k)(\mathtt{T}_s A + I - \mathtt{T}_s BP_\infty)^T Z_\infty (\mathtt{T}_s A + I - \mathtt{T}_s BP_\infty) x(t_k) \\ &-x^T(t_k) Z_\infty x(t_k) \\ =&x^T(t_k)\{(\mathtt{T}_s A + I - \mathtt{T}_s BP_\infty)^T Z_\infty (\mathtt{T}_s A + I - \mathtt{T}_s BP_\infty) \\ &-Z_\infty\} x(t_k) \end{aligned} \tag{3.16}$$

Substituting (3.15) into (3.16), we have

$$\begin{aligned} &x^T(t_{k+1}) Z_\infty x(t_{k+1}) \\ &= x^T(t_0) Z_\infty x(t_0) - \mathtt{T}_s \sum_{l=1}^{k} \{x^T(t_l) \mathcal{W}_\infty x(t_l)\}, \end{aligned} \tag{3.17}$$

where $\mathcal{W}_\infty = Q_\delta + P_\infty^T R_\delta P_\infty$, which further implies that

$$\lim_{k \to \infty} \{x^T(t_k)(\mathtt{T}_s Q_\delta + \mathtt{T}_s P_\infty^T R_\delta P_\infty) x(t_k)\} = 0.$$

Since $(\mathtt{T}_s A + I, (\mathtt{T}_s Q_\delta)^{1/2})$ is observable and $R_\delta > 0$, we have $\|x(t_k)\|^2 \to 0$ as $t_k \to \infty$. The proof is completed.

3.4 Robustness Analysis of ϵ-Optimum

In this section, the upper bound for the epsilon level of ϵ-optimum will be provided explicitly based on the obtained feedback gain $P(t_k)$ (respectively, P_∞) in finite-horizon (respectively, infinite-horizon). Furthermore, the algorithms to compute such epsilon level will be given.

3.4.1 Finite-Time Horizon Case

We first provide the definitions of the following dynamic process

1. $x^*(k)$ is the state vector for the system (Σ_n) if the optimal control strategy $u^*(t_k)$ is used.
2. $x(k)$ is the state vector for the system (Σ_e) if the optimal control strategy $u^*(t_k)$ is used.

Some notations are presented as follows

$$
\begin{aligned}
&\bar{e}(t_k) = x(t_k) - x^*(t_k), && (3.18)\\
&\bar{Q}_\delta(t_K) = \mathtt{T}_s Q_\delta^K,\ \bar{Q}_\delta(t_k) = \mathtt{T}_s Q_\delta + \mathtt{T}_s P^T(t_k) R_\delta P(t_k), && (3.19)\\
&\bar{A}(t_k) = A - \bar{\alpha} B P(t_k),\ V_1(t_k) = \bar{e}^T(t_k) S_1(t_k) \bar{e}(t_k), && (3.20)\\
&\lambda_{d_m}(t_k) = \lambda_{\max}\{\mathtt{T}_s B^T(\bar{\alpha} S_1(t_{k+1}) + \bar{\alpha}^2 S_1(t_{k+1})\\
&\qquad \Lambda_1^{-1} S_1(t_{k+1}) - \bar{\alpha}(1-\bar{\alpha}) S_1(t_{k+1}) \Lambda_2^{-1} S_1(t_{k+1})\\
&\qquad +\bar{\alpha}^2 S_1(t_{k+1}) \Lambda_4^{-1} S_1(t_{k+1}))B\}, && (3.21)\\
&\lambda_{d_u}(t_k) = \lambda_{\max}\{\mathtt{T}_s H^T(S_1(t_{k+1}) + S_1(t_{k+1}) \Lambda_3^{-1}\\
&\qquad \times S_1(t_{k+1}) + \Lambda_4 + S_1(t_{k+1}) \Lambda_7 S_1(t_{k+1}))H\}, && (3.22)\\
&\lambda_{x^*}(t_k) = \lambda_{\max}\{\mathtt{T}_s P^T(t_k) B^T(S_1(t_{k+1}) \Lambda_5^{-1} S_1(t_{k+1})\\
&\qquad +\bar{\alpha}(1-\bar{\alpha}) S_1(t_{k+1}) \Lambda_6^{-1} S_1(t_{k+1})\\
&\qquad +(1-\bar{\alpha})^2 \Lambda_7^{-1}) B P(t_k)\}, && (3.23)
\end{aligned}
$$

In the next stage, we shall proceed to tackle the estimation problem of the epsilon level such that, for the obtained composite control strategy $u^*(t_k)$, the system (Σ_e) satisfies the ϵ-optimum (3.5).

Theorem 3.8. *Consider the system Σ_e with the cost function (3.5). For given optimal control strategies $u^*(t_k)$ obtained from Theorem 3.5 and positive definite matrices $\{\Lambda_i\}_{i=1}^{7}$ and $\mathcal{L}_1$, if there exists positive definite matrix $S_1(t_k)$, $\forall k \in \boldsymbol{K}$ such that the following recursion holds*

$$\begin{aligned}\mathcal{F}(S_1(t_k)) := & \,\delta S_1(t_k) + T_s\bar{A}^T(t_k)S_1(t_{k+1})\bar{A}(t_k) \\ & + S_1(t_{k+1})\bar{A}(t_k) + \bar{A}^T(t_k)S_1(t_{k+1}) + \frac{1}{T_s}\mathcal{L}_1 \\ & + T_s\bar{\alpha}(1-\bar{\alpha})P^T(t_k)B^T(S_1(t_{k+1}) + \Lambda_6 - \Lambda_2)BP(t_k) \\ & + \frac{1}{T_s}(T_s\bar{A}(t_k)+I)^T(\Lambda_1 + \Lambda_3 + \Lambda_5)(T_s\bar{A}(t_k)+I) = 0,\end{aligned} \tag{3.24}$$

$$S_1(t_k) = S_1(t_{k+1}) - T_s\delta S_1(t_k), \; S_1(t_K) = I, \tag{3.25}$$

then the optimal control strategy in Theorem 3.5 provides the ϵ-optimum, i.e.,

$$\hat{J}^*(u^*) \leq J(u^*) + \epsilon_K, \tag{3.26}$$

where ϵ_K is given as inequality (3.27).

$$\begin{aligned}\epsilon_K \leq & \max_{\forall k \in \mathbf{K}} \lambda_{\max}\{\bar{Q}_\delta(t_k)\} \Big(\mathrm{T}_s\lambda_{\min}^{-1}\{\mathcal{L}_1\}S + \lambda_{\min}^{-1}\{\mathcal{L}_1\}V_1(t_0) \\ & + \sqrt{\sum_{k=0}^{K} \|x^*(t_k)\|^2(\mathrm{T}_s\lambda_{\min}^{-1}\{\mathcal{L}_1\}S + \lambda_{\min}^{-1}\{\mathcal{L}_1\}V_1(t_0))}\Big).\end{aligned} \tag{3.27}$$

where $S = \sum_{k=0}^{K}(\lambda_{d_m}(t_k)\zeta + \lambda_{d_u}(t_k)\xi + \lambda_{x^*}(t_k)\|x^*(t_k)\|^2)$.

Proof. According to the optimal strategies without disturbances and packets dropout, we obtain

$$J^*(u^*) = \sum_{k=0}^{K} x^{*T}(t_k)\bar{Q}_\delta(t_k)x^*(t_k). \tag{3.28}$$

When the disturbances and packets dropout exist, one has

$$\hat{J}^*(u^*) = \mathbb{E}\left\{\sum_{k=0}^{K} x^T(t_k)\bar{Q}_\delta(t_k)x(t_k)\right\}. \tag{3.29}$$

Then, the scalar ϵ_K is calculated as

$$
\begin{aligned}
\epsilon_K^i &= \mathbb{E}\{\hat{J}^i(u^*) - J^i(u^*)\} \\
&= \mathbb{E}\{\sum_{k=0}^{K} x^T(t_k)\bar{Q}_\delta(t_k)x(t_k) - \sum_{k=0}^{K} x^{*T}(t_k)\bar{Q}_\delta(t_k)x^*(t_k)\} \\
&= \mathbb{E}\{\sum_{k=0}^{K}(x(t_k) - x^*(t_k))^T\bar{Q}_\delta(t_k)(x(t_k) - x^*(t_k) + 2x^*(t_k))\} \\
&\leq \mathbb{E}\{\sum_{k=0}^{K}\|x(t_k) - x^*(t_k)\|^2_{\bar{Q}_\delta(t_k)} \\
&\quad +2\|x(t_k) - x^*(t_k)\|\|\bar{Q}_\delta(t_k)x^*(t_k)\|\} \\
&\leq \mathbb{E}\left\{\max_{\forall k\in\mathbf{K}}\lambda_{\max}\{\bar{Q}_\delta(t_k)\}\left(\sum_{k=0}^{K}\|x(t_k) - x^*(t_k)\|^2\right.\right. \\
&\quad \left.\left. +2\sum_{k=0}^{K}\|x(t_k) - x^*(t_k)\|\|x^*(t_k)\|\right)\right\} \\
&\leq \max_{\forall k\in\mathbf{K}}\lambda_{\max}\{\bar{Q}_\delta(t_k)\}\left(\mathbb{E}\{\sum_{k=0}^{K}\|\bar{e}(t_k)\|^2\}\right. \\
&\quad \left. +2\sqrt{\mathbb{E}\{\sum_{k=0}^{K}\|\bar{e}(t_k)\|^2\}}\sqrt{\sum_{k=0}^{K}\|x^*(t_k)\|^2}\right)
\end{aligned} \tag{3.30}
$$

Notice that the last inequality holds according to Cauchy inequality and Lemma 3.4. The error equation yields

$$
\begin{aligned}
\delta\bar{e}(t_k) &= \bar{\mathcal{A}}(t_k)\bar{e}(t_k) + (1 - \alpha(t_k))BP(t_k)x^*(t_k) \\
&\quad +\alpha(t_k)Bd_m(t_k) + Hd_u(t_k),
\end{aligned} \tag{3.31}
$$

where $\bar{\mathcal{A}}(t_k) = A - \alpha(t_k)BP(t_k)$. Calculating the difference of $V_1(t_k)$ along (3.31) in the delta domain yields

$$
\begin{aligned}
&\mathbb{E}\{\delta V_1(t_k)\} \\
&= \frac{1}{\mathtt{T}_s}\mathbb{E}\left\{\bar{e}^T(t_{k+1})S_1(t_{k+1})\bar{e}(t_{k+1}) - \bar{e}^T(t_k)S_1(t_k)\bar{e}(t_k)\right\} \\
&= \frac{1}{\mathtt{T}_s}\mathbb{E}\left\{\left((\mathtt{T}_s\bar{A}(t_k)+I)\bar{e}(t_k) + (1-\alpha(t_k))\mathtt{T}_s BP(t_k)\right.\right. \\
&\quad x^*(t_k) + \mathtt{T}_s\alpha(t_k)Bd_m(t_k) + \mathtt{T}_s Hd_u(t_k)\big)^T S_1(t_{k+1}) \\
&\quad \times \left((\mathtt{T}_s\bar{A}(t_k)+I)\bar{e}(t_k) + (1-\alpha(t_k))\mathtt{T}_s BP(t_k)x^*(t_k)\right. \\
&\quad \left.\left. +\mathtt{T}_s\alpha(t_k)Bd_m(t_k) + \mathtt{T}_s Hd_u(t_k)\right) - \bar{e}^T(t_k)S_1(t_k)\bar{e}(t_k)\right\} \\
&\le \frac{1}{\mathtt{T}_s}\mathbb{E}\left\{\bar{e}^T(t_k)\left((\mathtt{T}_s\bar{A}(t_k)+I)^T S_1(t_{k+1})(\mathtt{T}_s\bar{A}(t_k)+I)\right.\right. \\
&\quad +\mathtt{T}_s^2\bar{\alpha}(1-\bar{\alpha})P^T(t_k)B^T S_1(t_{k+1})BP(t_k) + \mathcal{L}_1 \\
&\quad +(\mathtt{T}_s\bar{A}(t_k)+I)^T(\Lambda_1+\Lambda_3+\Lambda_5)(\mathtt{T}_s\bar{A}(t_k)+I) \\
&\quad -\mathtt{T}_s^2\bar{\alpha}(1-\bar{\alpha})P^T(t_k)B^T\Lambda_2 BP(t_k) \\
&\quad \left. +\mathtt{T}_s^2\bar{\alpha}(1-\bar{\alpha})P^T(t_k)B^T\Lambda_6 BP(t_k) - S_1(t_k)\right)\bar{e}(t_k) \\
&\quad +\lambda_{d_m}(t_k)\|d_m(t_k)\|^2 + \lambda_{d_u}(t_k)\|d_u(t_k)\|^2 \\
&\quad \left. +\lambda_{x^*}(t_k)\|x^*(t_k)\|^2 - \frac{1}{\mathtt{T}_s}\bar{e}^T(t_k)\mathcal{L}_1\bar{e}(t_k)\right\},
\end{aligned} \tag{3.32}
$$

where$\lambda_{d_m}(t_k)$, $\lambda_{d_u}(t_k)$ and $\lambda_{x^*}(t_k)$ are given in (3.18). Note that the last inequality holds according to Lemma 3.3. Select positive definite matrix $S_1(t_k)$, $\forall k \in \mathbf{K}$ such that

$$
\begin{aligned}
&\mathtt{T}_s\bar{A}^T(t_k)S_1(t_{k+1})\bar{A}(t_k) + \frac{1}{\mathtt{T}_s}\mathcal{L}_1 - \delta S_1(t_k) \\
&+\mathtt{T}_s\bar{\alpha}(1-\bar{\alpha})P^T(t_k)B^T(S_1(t_{k+1})+\Lambda_6-\Lambda_2)BP(t_k) \\
&+\frac{1}{\mathtt{T}_s}(\mathtt{T}_s\bar{A}(t_k)+I)^T(\Lambda_1+\Lambda_3+\Lambda_5)(\mathtt{T}_s\bar{A}(t_k)+I) = 0
\end{aligned} \tag{3.33}
$$

Then

$$
\begin{aligned}
\sum_{k=0}^{K}\mathbb{E}\{\delta V_1(t_k)\} \le &\sum_{k=0}^{K}\{\lambda_{d_m}(t_k)\|d_m(t_k)\|^2 + \lambda_{d_u}(t_k)\|d_u(t_k)\|^2 \\
&+\lambda_{x^*}(t_k)\|x^*(t_k)\|^2\} - \frac{1}{\mathtt{T}_s}\mathbb{E}\left\{\sum_{k=0}^{K}\bar{e}^T(t_k)\mathcal{L}_1\bar{e}(t_k)\right\},
\end{aligned} \tag{3.34}
$$

Furthermore, we have

$$
\begin{aligned}
\lambda_{\min}\{\mathcal{L}_1\}\mathbb{E}\left\{\sum_{k=0}^{K}\|\bar{e}(t_k)\|^2\right\} \le \mathtt{T}_s &\sum_{k=0}^{K}\{\lambda_{d_m}(t_k)\|d_m(t_k)\|^2 + \lambda_{d_u}(t_k)\|d_u(t_k)\|^2 \\
&+\lambda_{x^*}(t_k)\|x^*(t_k)\|^2\} + V_1(t_0)
\end{aligned} \tag{3.35}
$$

Considering the upper bound of the parameters, we have

$$\mathbb{E}\left\{\sum_{k=0}^{K}\|\bar{e}(t_k)\|^2\right\} \leq \mathrm{T}_s\lambda_{\min}^{-1}\{\mathcal{L}_1\}\sum_{k=0}^{K}(\lambda_{d_m}(t_k)\zeta + \lambda_{d_u}(t_k)\xi$$
$$+\lambda_{x^*}(t_k)\|x^*(t_k)\|^2) + \lambda_{\min}^{-1}\{\mathcal{L}_1\}V_1(t_0). \tag{3.36}$$

Substituting (3.36) into (3.30), we arrive at the conclusion that the estimation of upper bound satisfies (3.27). The proof is completed.

It is not difficult to see that the inequalities $\mathcal{F}(S_1(t_k)) < 0$ also lead to (3.34). Thus, the equality (3.24) in Theorem 3.8 can be replaced by $\mathcal{F}(S_1(t_k)) < 0$. In the following, Algorithm 2 is proposed in terms of iterative LMIs to compute ϵ_K, which is less conservative than Theorem 3.8.

Algorithm 2 Calculation of the epsilon level ϵ_K

1: Input K, system matrix parameters in (3.1) and the optimal feedback gain $P(t_k)$, state system $x^*(t_k)$, $k \in \mathbf{K}$.
2: Select positive definite matrices $\{\Lambda_i\}_{i=1}^{7}$ and $\mathcal{L}_1$. Set that $S_1(t_K) = I$ and $k = K$.
3: Solve LMI problems $\mathcal{F}(S_1(t_k)) < 0$ to obtain $S_1(t_{k-1})$ with the known $S_1(t_k)$.
4: If the positive definite solution $S_1(t_{k-1})$ exists, and $k \geq 0$, set $k = k-1$ and go to *Step* 3. Else, if the positive definite solutions $S_1(t_{k-1})$ does not exist and $k \geq 0$, go to *Step* 2 to reselect matrices $\{\Lambda_i\}_{i=1}^{7}$ and $\mathcal{L}_1$. Else, output the feasible solutions $\{S_1(t_k),\ k \in \mathbf{K}\}$, and exit.
5: Compute $\lambda_{d_m}(t_k)$, $\lambda_{d_u}(t_k)$, $\lambda_{x^*}(t_k)$ the maximum eigenvalues of $\bar{Q}_\delta(t_k)$, $\forall k \in \mathbf{K}$, and the minimum eigenvalues of $\mathcal{L}_1$. Substitute these parameters into (3.27) and the scalar ϵ_K can be obtained.

Remark 3.9. It follows from Theorem 3.8 that the following argument can be easily verified: if $d_m(t_k) = d_u(t_k) \equiv 0, \alpha \equiv 0$ and $\bar{e}(0) = 0$, the so-called ϵ-optimum will reduce to the traditional optimum with $\epsilon_K \equiv 0$.

3.4.2 Infinite-Time Horizon Case

In this part, the upper bound for ϵ-optimum over the infinite-horizon will be explored. We assume here that $d_m(t_k)$ $d_u(t_k)$ and $x^*(t_k)$ here all belong to $l_2[0,\infty)$, and the corresponding upper bounds are redefined, i.e. $\|d_u(t_k)\|_2^2 \leq \xi^\infty$, $\|x^*(t_k)\|_2^2 \leq \beta_1^\infty$ and $\|d_m(t_k)\|_2^2 \leq \zeta^\infty$. Before presenting the main results, we provide the following definitions first.

$$\bar{Q}_\delta^\infty = \mathtt{T}_s Q_\delta + \mathtt{T}_s P_\infty^T R_\delta P_\infty,\ \bar{A}_\infty = A - \bar{\alpha} B P_\infty, \tag{3.37}$$
$$V_1^\infty(t_k) = \bar{e}^T(t_k) S_1^\infty \bar{e}(t_k),\ \bar{e}(t_k) = x(t_k) - x^*(t_k), \tag{3.38}$$
$$\lambda_{d_m}^\infty = \lambda_{\max}\{\mathtt{T}_s B^T(\bar{\alpha} S_1^\infty + \bar{\alpha}^2 S_1^\infty \Upsilon_1^{-1} S_1^\infty - \bar{\alpha}(1-\bar{\alpha}) S_1^\infty \Upsilon_2^{-1} S_1^\infty + \bar{\alpha}^2 S_1^\infty \Upsilon_4^{-1} S_1^\infty) B\}, \tag{3.39}$$
$$\lambda_{d_u}^\infty = \lambda_{\max}\{\mathtt{T}_s H^T(S_1^\infty + S_1^\infty \Upsilon_3^{-1} S_1^\infty + \Upsilon_4 + S_1^\infty \Upsilon_7 S_1^\infty) H\}, \tag{3.40}$$
$$\lambda_{x^*}^\infty = \lambda_{\max}\{\mathtt{T}_s P_\infty^T B^T(S_1^\infty \Upsilon_5^{-1} S_1^\infty + \bar{\alpha}(1-\bar{\alpha}) S_1^\infty \Upsilon_6^{-1} S_1^\infty + (1-\bar{\alpha})^2 \Upsilon_7^{-1}) B P_\infty\} \tag{3.41}$$

Theorem 3.10. *Consider the system Σ_e with the cost function (3.5) and $K \to \infty$. For given optimal control strategies $u^*(t_k) = -P_\infty x(t_k)$ obtained from Theorem 3.7 and positive definite matrices $\{\Upsilon_i\}_{i=1}^7$ and $\bar{\mathcal{L}}_1$, if there exists positive definite matrix S_1^∞ such that the following equation holds*

$$\mathcal{G}(S_1^\infty) := T_s \bar{A}_\infty^T S_1^\infty \bar{A}_\infty + S_1^\infty \bar{A}_\infty + \bar{A}_\infty^T S_1^\infty + T_s \bar{\alpha}(1-\bar{\alpha}) P_\infty^T B^T (S_1^\infty + \Upsilon_6 - \Upsilon_2) B P_\infty + \frac{1}{T_s}\bar{\mathcal{L}}_1 + \frac{1}{T_s}(T_s \bar{A}_\infty + I)^T(\Upsilon_1 + \Upsilon_3 + \Upsilon_5)(T_s \bar{A}_\infty + I) = 0, \tag{3.42}$$

then the optimal control strategy in Theorem 3.5 provides the ϵ-optimum, i.e.,

$$\hat{J}_\infty^*(u^*) \le J_\infty(u^*) + \epsilon_\infty, \tag{3.43}$$

where ϵ_∞ please refer to (3.44) on next page.

$$\epsilon_\infty = \lambda_{\max}\{\bar{Q}_\delta^\infty\}\left(T_s \lambda_{\min}^{-1}\{\bar{\mathcal{L}}_1\}(\lambda_{d_m}^\infty \zeta^\infty + \lambda_{d_u}^\infty \xi^\infty + \lambda_{x^*}^\infty \beta_1^\infty) + \sqrt{\beta_1^\infty (T_s \lambda_{\min}^{-1}\{\bar{\mathcal{L}}_1\}(\lambda_{d_m}^\infty \zeta^\infty + \lambda_{d_u}^\infty \xi^\infty + \lambda_{x^*}^\infty \beta_1^\infty) + \lambda_{\min}^{-1}\{\bar{\mathcal{L}}_1\} V_1^\infty(t_0))} + \lambda_{\min}^{-1}\{\bar{\mathcal{L}}_1\} V_1^\infty(t_0)\right) \tag{3.44}$$

Proof. The proof follows directly from Theorem 3.8 and is therefore omitted here.

Remark 3.11. It should be pointed out that as $\bar{\alpha} = 1$, the Riccati equation (3.42) will be reduced to the standard Lyapunov equation in the delta domain [97] as follows:

$$\mathtt{T}_s \bar{A}_\infty^T S_1^\infty \bar{A}_\infty + S_1^\infty \bar{A}_\infty + \bar{A}_\infty^{\mathtt{T}} S_1^\infty + \frac{1}{\mathtt{T}_s}\bar{\mathcal{L}}_1 + \frac{1}{\mathtt{T}_s}(\mathtt{T}_s \bar{A}_\infty + I)^T(\Upsilon_1 + \Upsilon_3 + \Upsilon_5)(\mathtt{T}_s \bar{A}_\infty + I) = 0, \tag{3.45}$$

According to [97], the existence and uniqueness condition for positive definite solution of (3.45) is that $\bar{A}$ lies within the stability boundary, which is a circle with the center $1/\mathtt{T}_s$ and radius $1/\mathtt{T}_s$ in the δ plane. It has been well known that the stability region of the classic shift implementation is fixed, causing clustering at 1 as the sample rate increases, and further leading to the insolubility of the corresponding Riccati equation. By contrast, the stability region for the delta implementation will approach that of the Laplace domain (i.e., the whole left hand plane) with a high sample rate and the corresponding Lyapunov equation (e.g., (3.45)) remains solvable.

It has been mentioned in Remark 3.11 that the Riccati equation (3.42) is not the standard delta domain Lyapunov equation, and hence we propose the following LMI method to compute the value of ϵ_∞.

3.5 Illustrate Examples

In this section, we aim to demonstrate the validity and applicability of the proposed method. For this purpose, we conduct the numerical simulation on the power system. Furthermore, the experimental verification on the ball and beam system is provided.

3.5.1 Numerical Simulation

The proposed methodology is applied to a multi-area Load Frequency Control (LFC) problem for power control systems [96]. Our main objective is to control the LFC system such that the outputs are kept at the desired setting, while maintaining robustness against load disturbances and packets dropout. Consider the two-area interconnected power system as follows:

$$\dot{x}(t) = Ax(t) + B(u(t) + P_{d_m}(t)) + H\Delta P_d(t), \tag{3.46}$$

where

$$x(t) = \left[x^{1T}(t)\ x^{2T}(t) \right]^T,\ u(t) = \left[u^{1T}(t)\ u^{2T}(t) \right]^T,$$
$$\Delta P_{d_m}(t) = \left[\Delta P_{d_{m1}}^T(t)\ \Delta P_{d_{m2}}^T(t) \right]^T,$$
$$\Delta P_d(t) = \left[\Delta P_{d_1}^T(t)\ \Delta P_{d_2}^T(t) \right]^T,\ A = \begin{bmatrix} A^{11} & A^{12} \\ A^{21} & A^{22} \end{bmatrix},$$
$$B = \mathrm{diag}\left[B^1\ B^2 \right],\ H = \mathrm{diag}\left[H^1\ H^2 \right],$$
$$A^{ii} = \begin{bmatrix} -\frac{1}{\mathrm{T}_{p_i}} & \frac{K_{p_i}}{\mathrm{T}_{p_i}} & 0 & 0 & A_{15}^{ii} \\ 0 & -\frac{1}{\mathrm{T}_{\mathrm{T}_i}} & \frac{1}{\mathrm{T}_{\mathrm{T}_i}} & 0 & 0 \\ -\frac{1}{R_i \mathrm{T}_{G_i}} & 0 & -\frac{1}{\mathrm{T}_{G_i}} & \frac{1}{\mathrm{T}_{G_i}} & 0 \\ K_{E_i} K_{B_i} & 0 & 0 & 0 & A_{45}^{ii} \\ 2\pi & 0 & 0 & 0 & 0 \end{bmatrix},$$
$$A_{15}^{ii} = -\frac{K_{p_i}}{2\pi \mathrm{T}_{p_i}} \sum_{j \in \mathbf{S}, j \neq i} K_{s_{ij}},\ A_{45}^{ii} = \frac{K_{E_i}}{2\pi} \sum_{j \in \mathbf{S}, j \neq i} K_{s_{ij}},$$
$$B^i = \left[0\ 0\ \tfrac{1}{\mathrm{T}_{G_i}}\ 0\ 0 \right]^T,\ H^i = \left[\tfrac{K_{p_i}}{\mathrm{T}_{p_i}}\ 0\ 0\ 0\ 0 \right]^T,$$
$$A^{ij} = \begin{bmatrix} 0\ 0\ 0\ 0 & -\frac{K_{p_i}}{2\pi \mathrm{T}_{p_i}} K_{s_{ij}} \\ 0\ 0\ 0\ 0 & 0 \\ 0\ 0\ 0\ 0 & 0 \\ 0\ 0\ 0\ 0 & \frac{K_{E_i}}{2\pi} K_{s_{ij}} \\ 0\ 0\ 0\ 0 & 0 \end{bmatrix},\ x^i(t) = \begin{bmatrix} \Delta f_i(t) \\ \Delta P_{g_i}(t) \\ \Delta X_{g_i}(t) \\ \Delta E_i(t) \\ \Delta \delta_i(t) \end{bmatrix}.$$

The state vector $x^i(t) \subset \mathbb{R}^n$ consists of variables $\Delta f_i(t)$, $\Delta P_{g_i}(t)$, $\Delta X_{g_i}(t)$, $\Delta E_i(t)$, and $\Delta \delta_i(t)$, which are the changes of frequency, power output, governor valve position, integral control, and rotor angle deviation, respectively. Vector $x^j(t) \in \mathbb{R}^n$ is the neighboring state vector of $x^i(t)$. $u^i(t) \in \mathbb{R}^m$ is the control input vector, $\Delta P_{d_i}(t) \in \mathbb{R}^k$ is the vector of the load disturbance. Parameters K_{p_i}, K_{E_i}, and K_{B_i} are power system gain, integral control gain, and frequency bias factor, and $K_{s_{ij}}$ is the interconnection gain between area i and j $(i \neq j)$. Parameters T_{p_i}, $\mathrm{T}_{\mathrm{T}_i}$, and T_{G_i} are time constants of power system, turbine, and governor, respectively. Parameter R_i is the speed regulation coefficient. The specific values of the parameters can be found in TABLE I in [96].

Firstly, we choose the initial value of state as $x(t_0) = \left[0\ 5\ 0\ 0\ 0\ 0\ 0\ 0\ 0\ 5 \right]^T$. Then, the matched and unmatched disturbances are given as $P_{d_m}(t) = 0.01 * \sin(t)$ and $\Delta P_d(t) = 1/10 * (10 + t)$, respectively. In the sequel, we estimate an upper bound for the epsilon level of the ϵ-optimum, and the specific parameters are chosen as sampling period $\mathrm{T}_s = 0.05s$, packet delivery rate $\bar{\alpha} = 0.9$, and finite time $K = 200$. The other parameters are given as $Q_\delta = I_{10 \times 10}$, $R_\delta = I_{2 \times 2}$, $\Lambda_j = 0.03 * I_{10 \times 10}$, for all $j = \{1, 2, 3, 4\}$ and $\mathcal{L}_1 = 0.03 * I_{10 \times 10}$. The specific values of the bounds can be obtained as

$\zeta = 0.01$, $\xi = 0.01$, $\lambda_{\max}\{\bar{Q}_\delta(t_k)\} = 5.3846$. It follows from (3.27) in Theorem 3.8 that $\epsilon_K = 53.8$. Since the value of the cost function is $\hat{J} = 1.4359 * 10^3$, the upper bound for a possible deviation of the ϵ-optimum is 3.75%.

Based on the setup of the above parameters, the resultant curves of the epsilon level ϵ_K versus finite time K can be obtained in Figure 3.1, where we can see that ϵ_K increase with K but decrease with packet delivery rate $\bar{\alpha}$. On the other hand, if the upper bounds for the disturbances change, the experimental results in Figure 3.2 can be obtained, where we can see that larger upper bounds of disturbances lead to larger ϵ_K.

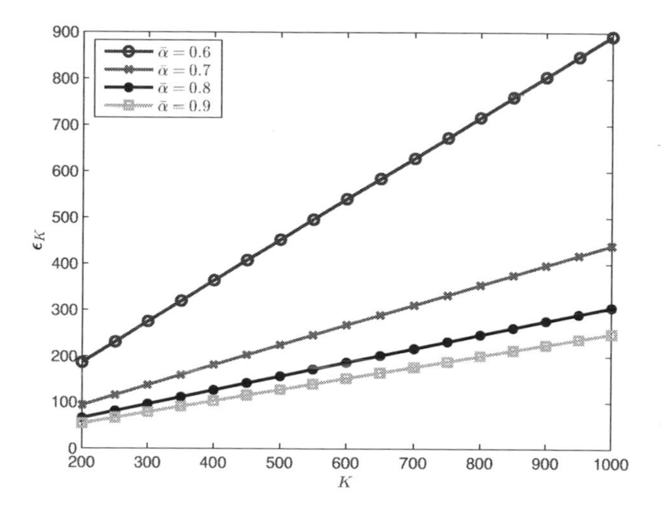

Figure 3.1 The relation between the epsilon level ϵ_K and finite time K with different packet delivery rates.

3.5.2 Experimental Verification

In this part, the methodology proposed in this chapter is applied to the ball and beam platform to illustrate the effectiveness and applicability. The ball and beam platform is shown in Figure 3.3(a), where the ball rolls back and forth on the groove of the beam. The physical structure of the ball and beam system is illustrated in Figure 3.3(b), where γ and θ are the ball position and beam angle, respectively. Let $x = [\gamma\ \dot{\gamma}\ \theta\ \dot{\theta}]$, and then the ball and beam system with added matched disturbance can be expressed in the continuous domain as [162]

$$\dot{x}(t) = A_s x(t) + B_s(u(t) + d_m(t)),$$

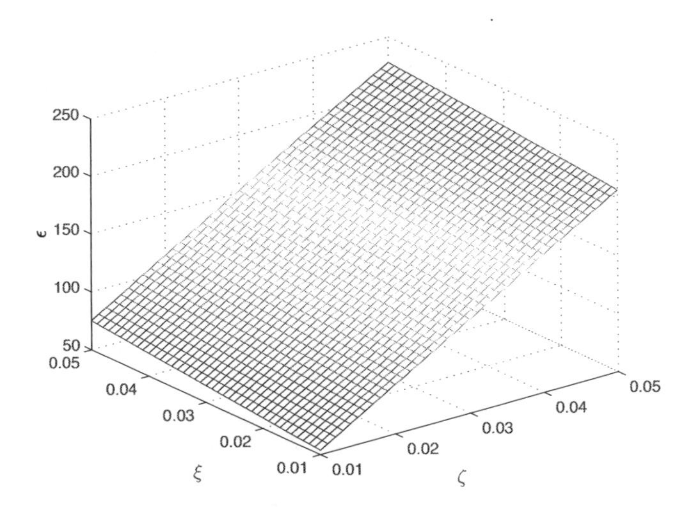

Figure 3.2 The relation between the epsilon level ϵ_K and upperbound of disturbances.

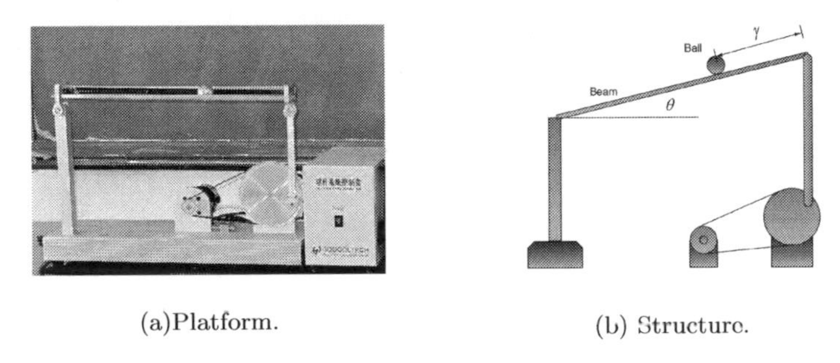

(a)Platform. (b) Structure.

Figure 3.3 Ball and beam system.

where

$$A_s = \begin{bmatrix} 0 & 1 & 0 & 0 \\ 0 & 0 & -7.007 & 0 \\ 0 & 0 & 0 & 1 \\ 0 & 0 & 0 & 0 \end{bmatrix}, \ B_s = \begin{bmatrix} 0 \\ 0 \\ 0 \\ 1 \end{bmatrix}.$$

The disturbance from the input channel is bounded by 0.05. The initial state is $x(t_0) = [-0.2\ 0\ 0\ 0]^T$, the packets dropout rate is chosen to be 1%, i.e., $\alpha = 0.01$, the sampling period is $\mathtt{T}_s = 0.02s$ and the finite time is $K = 1000$. It follows from Theorem 3.5 that the optimal controller yields $u^*(t_k) = [-9.2713\ \ -8.4746\ 27.1400\ 7.2887]x(t_k)$ with weighting matrices $Q_\delta = \mathrm{diag}\{1000, 0, 10, 0\}$ and $R_\delta = 10$. With Theorem 3.8, it is obtained that $\hat{J} = 1.8365{*}10^3$ and then we get $\epsilon_K = 319.2350$. It is concluded that the upper bound for a possible deviation of the ϵ-optimum is 17.38%. The experimental results with and without matched disturbance are shown in Figure 3.4. According to experimental results, the cost value is calculated as $\hat{J} = 2.0883 * 10^3$ with matched disturbance $d_m(t_k)$, and $J = 1.8155 * 10^3$ otherwise. In other words, in practical experiments, the difference between the cost value with and without matched disturbance is 13.06% of $\hat{J}$. As such, we could arrive at the conclusion that the proposed approach is applicable in the sense that it could estimate the actual deviation from the optimum in engineering practice.

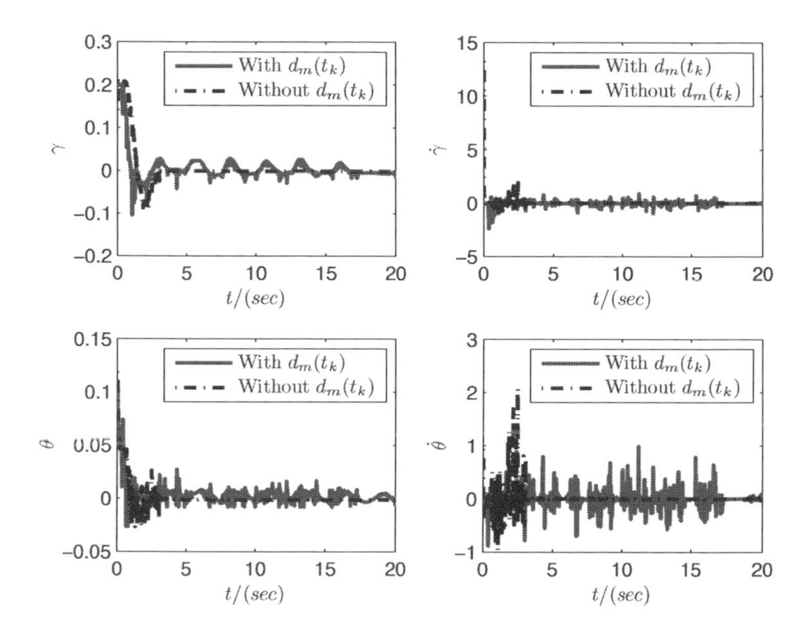

Figure 3.4 State responses of the ball and beam system with and without $d_m(t_k)$.

3.6 Conclusion

In this chapter, a novel optimal control problem has been addressed for a class of NCSs with missing data and disturbances. The so-called ϵ-optimum has been proposed as a performance index to describe the impact from the disturbances. The delta domain optimal controller has been designed over finite-/infinite-horizon, and an upper bound for the epsilon level of the ϵ-optimum has been provided explicitly. It has been shown that such upper bound can be obtained if the corresponding LMIs are feasible. Simulations and experimental verifications have demonstrated the feasibility and applicability of the proposed method.

Chapter 4
Resilient NPC for NCSs against DoS Attack

4.1 Introduction

In recent years, security problems on NCSs have received extensive attention for the reason that NCSs are vulnerable to malicious attacks. So far, researchers have done many researches on security issues and considered various forms of attackers, e.g., DoS attacks [172, 170], deception attacks [103, 98, 9], replay attacks [188] and so on. Though various attack schemes have been researched, optimal DoS attack schemes that are dangerous to NCSs have not been studied in depth yet. Moreover, it is very interesting to analyze optimal attack schemes for their serious harms on NCSs. On the other hand, NPC scheme is capable of dealing with dropped packet problems. Therefore, NPC schemes can be used to compensate random delays and consecutive packets dropout [104, 165, 35]. In practice, actuator saturation limitations generally exist in many practical control systems [146, 187, 123]. When control inputs are beyond certain values, actuators enter saturated situations, which may lead to instability of control systems. To the best of the authors' knowledge, NCSs under DoS attacks with NPC schemes and actuator saturation have not been studied adequately so far.

In this chapter, a defense strategy with NPC is proposed against optimal DoS attack schemes on NCSs. For reducing attack costs and achieving attack objectives, the least attack steps are estimated with properties for the domain of attraction. Optimal attack schemes are given as consecutive DoS attacks with the least steps in this chapter. Then random packets dropout are considered in backward channels, a Kalman filter is used at controller sides. Moreover, NPC strategies are proposed for coping with optimal DoS attack schemes. Then stability criteria for NCSs with NPC are presented in main results. Furthermore, offense and defense algorithms are shown for describing dynamic actions in detail. Finally, a numerical example is provided to illustrate the effectiveness of the proposed design techniques.

In Section 4.2, a specific formulation for optimal DoS attack schemes is presented and the domain of attraction for NCSs is obtained. The least attack steps are obtained, and meanwhile stability criteria for NCSs with NPC are given in Section 4.3. Then detailed attack and defense strategies are also formulated in forms of algorithms in this section. In Section 4.4, a numerical example is shown to illustrate the effectiveness of theoretical results. Conclusions of this chapter are drawn in Section 4.5.

4.2 Problem Formulation and Preliminaries

4.2.1 Optimal DoS Attack Scheme

Note that cyber attackers are inclined to launch effective attacks on vulnerable NCSs. As depicted in Figure4.1, a NCS controller sends data packets to a saturation actuator through an uncertain channel. When DoS attacks happen, the actuator does not obtain updating control signals, which reduces system performances and even makes the plant out of control. Then random packets dropout induced by networks are considered in the backward channel, and a Kalman filter is deployed at the controller side.

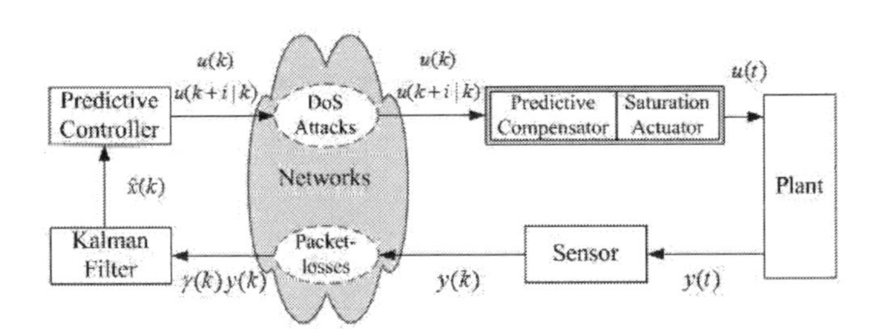

Figure 4.1 NPC system with actuator saturation under DoS attacks and packets dropout.

The plant is modeled as the following linear discrete stochastic system with actuator saturation

$$x(k+1) = Ax(k) + B\text{sat}(u(k)) + w(k), \tag{4.1}$$

$$y(k) = Cx(k) + v(k), \tag{4.2}$$

where $x(k) \in \mathbf{R}^n$, $u(k) \in \mathbf{R}^l$ and $y(k) \in \mathbf{R}^m$ are the system state, control input and plant output, respectively, $w(k) \in \mathbf{R}^n$ denotes the process noise and $v(k) \in \mathbf{R}^m$ is the measurement noise. Letting x_0 be the initial system state, x_0, $w(k)$ and $v(k)$ are uncorrelated random variables. Denote $x_0 \sim \mathcal{N}(\bar{x}_0, \Sigma)$, $w(k) \sim \mathcal{N}(0, Q)$ and $v(k) \sim \mathcal{N}(0, R)$, where $\mathcal{N}(m, M)$ represents a

Guassian distribution with mean m and covariance matrix M. Define sat$(\cdot)$ as a saturation function and $\text{sat}(u(k)) = \text{sgn}(u(k))\min\{1,\ |u(k)|\}$. A, B, and C are system constant matrices with appropriate dimensions. The pair $(A,\ C)$ is observable and $(A,\ B)$ is stabilizable in this chapter. Then a state feedback saturation controller of system (4.1) is designed as

$$\text{sat}(u(k)) = \text{sat}(Fx(k)), \tag{4.3}$$

where $F \in \mathbf{R}^{m\times n}$ is the static feedback gain.

As shown in [177], the more that attack times are grouped together, the larger system costs become. Furthermore, consecutive attacks are better than scatter attacks for attackers. Thus, consecutive DoS attacks on the forward communication channel are studied in this chapter. From the viewpoints of attackers, there exist two objectives that NCSs are unable to be stabilized with initial static saturation controllers and attack costs are reduced as low as possible. If controllers are valid on system (4.1) after attacks, then these attacks are failed, and vice versa. Moreover, when DoS attacks are considered on system (4.1), the system is written as

$$x(k+1) = Ax(k) + \nu(k)B\text{sat}(u(k)) + w(k), \tag{4.4}$$

where $\nu(k) \in \mathbf{I}[0,1]$ is a sign whether data packets dropout or not in forward channels and $\nu(k) = 1$ means that the data packets is transmitted successfully at time k.

Due to consecutive DoS attacks, any data packets do not arrive at actuators in a time interval T at time k and system (4.4) becomes

$$x(k+1) = Ax(k) + w(k). \tag{4.5}$$

Let the time interval T contain p consecutive DoS attack steps. When any defense measures are not taken, state $x(k+p)$ is obtained as

$$x(k+p) = A^p x(k) + \sum_{i=1}^{p} A^{p-i} w(k+i). \tag{4.6}$$

If system (4.5) is not self-stabilizing, then state $x(k+p)$ diverges to an uncertain point. When attacks end at step $k+p$, data packets arrive at the saturation actuator. At this time, $x(k+p)$ is regarded as a new initial state and controller (4.3) is reapplied. If state $x(k+p)$ is outside the domain of attraction of system (4.1), then the previous controller (4.3) is invalid on this NCS. That is, objectives of attackers are achieved in this case.

On the viewpoints of energy-constrained attackers, problems on minimizing attack costs are analyzed valuablly in the following. Then attack costs on system (4.1) are given as

$$J_A(r) = \mathcal{P}(r) + J_{\mathcal{G}}(r),\ \ r \in \mathbf{Z}^+, \tag{4.7}$$

where $J_A(r)$ denotes the cost of r consecutive attack steps, $\mathcal{P}(r)$ represents the fixed cost to launch r attacks. Let $\mathcal{G}(r) = (1-\epsilon)^r$ be a successful probability for r consecutive attacks in the case of accidents, where the positive constant $\epsilon \in \mathbf{I}[0,1]$ is the occurrence probability of unpredictable events. Then $J_{\mathcal{G}}(r)$ denotes the cost of preventing accidents. Further a problem on reducing attack costs is formulated as

$$\min_{\epsilon \to 0} J_A(r) \quad s.t.\ r \geq p. \tag{4.8}$$

For facilitating analysis on attacks costs, some properties for cost variables are given in the following. Considering equation (4.7), inequalities both $\mathcal{P}(r_1) \leq \mathcal{P}(r_2)$ and $\mathcal{G}(r_1) \geq \mathcal{G}(r_2)$ hold with $r_1 \leq r_2$. When $\mathcal{G}(r_1) \geq \mathcal{G}(r_2)$, there exists $J_{\mathcal{G}}(r_1) \leq J_{\mathcal{G}}(r_2)$.

In [177], an energy constraint is discussed in optimal DoS attack schedules. Not only sensors need to save transmission energies, but attackers always have finite energies to launch effective attacks. Therefore, it is very valuable to study the minimal attack costs with which attackers are sure to destroy stability of system (4.1). By cost properties, the optimization problem (4.8) is transformed into obtaining the minimal r which represents consecutive attack steps. In this chapter, optimal DoS attack schemes are given as consecutive DoS attacks with the least attack steps, which not only realize attack objectives but also minimize attack costs. Furthermore, this chapter aims to present a defense strategy against the optimal DoS attack schemes based on NPC.

4.2.2 The Domain of Attraction

Consider constraints of actuator saturation for system (4.1). When state $x(k)$ is outside the domain of attraction, controller (4.3) is invalid to stabilize system (4.1). Then attackers are likely to launch effective consecutive DoS attacks on this system with saturation constraints. Since the process noise $w(k)$ randomly exists in system (4.1) and is hardly obtained, it is conservative for attackers to study the domain of attraction for system (4.1) without noises. Furthermore, some results on the domain of attraction for system (4.1) without noises are shown in the following.

A nominal form of system (4.1) is given as

$$x(k+1) = Ax(k) + B\text{sat}(Fx(k)) \tag{4.9}$$

and the domain of attraction for system (4.9) is defined as

$$\mathcal{S} := \left\{ x_0 \in \mathbf{R}^n : \lim_{k \to \infty} \psi(k,\ x_0) = 0 \right\}, \tag{4.10}$$

where $\psi(k,\ x_0)$ denotes state trajectories starting from $x_0 \in \mathbf{R}^n$. Note that $\mathcal{S}$ is an invariant set that all trajectories starting from it will remain in it.

Denoting f_i as the i-th row of F, then define

$$\mathcal{L}(F) := \{x \in \mathbf{R}^n :\ |f_i x| \leq 1,\ i \in \mathbf{I}[1,\ m]\}, \tag{4.11}$$

where $\mathcal{L}(F)$ is the region in which controller (4.3) is linear.

Letting $\mathcal{N}$ be a set of $m \times m$ diagonal matrices whose diagonal elements are either 1 or 0, there exist 2^m elements in $\mathcal{N}$. In the case of $m = 2$, one has that

$$\mathcal{N} = \left\{ \begin{bmatrix} 0 & 0 \\ 0 & 0 \end{bmatrix}, \begin{bmatrix} 0 & 0 \\ 0 & 1 \end{bmatrix}, \begin{bmatrix} 1 & 0 \\ 0 & 0 \end{bmatrix}, \begin{bmatrix} 1 & 0 \\ 0 & 1 \end{bmatrix} \right\}.$$

Denote each element of $\mathcal{N}$ as N_i and $i \in \mathbf{I}[1,\ 2^m]$, then let $N_i^- = I - N_i$. Note that N_i^- is also an element of $\mathcal{N}$ if $N_i \in \mathcal{N}$.

Giving matrices $F, G \in R^{m \times n}$, when $x(k) \in \mathcal{L}(G)$, for all $i \in \mathbf{I}[1,\ 2^m]$, there exists

$$\text{sat}(Fx(k)) \in \text{co}\{N_i Fx(k) + N_i^- Gx(k)\},$$

where co$\{\cdot\}$ denotes a convex hull for a group of vectors. Therefore, sat$(Fx(k))$ is expressed as

$$\text{sat}(Fx(k)) = \sum_{i=1}^{2^m} \delta_i (N_i F + N_i^- G)x(k), \tag{4.12}$$

where $\sum_{i=1}^{2^m} \delta_i = 1$ and $\delta_i \geq 0$.

Let $P \in \mathbf{R}^{n \times n}$ be a positive definite matrix. Denote a state set as

$$\varepsilon(P,\ \rho) = \left\{x \in \mathbf{R}^n : x^T P x \leq \rho\right\}. \tag{4.13}$$

Defining a Lyapunov function as $W(x(k)) = x(k)^T P x(k)$, if there exists a positive diagonal matrix $E \in \mathbf{R}^{m \times m}$ satisfying $E < I$ and $G = EF$ such that

$$\begin{aligned} &(A + B(N_i F + N_i^- G))^T P (A + B(N_i F + N_i^- G)) \\ &-P < 0, i \in \mathbf{I}[1, 2^m], \end{aligned} \tag{4.14}$$

and $\varepsilon(P, \rho) \subset \mathcal{L}(G)$ hold, then $\varepsilon(P,\ \rho)$ is a contractive invariant set [60].

Let $\mathcal{X}_R \subset \mathbf{R}^n$ be a bounded convex set of some desired shapes and $0 \in \mathcal{X}_R$. For a positive real number β, denote

$$\beta \mathcal{X}_R = \{\alpha x : x \in \mathcal{X}_R\}.$$

Take the size of the domain of attraction $\mathcal{S}$ with respect to $\mathcal{X}_R$ as

$$\beta_R(\mathcal{S}) := \sup\{\beta > 0 : \beta\mathcal{X}_R \subset \mathcal{S}\}.$$

Then a problem on estimating the domain of attraction for system (4.9) is transformed into

$$\sup_{P>0,\rho,H} \quad \alpha \tag{4.15}$$
$$s.t. \begin{cases} \text{(i) Inequality (4.14)}, \\ \text{(ii)} \quad \beta\mathcal{X}_R \subset \varepsilon(P,\rho), \\ \text{(iii)} \quad \varepsilon(P,\rho) \subset \mathcal{L}(G). \end{cases}$$

When $\mathcal{X}_R$ is a polyhedron, the optimization problem (4.15) is changed to

$$\inf_{H>0,Z} \quad \varpi \tag{4.16}$$
$$s.t. \begin{cases} \text{(i)} & \begin{bmatrix} \varpi & x_i^{\mathrm{T}} \\ x_i & H \end{bmatrix} \geq 0, \; i \in \mathbf{I}[1,l], \\ \text{(ii)} & \begin{bmatrix} 1 & z_i \\ z_i^T & H \end{bmatrix} \geq 0, \; i \in \mathbf{I}[1,m], \\ \text{(iii)} & \begin{bmatrix} H & \Phi_i^T \\ \Phi_i & H \end{bmatrix} \geq 0, \; i \in \mathbf{I}[1,2^m], \end{cases}$$

where z_i and g_i are the i-th row of Z and G, respectively, and

$$\varpi = \frac{1}{\beta^2}, \; H = \left(\frac{P}{\rho}\right)^{-1}, \; Z = G\left(\frac{P}{\rho}\right)^{-1},$$
$$\Phi_i = AH + B(N_iFH + N_i^- Z).$$

A detailed transformation process for the optimization problem (4.16) is studied in monograph [60]. By solving the optimization problem (4.16), optimal values of matrices P and F are obtained. Then an estimation of the largest domain of attraction and a corresponding controller are obtained as $x(k)^T Px(k) < 1$ and $u = Fx(k)$, respectively.

For obtaining main results of this chapter, two lemmas are given in the following.

Lemma 4.1. *[71] If matrices $\mathcal{A}$, $\mathcal{B} \in \mathbf{R}^n$ are symmetric positive definite, then*

$$(\mathcal{A} + \mathcal{B})^{-1} > \mathcal{A}^{-1} - \mathcal{A}^{-1}\mathcal{B}\mathcal{A}^{-1}. \tag{4.17}$$

Lemma 4.2. *For any positive number η and for all $a, b \in \mathbf{R}^n$, there exists*

$$2a^T b \leq \frac{1}{\eta}a^T a + \eta b^T b. \tag{4.18}$$

4.3 Main Results

4.3.1 Least Attack Steps

In this subsection, the least consecutive DoS attack steps are obtained with properties for the domain of attraction on system (4.1) without noises. On the viewpoints of attackers, consecutive DoS attacks with the least steps make system (4.9) unable to be stabilized by the initial controller $\text{sat}(u(k))$. Then a solution for the least attack steps p is given in the following theorem.

Theorem 4.3. *Consider that attackers launch consecutive DoS attacks on system (4.9) at time k for p updating steps. System (4.9) is unable to be stabilized by initial controller $sat(u(k))$ if the least attack steps p satisfy*

$$p \geq log_{|\lambda(\tilde{\Lambda})|_{\min}} \left(\frac{1}{\lambda_{\min}(\tilde{P}) \|\tilde{\eta}(k)\|} \right), \tag{4.19}$$

where $\tilde{\Lambda} = diag\{\tilde{\lambda}_1, \ \tilde{\lambda}_2, \ \cdots, \ \tilde{\lambda}_s\} (s \leq m)$ and $\tilde{\lambda}_i > 1 (i \in \mathbf{I}[1, s])$ are eigenvalues of matrix A. Note that λ_j and R_j are eigenvalues and eigenvectors of matrix A, respectively, and $\lambda_j R_j = AR_j \ (j \in \mathbf{I}[1, m])$. Denote $\tilde{P} = R^T PR$, where $R = [R_1 \ R_2 \ \cdots \ R_m]$.

Proof. For attackers, consecutive DoS attacks happen on system (4.9) at time k for p updating steps, then state $x(k+p)$ diverges to $x(k+p) = A^p x(k)$. From the optimization problem (4.16), the domain of attraction is obtained as $x^T(k+p)Px(k+p) < 1$. When state $x(k+p)$ is outside of the domain of attraction, system (4.9) is unable to be recontrolled by the initial saturation controller based on properties of the domain of attraction. In order to destroy stability of system (4.9) for attackers, state $x(k+p)$ should satisfy

$$x^T(k+p)Px(k+p) \geq 1. \tag{4.20}$$

Since $A^p = R\Lambda^p R^{-1}$, inequality (4.20) is changed to

$$x^T(k)(R^{-1})^T \Lambda^p R^T PR\Lambda^p R^{-1} x(k) \geq 1. \tag{4.21}$$

Letting $\eta(k) = R^{-1}x(k)$ and $\tilde{P} = R^T PR$, inequality (4.21) is rewritten as

$$\eta^T(k)\Lambda^p \tilde{P} \Lambda^p \eta(k) \geq 1. \tag{4.22}$$

Moreover, with properties of matrix quadratic forms, inequality (4.22) is changed to

$$\eta^T(k)\Lambda^p \tilde{P} \Lambda^p \eta(k) \geq \lambda_{\min}(\tilde{P}) \|\Lambda^p \eta(k)\|. \tag{4.23}$$

Letting the j-th element of vector $\eta(k)$ be $\eta_j(k)$, there exists

$$\|\Lambda^p \eta(k)\| = \left(\sum_{j=1}^{m} |\lambda_j^p \eta_j(k)|^q \right)^{\frac{1}{q}}. \tag{4.24}$$

Note that $\lim_{p\to\infty} \lambda_i^p = 0$ holds when $\lambda_i < 1$. If $\lambda_i > 1$, then $\eta_i(k)$ is chosen from vector $\eta(k)$ to form a s dimensional vector $\tilde{\eta}(k)$. Considering equation (4.24), it is obtained that

$$\begin{aligned} \|\Lambda^p \eta(k)\| &\geq \left(\sum_{i=1}^{s} |\tilde{\lambda}_i^p \tilde{\eta}_i(k)|^q \right)^{\frac{1}{q}} \\ &= \left(\sum_{i=1}^{s} |\tilde{\lambda}_i^p|^q |\tilde{\eta}_i(k)|^q \right)^{\frac{1}{q}} \\ &\geq \left(\sum_{i=1}^{s} |\lambda^p(\tilde{\Lambda})|_{\min}^q |\tilde{\eta}_i(k)|^q \right)^{\frac{1}{q}} \\ &= |\lambda(\tilde{\Lambda})|_{\min}^p \|\tilde{\eta}(k)\|. \end{aligned} \tag{4.25}$$

Combining inequalities (4.22), (4.23) and (4.25), the least attack steps p are obtained as inequality (4.19). This completes the proof.

Note that stabilized controllers make state $x(k)$ obey $\lim_{k\to\infty} x(k) \to 0$, then a vector variable ε is defined as $\lim_{k\to\infty} x(k) = \varepsilon$. To destroy stability of system (4.9), attack steps p satisfy

$$\begin{aligned} &\min_{p\in\mathbf{Z}} p \\ &s.t.\ (A^p\varepsilon)^T P(A^p\varepsilon) \geq 1. \end{aligned} \tag{4.26}$$

Due to an effect of process noise $w(k)$ in system (4.1), state $x(k)$ is hardly kept at the zero point. Therefore, it is reasonable to give the vector variable ε based on system disturbances in practice.

Inequality (4.19) shows that p is a function of $x(k)$, and p changes in different updating steps. In [177], it is mentioned that attackers are able to eavesdrop on transmission channels over a long period before starting their attack actions. This assumption is also provided in this chapter; thus attackers tend to choose the best attack points after considering all conditions. When energy constraints are considered for attackers, attackable ranges are obtained through inequality (4.19).

4.3.2 Design of Kalman Filter

In this subsection, a Kalman filter with actuator saturation is designed to estimate system states at the controller side as shown in Figure 4.1. Moreover, packets dropout are considered in the backward channel in Figure 4.1. Then the Kalman filter is designed as

$$\hat{x}(k+1|k) = A\hat{x}(k) + B\text{sat}(F\hat{x}(k)), \tag{4.27}$$

$$\begin{aligned}\hat{x}(k) = {} & \hat{x}(k|k-1) + \gamma(k)K(k)(y(k) \\ & -C\hat{x}(k|k-1)),\end{aligned} \tag{4.28}$$

where $\gamma(k) \in \mathbf{I}[0,1]$ is a sign whether the Kalman filter obtains data packets or not. Note that $\gamma(k) = 1$ means this filter gets data packets at time k. Furthermore, a Kalman filter gain $K(k)$ is also used to predict the first step of inputs $\text{sat}(u(k))$ in defense strategies.

The following theorem is presented to ensure effectiveness of the Kalman filter with actuator saturation in networks.

Theorem 4.4. *Consider system (4.1)-(4.2). Let* $\|A\| \le \bar{a}$, $\|C\| \ge \underline{c}$, $\mathbf{E}(\gamma_k) = \gamma$, $Q \le \bar{q}I_n$, $R \le \bar{r}I_q$ *and* $P(1|0) = \Sigma > 0$. *If arrival rates of output data packets satisfy*

$$\gamma > 1 - \bar{a}^{-2}, \tag{4.29}$$

then the Kalman filter is effective for system (4.1) with packets dropout in the backward channel.

Proof. Combining system (4.1) and the Kalman filter (4.27)-(4.28), an error system is given as follows:

$$e(k+1|k) = Ae(k) + w(k), \tag{4.30}$$

$$\begin{aligned}P(k+1|k) &= \mathbf{E}\left[e^T(k+1|k)e(k|k+1)\right] \\ &= AP(k|k)A^T + Q,\end{aligned} \tag{4.31}$$

where $e(k+1|k) = x(k+1) - \hat{x}(k+1|k)$ and $e(k) = x(k+1) - \hat{x}(k)$.

By the same way, it is also obtained that

$$P(k|k) = P(k|k-1) - \gamma(k)K(k)CP(k|k-1), \tag{4.32}$$

$$K(k) = P(k|k-1)C^T(CP(k|k-1)C^T + R)^{-1}. \tag{4.33}$$

Combining equations (4.31), (4.32), and (4.33), an iteration of $P(k+1|k)$ is obtained as

$$\begin{aligned} P(k+1|k) &= A(P(k|k-1) - \gamma(k)P(k|k-1)C^T \\ &\quad \times(CP(k|k-1)C^T + R)^{-1}C \\ &\quad \times P(k|k-1))A^T + Q \\ &= A(P(k|k-1) - \gamma(k)P(k|k-1) \\ &\quad -\gamma(k)P(k|k-1)C^T R^{-1}C \\ &\quad \times P(k|k-1))A^T + Q. \end{aligned} \tag{4.34}$$

Then applying Lemma 4.1, equation (4.34) turns into

$$\begin{aligned} P(k+1|k) &\leq (1-\gamma(k))AP(k|k-1)A^T \\ &\quad +\gamma(k)A(C^T R^{-1}C)^{-1}A^T + Q. \end{aligned}$$

Furthermore, it is obtained that

$$\begin{aligned} \mathbf{E}[P(k+1|k)] &\leq \mathbf{E}[(1-\gamma(k))AP(k|k-1)A^T \\ &\quad +\gamma(k)A(C^T R^{-1}C)^{-1}A^T + Q] \\ &\leq (1-\gamma)\bar{a}^2\mathbf{E}[P(k|k-1)] \\ &\quad +\gamma\bar{a}^2\bar{r}\underline{c}^{-2}I_n + \bar{q}I_n. \end{aligned} \tag{4.35}$$

Moreover, inequality (4.35) is derived with mathematical induction in the following.

Firstly, let

$$\mathbf{E}[P(k|k-1)] \leq \varphi\sum_{j=0}^{k-1}[(1-\gamma)\bar{a}^2]^j I_n, \ \forall k \geq 1, \tag{4.36}$$

where $\varphi = \max(\|P(1|0)\|, \bar{a}^2\gamma\bar{r}\underline{c}^{-2} + \bar{q})$.

Secondly, when $k = 1$, inequality (4.35) is changed to

$$\begin{aligned} \mathbf{E}[P(2|1)] &\leq \mathbf{E}[(1-\gamma_0)AP(1|0)A^T \\ &\quad +\gamma_0 A(C^T R^{-1}C)^{-1}A^T + Q] \\ &\leq (1-\gamma)\bar{a}^2P(1|0) + \gamma\bar{a}^2\bar{r}\underline{c}^{-2}I_n + \bar{q}I_n \\ &\leq (1-\gamma)\bar{a}^2\varphi I_n + \varphi I_n. \end{aligned} \tag{4.37}$$

Lastly, it is obtained from inequality (4.36) that

$$\begin{aligned} \mathbf{E}[P(k+1|k)] &\leq (1-\gamma)\bar{a}^2\varphi\sum_{j=0}^{k-1}[(1-\gamma)\bar{a}^2]^j I_n \\ &\quad +\gamma\bar{a}^2\bar{r}\underline{c}^{-2}I_n + \bar{q}I_n \\ &= \varphi\sum_{j=0}^{k}[(1-\gamma)\bar{a}^2]^j I_n. \end{aligned} \tag{4.38}$$

Combining inequalities (4.36), (4.37) and (4.38), $\mathbf{E}[P(k+1|k)]$ has upper bounds when $\gamma > 1 - \bar{a}^{-2}$. That is, the Kalman filter is effective for system (4.1) with packets dropout in the backward channel.

This completes the proof.

4.3.3 Defense Strategy

In this subsection, a defense strategy based on NPC is presented to cope with optimal DoS attack schemes on system (4.1). When attackers launch consecutive DoS attacks on system (4.1) in a time interval T, control packets do not arrive at the actuator. At this time, the defense strategy is used to resist these attacks. That is, predictive inputs are applied to stabilize system (4.1) under DoS attacks. Moreover, NPC schemes are designed with the Kalman gain $K(k)$, which usually converges in a few steps. Hence, $\bar{P}$ and K are shown as follows:

$$\bar{P} \triangleq \lim_{k\to\infty} P(k|k-1),$$
$$K \triangleq \bar{P}C^T(C\bar{P}C^T + R)^{-1}.$$

An initial condition $\Sigma = \bar{P}$ reduces the Kalman filter to a fixed gain estimator [98].

As mentioned in equation (4.12), it is shown that

$$\text{sat}(Fx(k)) = \sum_{i=1}^{2^m} \delta_i(N_iF + N_i^- G)x(k)$$

for all $x(k) \in \varepsilon(P, \rho)$. For simplification, denote $\phi_i(F, G) = \sum_{i=1}^{2^m} \delta_i(N_iF + N_i^- G), i \in \mathbf{I}[1, 2^m]$. Then a predictive control model is designed as

$$\begin{aligned} \hat{x}(k+d|k) &= A\hat{x}(k+d-1|k) \\ &\quad + B\text{sat}(F\hat{x}(k+d-1|k)),\ d \in \mathbf{I}[1,p]. \end{aligned} \tag{4.39}$$

When a NPC scheme is applied in system (4.1) as the defense strategy, a system with predictive inputs is given as

$$\begin{aligned} x(k+d) &= Ax(k+d-1) + B\text{sat}(F\hat{x}(k+d-1|k)) \\ &\quad + w(k+d-1),\ d \in \mathbf{I}[1,p]. \end{aligned} \tag{4.40}$$

Thereby, stability of system (4.40) should be ensured. Moreover, the following theorem is given to illustrate stability of system (4.40) without disturbances.

Theorem 4.5. *Let the arrival packet rate be satisfied with $\gamma > 1 - \bar{a}^{-2}$ in the backward channel for system (4.40). Then system (4.40) free of disturbances* *is asymptotically stable, if there exist* $\mathcal{P} > 0$, $\mathcal{Q} > 0$, $\mathcal{P}, \mathcal{Q} \in \mathbf{R}^{n \times n}$ *satisfying*

$$[A^p(I - \gamma KC)]^T \mathcal{P} A^p(I - \gamma KC) - \mathcal{P} < 0, \tag{4.41}$$

$$[A + B\phi_i(F, G)]^T \mathcal{Q}[A + B\phi_i(F, G)] - \mathcal{Q} < 0, \tag{4.42}$$

where $\phi_i(F, G) = \sum_{i=1}^{2^m} \delta_i(N_i F + N_i^- G)$, $i \in \mathbf{I}[1, 2^m]$.

Proof. For the Kalman filter (4.27)-(4.28), Theorem 4.4 has presented a stability criterion that the arrival packet rate satisfy $\gamma > 1 - \bar{a}^{-2}$ in the backward channel for system (4.40). Moreover, combining system (4.1) and the Kalman filter (4.27)-(4.28), it is obtained that

$$\begin{aligned} e(k+1|k) = {} & A(I - \gamma(k)KC)e(k|k-1) \\ & -\gamma(k)AKv(k) + w(k). \end{aligned} \tag{4.43}$$

Subtracting equation (4.40) to (4.39), it is given that

$$e(k+d|k) = Ae(k+d-1|k) + w(k+d-1), \tag{4.44}$$

where $d \in \mathbf{I}[1, p]$. When consecutive attacks end at time $k+p$, error $e(k+p|k)$ is changed to

$$\begin{aligned} e(k+p|k) = {} & A^p(I - \gamma(k)KC)e(k|k-1) \\ & -\gamma(k)A^pKv(k) + \sum_{i=0}^{p-1} A^{p-1-i}w(k+i) \end{aligned} \tag{4.45}$$

and state $x(k+p)$ is given as

$$\begin{aligned} x(k+p) = {} & Ax(k+p-1) + B\phi_i(F, G)\hat{x}(k+p-1|k) \\ & +w(k+p-1) \\ = {} & (A + B\phi_i(F, G))x(k+p-1) \\ & -B\phi_i(F, G)A^{p-1}(I - \gamma(k)KC)e(k|k-1) \\ & +\gamma(k)B\phi_i(F, G)A^{p-1}Kv(k) \\ & -\sum_{j=0}^{p-2} B\phi_i(F, G)A^{p-2-j}w(k+j) \\ & +w(k+p-1), \end{aligned} \tag{4.46}$$

where $i \in \mathbf{I}[1, 2^m]$. Without effects of disturbances, equations (4.45) and (4.46) are reduced to

$$
\begin{aligned}
e(k+p|k) &= A^p(I-\gamma(k)KC)e(k|k-1), \qquad (4.47)\\
x(k+p) &= (A+B\phi_i(F,G))x(k+p-1)\\
&\quad -B\phi_i(F,G)A^{p-1}(I-\gamma(k)KC)\\
&\quad \times e(k|k-1),\ i\in\mathbf{I}[1,2^m]. \qquad (4.48)
\end{aligned}
$$

Then it is shown that

$$
\mathbf{E}[e(k+p|k)] = A^p(I-\gamma KC)e(k|k-1). \qquad (4.49)
$$

When condition (4.41) holds, system (4.49) is asymptotically stable as shown in [124].

Note that error $e(k+p|k)$ of subsystems in system (4.48) converges to zero when system (4.49) is stable. If inequality (4.42) holds, then the following system

$$
x(k+p) = (A+B\phi_i(F,G))x(k+p-1), i\in\mathbf{I}[1,2^m]
$$

is asymptotically stable. Therefore, system (4.48) is asymptotically stable based on small gain theorem referring to [67]. This completes the proof.

Furthermore, stability of system (4.40) with disturbances is presented in the following.

Theorem 4.6. *Considering system (4.1) under DoS attacks, system (4.40) possesses an input-to-state stable property with respect to disturbances $w(k)$ and $v(k)$ if conditions (4.29), (4.41), (4.42), and the following inequality hold*

$$
\begin{aligned}
&\gamma^2\lambda_{\max}(\mathbf{P})Tr(R)+\lambda_{\max}(\mathbf{S})(p-1)Tr(Q)\\
&+\lambda_{\max}(P)Tr(Q) < \frac{1}{4}-\frac{\varsigma}{2} \qquad (4.50)
\end{aligned}
$$

with

$$
\begin{aligned}
\mathbf{P} &= (B\phi_i(F,G)A^pK)^T PB\phi_i(F,G)A^pK,\\
\mathbf{S} &= [\Psi(p-2)\ \Psi(p-3)\ \cdots\ I]^T P\\
&\quad\times[\Psi(p-2)\ \Psi(p-3)\ \cdots\ I],
\end{aligned}
$$

where p is obtained from Theorem 4.3 and $\Psi(j) = B\phi_i(F,G)A^j$, $i\in\mathbf{I}[1,2^m]$, $j\in\mathbf{I}[1,p-2]$.

Proof. For simplicity, denote

$$
\begin{aligned}
\mathbf{A}_1 &= (A + B\phi_i(F,G))x(k+p-1), \\
\mathbf{A}_2 &= B\phi_i(F,G)A^{p-1}(I - \gamma(k)KC)e(k|k-1), \\
\mathbf{B}_1 &= \gamma(k)B\phi_i(F,G)A^{p-1}Kv(k), \\
\mathbf{B}_2 &= \sum_{j=0}^{p-2} B\phi_i(F,G)A^{p-2-j}w(k+j).
\end{aligned}
$$

Then equation (4.46) is written as

$$
x(k+p) = \mathbf{A}_1 - \mathbf{A}_2 + \mathbf{B}_1 - \mathbf{B}_2 + w(k+p-1).
$$

Consider actuator saturation in system (4.1). Furthermore, constraint $x^T(k+p)Px(k+p) < 1$ holds, i.e.,

$$
\begin{aligned}
&(\mathbf{A}_1 - \mathbf{A}_2 + \mathbf{B}_1 - \mathbf{B}_2 + w(k+p-1))^T P \\
&\times (\mathbf{A}_1 - \mathbf{A}_2 + \mathbf{B}_1 - \mathbf{B}_2 + w(k+p-1)) < 1.
\end{aligned} \tag{4.51}
$$

Applying Lemma 4.2, it is obtained from inequality (4.51) that

$$
\begin{aligned}
&(\mathbf{B}_1 - \mathbf{B}_2 + w(k+p-1))^T P(\mathbf{B}_1 - \mathbf{B}_2 \\
&+ w(k+p-1)) + (\mathbf{A}_1 - \mathbf{A}_2)^T P(\mathbf{A}_1 - \mathbf{A}_2) < \frac{1}{2}.
\end{aligned} \tag{4.52}
$$

Stability conditions have been presented in Theorem 4.5 for system (4.48). Denoting $\varsigma = (\mathbf{A}_1 - \mathbf{A}_2)^T P(\mathbf{A}_1 - \mathbf{A}_2)$, inequality (4.52) turns into

$$
\begin{aligned}
&(\mathbf{B}_1 - \mathbf{B}_2 + w(k+p-1))^T P \\
&\times (\mathbf{B}_1 - \mathbf{B}_2 + w(k+p-1)) < \frac{1}{2} - \varsigma.
\end{aligned} \tag{4.53}
$$

Using Lemma 4.2 in inequality (4.53), one has that

$$
\begin{aligned}
&(\mathbf{B}_1 - \mathbf{B}_2 + w(k+p-1))^T P \\
&\times (\mathbf{B}_1 - \mathbf{B}_2 + w(k+p-1)) \\
&< 2(\mathbf{B}_1^T P \mathbf{B}_1 + \mathbf{B}_2^T P \mathbf{B}_2 \\
&+ w^T(k+p-1)Pw(k+p-1)).
\end{aligned} \tag{4.54}
$$

With properties of matrix quadratic forms, it is obtained that

$$
\begin{aligned}
&\mathbf{E}\left[\mathbf{B}_1^T P \mathbf{B}_1 + \mathbf{B}_2^T P \mathbf{B}_2 \right. \\
&\left. + w^T(k+p-1)Pw(k+p-1)\right] \\
&\leq \gamma^2 \lambda_{\max}(\mathbf{P}) Tr(R) \\
&+ \left[\lambda_{\max}(\mathbf{S})(p-1) + \lambda_{\max}(P)\right] Tr(Q).
\end{aligned} \tag{4.55}
$$

Combining inequalities (4.54) and (4.55), inequality (4.53) is satisfied if inequality (4.50) holds. Moreover, inequality (4.51) is achieved from inequality (4.53), and stability for system (4.40) with actuator saturation is guaranteed.

Thus, combining Theorem 4.5 and small gain theorem, the proof is completed.

4.3.4 Algorithms of Attacks and Defenses

In this subsection, both offensive and defensive algorithms are presented based on the obtained results in this chapter. When the controller of system (4.1) sends data packets to the saturation actuator through a vulnerable communication channel, attackers launch consecutive DoS attacks with an optimal attack scheme. In order to destroy stability of system (4.1) and make it unable to recontrol by the initial controller, attackers use the following algorithm.

Algorithm 3 The algorithm for optimal DoS attack strategies

Require: Steal continuous system initial state x_0, matrices A and B. Eavesdrop running time tamp k and sampling period $\mathtt{T}$.

1: Discretize the continuous system with sampling period $\mathtt{T}$.
2: Calculate the optimal domain of attraction $x(k)^T Px(k) < 1$ and the corresponding controller $\text{sat}(Fx(k))$.
3: Calculate the least consecutive DoS attack steps $p(k+i)$, $i \in \mathbf{I}[0, l]$ for each step.
4: Choose the best attack time point $k+j$ between time k and $k+l$.
5: Launch $p(k+j)$ steps consecutive DoS attacks at time $k+j$.

Normally, main constraints for attackers are limitations of energy and concealment as mentioned in [9] and [177]. The two constraints make attackers unable to launch effective consecutive DoS attacks on any time points. If attackers have enough energies to launch the maximal steps $\bar{p}$ consecutive DoS attacks, then an attackable range $[0,\ t]$ is obtained. To guarantee high success rates, attackers choose a random attack time point $t^* \in [0, t]$ to launch $p(t^*)$ consecutive DoS attack steps.

For coping with consecutive DoS attacks as presented in Algorithm 3, a NPC defense strategy is described in the following algorithm.

From Algorithm 4, when packets dropout caused by DoS attacks happen in system (4.1), predictive inputs $\text{sat}(F\hat{x}(k))$ are used in actuators. Then system (4.1) remains stable according to Theorems 4.5-4.6. Due to effects of disturbances, state $x(k)$ of system (4.1) can diverge to outside of the domain of attraction with presented offense strategies. Therefore, for defending more

Algorithm 4 The algorithm for defense strategies

Require: Set $\hat{x}_0$, $\hat{x}_{0|-1} = \bar{x}_0$, Σ, Q, R and total running steps N.

1: Obtain the optimal domain of attraction $x(k)^T P x(k) < 1$ and saturation predictive controller $\text{sat}(u(k))$ under stability conditions.
2: **for** $k = 1 : 1 : N$ **do**
3: Calculate packets dropout upper bounds $p(k)$ as predictive upper bounds.
4: Calculate observer gain (4.33) and observe state $\hat{x}(k)$ at each step.
5: Predict the control inputs for $p(k)$ steps.
6: Package predictive values of each step in the predictive controller and send these packets to the actuator.
7: **if** $\nu(k) = 1$ **then**
8: The actuator selects $\text{sat}(u(k))$=$\text{sat}(F\hat{x}(k))$.
9: **else if** $\nu(k)$=0 **then**
10: The actuator selects $\text{sat}(u(k))$=$\text{sat}(F\hat{x}(k|k-i))$, $i \in \mathbf{I}[1, p]$.
11: **end if**
12: **end for**

intense DoS attacks than the presented scheme, this chapter has given a coping method by predictive control theories. The effectiveness of the presented NPC strategy is verified by the following numerical simulation.

4.4 Numerical Example

In this section, a single inverted pendulum (SIP) model system is given to demonstrate the effectiveness of the previously derived theorems. The SIP physical model is shown as Figure 4.2. In Figure 4.2, m and M are the mass

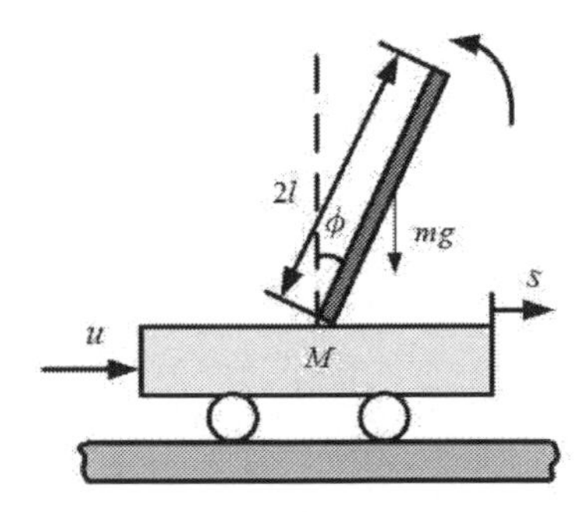

Figure 4.2 Physical model of the single inverted pendulum.

of the pendulum and the cart, respectively, l is the length between the center of the rotational shaft and the centroid of the pendulum, u is the force exerted on the cart, s is the displacement of the cart, ϕ is the pendulum angle from vertical. Furthermore, four states are given as $x = [s\ \dot{s}\ \phi\ \dot{\phi}]^T$. With sampling time T $= 0.005s$, the system model with actuator saturation is discretized as

$$x(k+1) = \begin{bmatrix} 1 & 0.005 & 0 & 0 \\ 0 & 0.9996 & 0.0031 & 0 \\ 0 & 0 & 1.0003 & 0.005 \\ 0 & -0.0012 & 0.1392 & 1.0003 \end{bmatrix} x(k) + \begin{bmatrix} 0 \\ 0.0044 \\ 0 \\ 0.0118 \end{bmatrix} \mathrm{sat}(u(k)) + w(k), \tag{4.56}$$

$$y(k) = \begin{bmatrix} 1\,0\,0\,0 \\ 0\,0\,1\,0 \end{bmatrix} x(k) + v(k). \tag{4.57}$$

Based on the optimization problem (4.16), an optimal estimation for the domain of attraction $x^T(k)Px(k) < 1$ and the corresponding controller $u(k) = Fx(k)$ are obtained, where

$$P = \begin{bmatrix} 3.564 & 6.823 & -14.759 & -5.381 \\ 6.823 & 14.4 & -35.912 & -11.102 \\ -14.759 & -35.912 & 262.833 & 53.776 \\ -5.381 & -11.102 & 53.776 & 15.303 \end{bmatrix},$$

$$F = \begin{bmatrix} 13.966\ 27.223\ -158.765\ -51.875 \end{bmatrix}.$$

An initial state and a referred shape are noted as $x_0 = [0\ 0.01\ 0.05\ 0]^T$ and $x_1 = [1.64\ -0.66\ -0.14\ 0.09]^T$, respectively. The optimal value of ϖ is obtained as $\varpi = 4.799$.

For destroying the stability of the SIP system (4.56), attackers choose the most cost-saving DoS attack schemes. Analysis on attack schemes has been presented in Sections II and III. Theorem 4.3 has given an approach to obtain the least consecutive DoS attack steps p for each time. Moreover, results on p for each step are shown in Figure 4.3. In order to make the Kalman filter (4.27)-(4.28) effective, arrival packet rate γ satisfies $\gamma \geq 0.0513$ referring to Theorem 4.4. Let $\gamma = 0.95$, $x_0 = [0\ 0.1\ 0.05\ 0]^T$, and $\hat{x}(0|-1) = [0.001\ 0.101\ 0.051\ 0.001]^T$. Covariance matrices of disturbances are chosen as $Q = 1 \times 10^{-8}I$, $R = 1 \times 10^{-8}I$ by inequality (4.50), then state $x(k)$ responses of SIP system (4.56) are shown as Figure 4.4. It is easy to find that the least attack steps p keep stable after $1s$ and $p = 336$ in Figure 4.3. Then from Figure 4.4, state $x(k)$ of SIP system (4.56) tends to zero after $1s$ with disturbances $w(k)$ and $v(k)$, that is, SIP system (4.56) has been stable. When SIP system (4.56) suffers consecutive DoS attacks at time $t = 1s$ for 336 steps, state $x(k)$ responses are given in Figure 4.5.

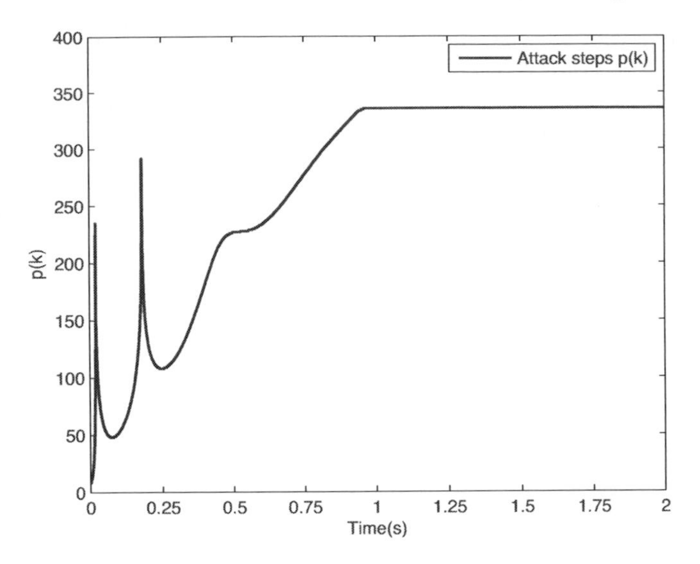

Figure 4.3 Least consecutive DoS attack steps for each step.

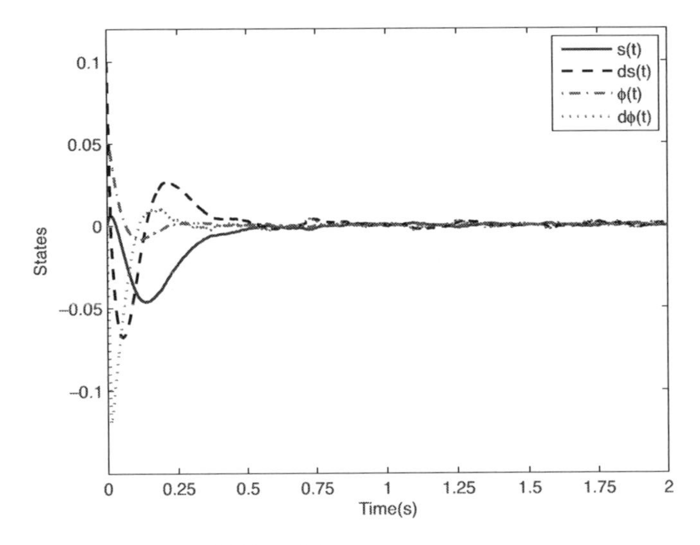

Figure 4.4 State responses of SIP model system.

From Figure 4.5, SIP system (4.56) is out of control after being attacked. When a NPC defense strategy is applied in SIP system (4.56), state $x(k)$ responses and input $\mathrm{sat}(u(k))$ responses are shown in Figure 4.6 and Figure 4.7, respectively.

In Theorem 4.6, stability conditions for SIP system (4.56) with NPC have been presented when disturbances exist. Comparing Figure 4.5 with Figure 4.6, the NPC scheme is able to resist DoS attacks which are under the optimal

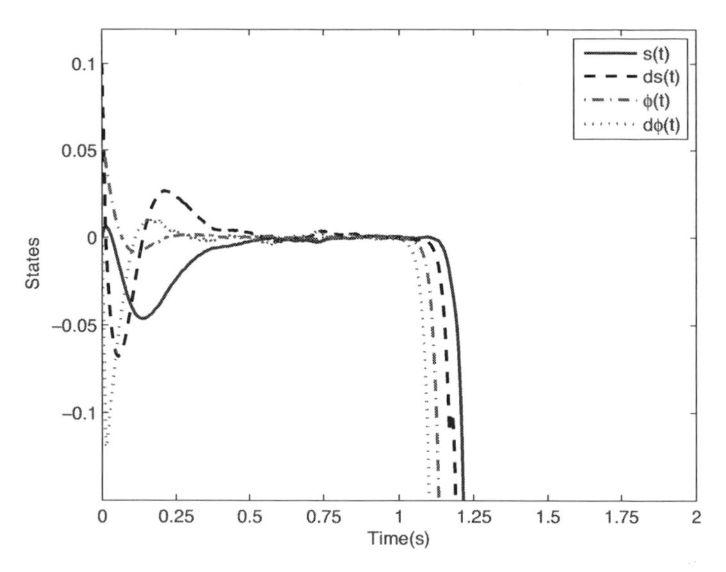

Figurc 4.5 State responses of SIP model system under optimal DoS attacks.

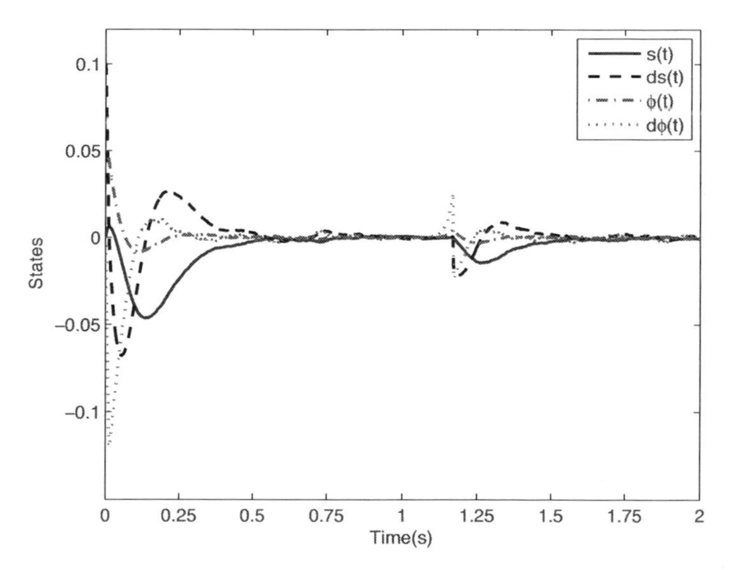

Figure 4.6 State responses of SIP model system under optimal DoS attacks with defense strategy.

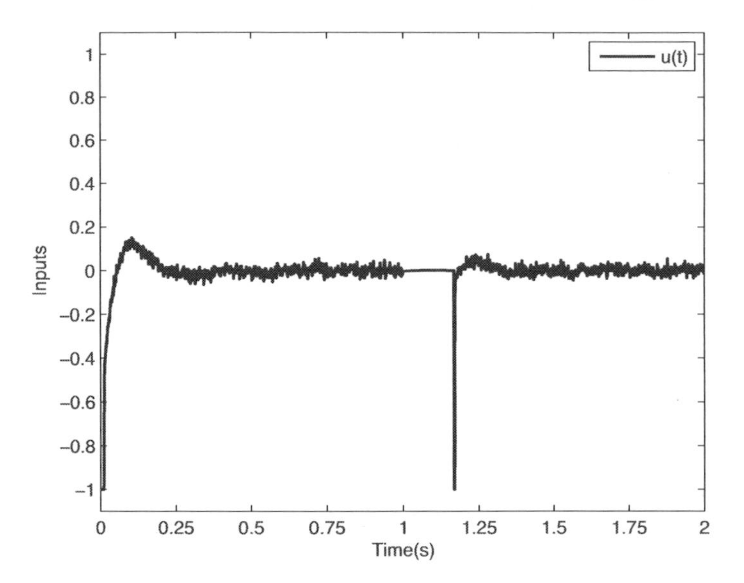

Figure 4.7 Input responses of SIP model system under optimal DoS attacks with defense strategy.

scheme. As shown in Figure 4.7, predictive control signals are smooth in attack periods, but real time control signals are rough when SIP system (4.56) is free of attacks. Then control errors are accumulated in every predictive step as presented in equation (4.45), which leads state $x(k)$ to be shifting in Figure 4.6. As a result, the effectiveness of the presented defense strategy has been shown.

4.5 Conclusion

This chapter has presented a defense strategy based on NPC schemes against an optimal DoS attack scheme on NCSs with actuator saturation. The least consecutive DoS attack steps p have been presented with properties for the domain of attraction, with which attackers realized attack objectives and minimized attack costs. Moreover, stability criteria of NCSs under NPC schemes have been presented for the defense strategy. Both attack and defense strategies have been shown in forms of algorithms with obtained results. Finally, a numerical simulation has illustrated the effectiveness and validity of the proposed schemes.

Part II
Resilient Control of WNCSs

Chapter 5
A Hierarchical Game Approach to Secure WNCSs

5.1 Introduction

In WNCSs, the sensor and actuator communicate with the controller through wireless networks. The availability attacks are also known as DoS attacks or jamming attacks, and they are realized by corrupting the communication network. Many control systems relying on real-time information become performance loss or even unstable under DoS attacks [193]. In[177], jamming attack scheduling schemes to optimize the linear quadratic Gaussian control cost function for single and multiple systems have been investigated. In practice, all the control systems operate in the presence of disturbances, which are caused by many factors such as load variation [38], friction [186], and measurement noise [168]. Therefore, considering the influence of disturbances on WNCSs is of great importance [45]. The H_∞ minmax control has advantages compared with the traditional observer-based disturbance control method when it is difficult to model the disturbances. Nowadays, the sampling intervals in industrial control systems are becoming smaller and smaller with the rapid development of sensing technology. The delta operator method is recommended as an effective way to solve the sampling issues for WNCSs [162]. In the last few decades, some inspiring control results have been reported for the control problems in the delta domain of NCSs [159].

In this chapter, a SINR-based DoS attack model under dynamic network environment is investigated for a WNCS, which is stratified as the cyber- and physical layer. In the presence of a DoS attack, a two player zero-sum game framework is introduced to model the interaction between transmitter and attacker in the cyber layer. Then, for the physical-layer, an H_∞ minmax control problem in delta domain is studied to guarantee the optimal system performance for high frequency sampled WNCS under disturbance. Iteration algorithms are given to solve the cross layer optimization problem.

The rest of this chapter is organized as follows. The problem formulation and design objectives are given in Section 5.2. In Section 5.3, the Markov zero-sum game in cyber layer is analysed when the transition probability of dynamic network is known and unknown. The H_∞ minimax controller is given in the form of saddle-point solution with dynamic programming. The value iteration and Q-learning algorithms are used for the coupled design of WNCS. In Section 5.4, numerical simulations on a two-area load frequency control system are shown to verify the effectiveness of the proposed method. Conclusions are drawn in Section 5.5.

5.2 Problem Setup

As shown in Figure 5.1, a WNCS under DoS attack is stratified as the cyber- and physical-layer, in which the controller and actuator are linked by wireless network as well as the sensor and controller communicating by reliable network. There exists a DoS attacker that aims at interfering with the transmission of the control signal between the transmitting and receiving modules. Under the DoS attacker and disturbance, system models are set up for cyber- and physical-layers, respectively.

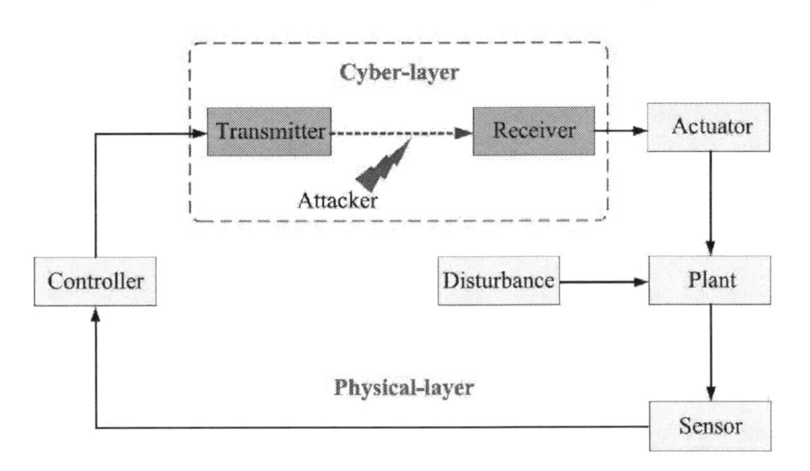

Figure 5.1 WNCS under DoS attack.

5.2.1 Transmit Model with SINR

In the cyber layer, the transmit power of transmitter p_m is chosen in M levels denoted as $\mathbf{p} = \{p_1, p_2, \cdots, p_M\}$, $\mathcal{M} = \{1, 2, \cdots, M\}$. The transmit strategy of attacker is denoted as w_l, which is chosen from $\mathbf{w} = \{w_1, w_2, \cdots, w_L\}$,

$\mathcal{L} = \{1, 2, \cdots, L\}$. Actually, the environment of a wireless communication network fluctuates over time. Assume that the wireless network is random, which subjects to a discrete-time Markov jump process in a finite set $\Pi = \{1, 2, \cdots, S\}$ [6]. The channel fading gain of transmitter in network mode $s, s \in \Pi$ is given as ζ_s, denoting that $\zeta_s \in \Xi = \{\zeta_1, \zeta_2, \cdots, \zeta_S\}$ and the channel gain of attacker is η_s, where $\eta_s \in \Gamma = \{\eta_1, \eta_2, \cdots, \eta_S\}$. The stationary transition probability from mode s at time n to s' at time $n+1$ is

$$\mathbb{P}\{s'(n+1)|s(n)\}, \; s'(n+1), s(n) \in \Pi \tag{5.1}$$

and one has that

$$\sum_{s' \in \Pi} \mathbb{P}\{s'(n+1)|s(n)\} = 1$$

In the presence of the transmitter and DoS attacker, the SINR at the receiver yields

$$\gamma_{T,s} = \frac{\zeta_s p_m}{\eta_s w_l + \sigma^2}, \; s \in \Pi, \; m \in \mathcal{M} \text{ and } l \in \mathcal{L} \tag{5.2}$$

where σ^2 is the additive white Gaussian noise. Assume that the communication between controller and actuator is modulated by quadrature amplitude technique. Based on digital communication theory [109], the relationship between Symbol Error Rate (SER) and SINR is shown as

$$\text{SER}(s) = 2\mathbf{Q}\left(\sqrt{\kappa \gamma_{T,s}}\right) \tag{5.3}$$

where

$$\mathbf{Q}(x) \triangleq \frac{1}{\sqrt{2\pi}} \int_x^{\infty} \exp(-\tau^2/2) \mathrm{d}\tau$$

and $\kappa > 0$ is a constant. Before proceeding, the following assumption is given.

Assumption 5.1 *The time scale on network environment varying which is usually based on order of hours is much larger than the physical systems which evolve in seconds.*

Function $r : \Pi \times \mathbf{p} \times \mathbf{w} \to \mathbb{R}$ is defined as the cost of possible action pair (p_m, w_l) in certain network mode s. Specially, the cost function is given as

$$r(s, p_m, w_l) = c_0 J_p(s) + c_1 p_m - c_2 w_l + C_{ml} \tag{5.4}$$

where $c_0 > 0$ is a weighting coefficient. $J_p(s)$ is the physical layer system performance. Parameters c_1 and c_2 are cost per unit energy consumption of transmitter and attacker. Scalar C_{ml} denotes the inherent cost when taking the action pair (p_m, w_l). The transmitter is a minimizer, that minimizes the cost function $r(s, p_m, w_l)$. In contrast, the attacker has an opposite objective, that wants to maximize the cost function. The transmitter and attacker are involved in a two player zero-sum game, and the relation is presented as

$$r(s, p_m, w_l) = r_T(s, p_m, w_l) = -r_J(s, p_m, w_l)$$

To give a formal model of the Markov dynamic jamming game, the distribution vectors are given subsequently. Denote that $f_m(s) \in [0,1]$ and $g_l(s) \in [0,1]$ are the probabilities that the transmitter and attacker choose action $p_m \in \mathbf{p}$ and $w_l \in \mathbf{w}$ in mode s. For specific mode s, we have $\sum_{m=1}^{M} f_m(s) = 1$ and $\sum_{l=1}^{L} g_l(s) = 1$. Then, denote that

$$\begin{aligned}
\mathbf{f}(s) &= [f_1(s), f_2(s), \cdots, f_M(s)],\ \forall s \in \Pi, \\
\mathbf{g}(s) &= [g_1(s), g_2(s), \cdots, g_L(s)],\ \forall s \in \Pi, \\
\mathbf{F} &= [\mathbf{f}(1), \mathbf{f}(2), \cdots, \mathbf{f}(S)], \\
\mathbf{G} &= [\mathbf{g}(1), \mathbf{g}(2), \cdots, \mathbf{g}(S)].
\end{aligned}$$

The following cost function J_c is introduced to quantify the discounted sum of the expected cost in cyber-layer

$$J_c(s) = \mathbb{E}_s^{\mathbf{f}(s),\mathbf{g}(s)} \left(\sum_{n=1}^{+\infty} \rho^n r_n(s, p_m, w_l) \right) \tag{5.5}$$

where s is the initial state choosing in Π and n is the time step, which is the time scale of the network environment varying. Parameter $\rho \in (0, 1)$ is the discount factor for discounting the future rewards.

5.2.2 Control Model under Disturbance

The control problem in the physical-layer of the WNCS is described in this subsection. Consider a general linear time-invariant system as follows

$$\dot{x}(t) = A_0 x(t) + B_0 u(t) + D_0 w(t) \tag{5.6}$$

where $x(t) \in \mathbb{R}^n$ is the state vector, $u(t) \in \mathbb{R}^m$ is the control signal and $\omega(t) \in \mathbb{R}^p$ is the disturbance. $A_0 \in \mathbb{R}^{n \times n}$, $B_0 \in \mathbb{R}^{n \times m}$, and $D_0 \in \mathbb{R}^{n \times p}$ are system matrices. By using the delta operator approach, system (5.6) is discretized as

$$\delta x(t_k) = A(t_k) x(t_k) + v(t_k) B(t_k) u(t_k) + D(t_k) \omega(t_k) \tag{5.7}$$

where

$$A(t_k) = \frac{e^{A_0 \mathtt{T}_k} - I}{\mathtt{T}_k},$$
$$B(t_k) = \frac{1}{\mathtt{T}_k} \int_0^{\mathtt{T}_k} e^{A_0(\mathtt{T}_k - \tau)} B_0 d\tau,$$
$$D(t_k) = \frac{1}{\mathtt{T}_k} \int_0^{\mathtt{T}_k} e^{A_0(\mathtt{T}_k - \tau)} D_0 d\tau.$$

In equation (5.7), the delta operator is given as follows

$$\delta x(t_k) = \frac{x(t_k + \mathtt{T}_k) - x(t_k)}{\mathtt{T}_k}$$

where $\mathtt{T}_k$ is sampling period, t is the continuous time, and k is the time step with $t_k = \sum_{l=0}^{k-1} \mathtt{T}_l$. The initial state $x(t_0)$ is assumed to be Gaussian random vector with mean 0 and covariance Σ. The attack-induced packets dropout is modeled as a Markov stochastic process. The process $\{v(t_k)\}$ obeys the following probability distribution

$$\begin{bmatrix} \mathbb{P}(v(t_{k+1}) = 0 \,|v(t_k) = 0) & \mathbb{P}(v(t_{k+1}) = 1 \,|v(t_k) = 0) \\ \mathbb{P}(v(t_{k+1}) = 0 \,|v(t_k) = 1) & \mathbb{P}(v(t_{k+1}) = 1 \,|v(t_k) = 1) \end{bmatrix} = \begin{bmatrix} 1-\alpha & \alpha \\ \beta & 1-\beta \end{bmatrix} \quad (5.8)$$

In the sequence, probabilities that a packet is received or not depend on the previous packet. The variables satisfy $0 < \alpha \leq 1$ and $0 < \beta \leq 1$. It is known that the Markov packets dropout model is a generalization of Bernoulli packets dropout. The average sojourn time in the mode of packets dropout is $(1-\alpha)/\alpha$ [50]. Scalar α, which is a critical condition for the convergence of the cost, is in a sense "more critical" than β. DoS attacker tends to destroy the system by making the system packets dropout consecutively [99]. Therefore, parameter α strictly connected with the probability to have a long "bursts" of packet losses is seen as the consequence of a game between the transmitter and DoS attacker. Then, based on equation (5.3) and the preceding discussion, we have

$$\alpha = 1 - \mathrm{SER} \quad (5.9)$$

For the stationary situation, in the absence of any past information, the probability is always the same for all steps, that is, $\mathbb{P}(v(t_0) = 0) = \cdots = \mathbb{P}(v(t_k) = 0) = \beta/(\alpha+\beta)$ and $\mathbb{P}(v(t_0) = 1) = \cdots = \mathbb{P}(v(t_k) = 1) = \alpha/(\alpha+\beta), \forall k > 0$.

Transmission control protocol (TCP) is applied in the WNCS and the information set $\mathcal{I}(t_k)$ is given as

$$\mathcal{I}(t_k) = \{x(t_0)\},\ \mathcal{I}(t_k) = \{x(t_0), \cdots, x(t_k), v(t_0), \cdots, v(t_{k-1})\} \quad (5.10)$$

and is acknowledged at each step k. Denote the control sequence $\{u(t_k)\}$ and disturbance sequence $\{\omega(t_k)\}$ as $\mu(\mathcal{I}(t_k))$ and $\nu(\mathcal{I}(t_k))$, respectively. The goal of this chapter is to determine the optimal action $\mu^*(\mathcal{I}(t_k))$ such that the following system performance

$$J_p(\mu, \nu, \mathbf{F}^*, \mathbf{G}^*) = \mathbb{E}_{v(t_k)}\{x^T(t_K)Q^K x(t_K) + \mathrm{T}_k \sum_{k=0}^{K-1} \{x^T(t_k)Qx(t_k) + v(t_k)u^T(t_k)Ru(t_k) - \gamma^2\omega^2(t_k)\}\} \tag{5.11}$$

is minimized with the worst case disturbance for a prescribed fixed value γ, where γ is an upper bound on the desired $\mathcal{L}_2$ gain disturbance attenuation. Above all, our design objectives are given in the following.

Problem 5.1. Let the strategy pair (μ, ν) be fixed. A pair $(\mathbf{F}^*, \mathbf{G}^*)$ is a saddle-point of the ρ-discount game if the following inequality

$$\mathcal{G}_1 : \mathbf{J}_c(\mathbf{F}^*, \mathbf{G}) \leq \mathbf{J}_c(\mathbf{F}^*, \mathbf{G}^*) \leq \mathbf{J}_c(\mathbf{F}, \mathbf{G}^*) \tag{5.12}$$

holds, where $\mathbf{J}_c = [J_c(1), J_c(2), \cdots, J_c(S)]$.

Problem 5.2. With given cyber layer periodical optimal strategy $(\mathbf{F}^*, \mathbf{G}^*)$, if for all feasible strategy (μ, ν), there exists

$$\mathcal{G}_2 : J_p(\mu^*, \nu, \mathbf{F}^*, \mathbf{G}^*) \leq J_p(\mu^*, \nu^*, \mathbf{F}^*, \mathbf{G}^*) \leq J_p(\mu, \nu^*, \mathbf{F}^*, \mathbf{G}^*) \tag{5.13}$$

then (μ^*, ν^*) is the saddle-point equilibrium, and the optimal system performance is $J_p(\mu^*, \nu^*, \mathbf{F}^*, \mathbf{G}^*)$.

A hierarchical game structure is introduced to characterize the cross layer relation of the whole WNCS. In the hierarchical game, the cyber layer game and physical layer game are updated crosswise. The optimal cyber-layer strategy $(\mathbf{F}^*, \mathbf{G}^*)$ leads to a specific Markov packets dropout transition probability α. On the other hand, the control system performance effected by Markov packets dropout is also considered as a part of the cost in the cyber-layer. Based on game theory, the optimal transmit strategy under attack and optimal H_∞ controller are given in the form of saddle-point equilibriums.

5.3 Main Results

In this section, solutions of the design objectives for WNCS are presented with saddle-point equilibriums for cyber- and physical-layer. The couple relationship between cyber and physical layers is captured as a hierarchical game. The value iteration and Q-learning methods are employed with known and unknown transition probability cases, respectively.

5.3.1 Strategy Design for $\mathcal{G}_1$

The mixed Nash equilibrium strategy for $\mathcal{G}_1$ is obtained when the transition probability matrix of network environment is known and unknown. The value iteration method and Q-learning method are exploited for the known and unknown transition probability matrix cases, respectively. The following lemma is provided for further analysis.

Lemma 5.3. *[87] The standard solution to problem (5.5) above is through an iterative search method that searches for a fixed point of the following Bellman equation*

$$\mathcal{Q}(s, p_m, w_l) = r_n(s, p_m, w_l) + \rho \sum_{s' \in \Pi} \mathbb{P}(s'|s) J_c(s', n) \tag{5.14}$$

Theorem 5.4. *The discounted, zero-sum, stochastic game has a value vector $\boldsymbol{J}_c^*$ that is the unique solution of equation*

$$\boldsymbol{J}_c^* = val([\boldsymbol{Q}(s)]_{s \in \Pi}) \tag{5.15}$$

where $val(\cdot)$ is a function yielding the game value of a zero-sum matrix game. $\boldsymbol{Q}(s)$ is an auxiliary matrix which is given as

$$\boldsymbol{Q}(s) = [\mathcal{Q}(s, p_m, w_l)]_{m \in \mathcal{M}, l \in \mathcal{L}} \tag{5.16}$$

where $\mathcal{Q}(s, p_m, w_l)$ is shown as equation (5.14) of Lemma 5.3.

Proof. The results are obtained by using Shapley's Theorem [120] and the proof is omitted.

In [120], the value iteration method is used for solving the zero-sum stochastic game (5.15). With an initial approximate game value, iterate according to

$$\mathbf{J}_c(n+1) = \mathrm{val}([\mathbf{Q}(s, \mathbf{J}_c(n))]_{s \in \Pi}) \tag{5.17}$$

Obviously, the transition probability matrix of the dynamic network environment must be known by transmitter and attacker when using the value iteration method in Theorem 5.4. However, the transmitter and attacker may not have access to such information. Therefore, Q-learning method is introduced to obtain the optimal value with limited information. The instrumental lemma is provided for the subsequent theorem as follows:

Lemma 5.5. *[58] Let $\mathbb{Q}$ be the space of all $\boldsymbol{Q}$ functions. Define a contraction operator $\mathcal{P}_n : \mathbb{Q} \to \mathbb{Q}$ that satisfies the following equation*

1. $\boldsymbol{Q}^*(s) = \mathbb{E}\{\mathcal{P}_n \boldsymbol{Q}^*(s)\}$.
2. *For all $s \in \Pi$, $\boldsymbol{Q}(s)$, $\boldsymbol{Q}^*(s) \in \mathbb{Q}$, there exists*

$$\|\mathcal{P}_n \boldsymbol{Q}(s) - \mathcal{P}_n \boldsymbol{Q}^*(s)\| \leq \varpi \|\boldsymbol{Q}(s) - \boldsymbol{Q}^*(s)\| + \lambda_n,$$

where $0 < \varpi < 1$ *and the sequence* $\lambda_n > 0$ *converges to 0 with probability 1.*

Then, the iteration

$$\boldsymbol{Q}_{n+1}(s) = (1 - \varsigma(n))\boldsymbol{Q}_n(s) + \varsigma(n)[\mathcal{P}_n \boldsymbol{Q}(s)]$$

converges to $\boldsymbol{Q}^*$ *with probability 1, if learning rate satisfies*

$$\varsigma(n) \in [0, 1), \sum_{n=0}^{+\infty} \varsigma(n) = +\infty, \sum_{n=0}^{+\infty} \varsigma^2(n) < +\infty.$$

Theorem 5.6. *If the transition probability is not known,* $\mathcal{Q}_{n+1}(s, p_m, w_l)$ *is updated as follows*

$$\begin{aligned} \mathcal{Q}_{n+1}(s, p_m, w_l) = & (1 - \theta(n))\mathcal{Q}_n(s, p_m, w_l) \\ & + \theta(n)\left(r_n(s, p_m, w_l) + \rho J_c^*(s^{'})\right) \end{aligned} \quad (5.18)$$

where $\mathcal{Q}_{n+1}(s, p_m, w_l)$ *is the* $(n+1)$*th iteration of (5.18). Scalar* $\theta(n)$ *is learning rate satisfies*

$$\theta(n) \in [0, 1], \sum_{n=0}^{+\infty} \theta(n) = +\infty, \sum_{n=0}^{+\infty} \theta^2(n) < +\infty,$$

The saddle value $J_c^*(s^{'})$ *is obtained as*

$$J_c^*(s^{'}) = \min_{f(s^{'})} \max_{g(s^{'})} \sum_{p_m, w_l} \mathcal{Q}(s^{'}, p_m, w_l) f_m(s^{'}) g_l(s^{'}). \quad (5.19)$$

Proof. It is known that $\mathcal{Q}_{n+1}(s, p_m, w_l)$ converges if Condition 1 and 2 hold in Lemma 5.5. Let the operator $\mathcal{P}_n$ be defined as

$$\mathcal{P}_n \mathbf{Q}(s) = r_n(s, p_m, w_l) + \rho \text{val}(\mathbf{Q}_n(s^{'})) \quad (5.20)$$

for all $\mathbf{Q} \in \mathbb{Q}$. Denote that $\mathbf{Q}^*(s)$ in the following is the steady state of equation (5.14) if it is converges.

$$\mathbf{Q}^*(s) = r_n(s, p_m, w_l) + \rho \sum_{s^{'} \in \Pi} \mathbb{P}(s^{'}|s)\mathbf{J}_c^*(s^{'}) \quad (5.21)$$

Based on (5.21), we have

$$\begin{aligned}\mathbf{Q}^*(s) &= \sum_{s' \in \Pi} \mathbb{P}(s'|s)(r_n(s, p_m, w_l) + \rho \mathbf{J}_c^*(s')) \\ &= r_n(s, p_m, w_l) + \rho \sum_{s' \in \Pi} \mathbb{P}(s'|s)\text{val}(\mathbf{Q}_n(s')) \\ &= \mathbb{E}\{\mathcal{P}_n \mathbf{Q}^*(s)\}\end{aligned} \tag{5.22}$$

It is concluded that Condition 1 sets up. Next we will show the contraction property. By matrix game theory, if $A \geq B$, we have $\text{val}(A) \geq \text{val}(B)$. It is obtained that

$$\begin{aligned}\text{val}(\mathbf{Q}_2(s)) &\geq \text{val}(\mathbf{Q}_1(s)) - \|\mathbf{Q}_1(s) - \mathbf{Q}_2(s)\|, \\ \text{val}(\mathbf{Q}_2(s)) &\leq \text{val}(\mathbf{Q}_1(s)) + \|\mathbf{Q}_1(s) - \mathbf{Q}_2(s)\|\end{aligned}$$

and then

$$|\text{val}(\mathbf{Q}_1(s)) - \text{val}(\mathbf{Q}_2(s))| \leq \|\mathbf{Q}_1(s) - \mathbf{Q}_2(s)\| \tag{5.23}$$

According to inequality (5.23), for all $s \in \Pi$, one has that

$$\begin{aligned}\|\mathcal{P}_n \mathbf{Q} - \mathcal{P}_n \mathbf{Q}^*\|_\infty &\leq \rho \max_{s \subset \Pi} \|\mathbf{Q}(s) - \mathbf{Q}^*(s)\| \\ &= \rho \|\mathbf{Q} - \mathbf{Q}^*\|_\infty\end{aligned} \tag{5.24}$$

By Lemma 5.5, it is concluded that $\mathbf{Q}_n \rightarrow \mathbf{Q}^*$. This completes the proof.

From Theorem 5.6, it is known that the discount factor ρ is related to convergence rate of Q-learning method. With the optimal auxiliary matrix $\mathbf{Q}^*$, the mixed saddle-point equilibrium strategies $\mathbf{f}^*(s)$, $\mathbf{g}^*(s)$ and optimal value $J_c^*(s)$, $s \in \Pi$ are obtained by equation (5.19).

5.3.2 Strategy Design for $\mathcal{G}_2$

With the periodical optimal transmit and attack strategies $(\mathbf{F}^*, \mathbf{G}^*)$, the following theorem is given to provide solutions for Problem 5.2. The preliminary notations and lemma for theorem are given.

$$\begin{aligned}
-\delta\mathcal{S}(t_k) &= Q + \alpha P_{u0}^T(t_k)RP_{u0}(t_k) - \gamma^2 P_{\omega 0}^T(t_k)P_{\omega 0}(t_k) \\
&\quad +(1-\alpha)(\mathtt{T}_k\mathcal{X}_{\mathcal{S}_0}^T(t_k)\mathcal{S}(t_{k+1})\mathcal{X}_{\mathcal{S}_0}(t_k) + \mathcal{S}(t_{k+1})\mathcal{X}_{\mathcal{S}_0}(t_k) \\
&\quad +\mathcal{X}_{\mathcal{S}_0}^T(t_k)\mathcal{S}(t_{k+1}))\alpha\left(\mathtt{T}_k\mathcal{X}_{\mathcal{R}_0}^T(t_k)\mathcal{R}(t_{k+1})\mathcal{X}_{\mathcal{R}_0}(t_k)\right. \\
&\quad \left.+\mathcal{R}(t_{k+1})\mathcal{X}_{\mathcal{R}_0}(t_k) + \mathcal{X}_{\mathcal{R}_0}^T(t_k)\mathcal{R}(t_{k+1})\right) \\
&\quad +\frac{\alpha}{\mathtt{T}_k}(\mathcal{R}(t_{k+1}) - \mathcal{S}(t_{k+1})) \qquad (5.25) \\
\mathcal{S}(t_k) &= \mathcal{S}(t_{k+1}) - \mathtt{T}_k\delta\mathcal{S}(t_k) \\
-\delta\mathcal{R}(t_k) &= Q + (1-\beta)P_{u1}^T(t_k)RP_{u1}(t_k) - \gamma^2 P_{\omega 1}^T(t_k)P_{\omega 1}(t_k) \\
&\quad +\alpha\left(\mathtt{T}_k\mathcal{X}_{\mathcal{R}_1}^T(t_k)\mathcal{R}(t_{k+1})\mathcal{X}_{\mathcal{R}_1}(t_k) + \beta\left(\mathtt{T}_k\mathcal{X}_{\mathcal{S}_1}^T(t_k)\mathcal{S}(t_{k+1})\mathcal{X}_{\mathcal{S}_1}(t_k)\right.\right. \\
&\quad \left.+\mathcal{S}(t_{k+1})\mathcal{X}_{\mathcal{S}_1}(t_k) + \mathcal{X}_{\mathcal{S}_1}^T(t_k)\mathcal{S}(t_{k+1})\right) + \mathcal{R}(t_{k+1})\mathcal{X}_{\mathcal{R}_1}(t_k) \\
&\quad \left.+\mathcal{X}_{\mathcal{R}_1}^T(t_k)\mathcal{R}(t_{k+1})\right) + \frac{\beta}{\mathtt{T}_k}(\mathcal{S}(t_{k+1}) - \mathcal{R}(t_{k+1})) \qquad (5.26) \\
\mathcal{R}(t_k) &= \mathcal{R}(t_{k+1}) - \mathtt{T}_k\delta\mathcal{R}(t_k)
\end{aligned}$$

where

$$\begin{aligned}
\Theta(t_k) &= R + \mathtt{T}_k B^T(t_k)\mathcal{R}(t_{k+1})B(t_k), \\
\Lambda_0(t_k) &= -\gamma^2 + (1-\alpha)\mathtt{T}_k D^T(t_k)\mathcal{S}(t_{k+1})D(t_k) \\
&\quad +\alpha\mathtt{T}_k D^T(t_k)\mathcal{R}(t_{k+1})D(t_k), \\
\Lambda_1(t_k) &= -\gamma^2 + \beta\mathtt{T}_k D^T(t_k)\mathcal{S}(t_{k+1})D(t_k) \\
&\quad +(1-\beta)\mathtt{T}_k D^T(t_k)\mathcal{R}(t_{k+1})D(t_k), \\
\Pi_{u0}(t_k) &= \Theta(t_k) - \alpha\mathtt{T}_k^2 B^T(t_k)\mathcal{R}(t_{k+1})D(t_k) \\
&\quad \times\Lambda_0^{-1}(t_k)D^T(t_k)\mathcal{R}(t_{k+1})B(t_k), \\
\Pi_{\omega 0}(t_k) &= \Lambda_0(t_k) - \alpha\mathtt{T}_k^2 D^T(t_k)\mathcal{R}(t_{k+1})B(t_k) \\
&\quad \times\Theta^{-1}(t_k)B^T(t_k)\mathcal{R}(t_{k+1})D(t_k), \\
\Pi_{u1}(t_k) &= \Theta(t_k) - (1-\beta)\mathtt{T}_k^2 B^T(t_k)\mathcal{R}(t_{k+1})D(t_k) \\
&\quad \times\Lambda_1^{-1}(t_k)D^T(t_k)\mathcal{R}(t_{k+1})B(t_k), \\
\Pi_{\omega 1}(t_k) &= \Lambda_1(t_k) - (1-\beta)\mathtt{T}_k^2 D^T(t_k)\mathcal{R}(k+1)B(t_k) \\
&\quad \times\Theta^{-1}(t_k)B^T(t_k)\mathcal{R}(t_{k+1})D(t_k), \\
\Xi_{u0}(t_k) &= I - \mathtt{T}_k D(t_k)\Lambda_0^{-1}(t_k)((1-\alpha)D^T\mathcal{S}(t_{k+1}) + \alpha D^T\mathcal{R}(t_{k+1})),
\end{aligned}$$

$$
\begin{aligned}
\Xi_{\omega 0}(t_k) &= (1-\alpha)\mathcal{S}(t_{k+1}) + \alpha\mathcal{R}(t_{k+1}) \\
&\quad -\alpha \mathrm{T}_k D^T \mathcal{R}(t_{k+1}) B \Theta^{-1}(k) B^T \mathcal{R}(t_{k+1}), \\
\Xi_{u1}(t_k) &= I - \mathrm{T}_k D \Lambda_1^{-1} D^T (\beta \mathcal{S}(t_{k+1}) + (1-\beta)\mathcal{R}(t_{k+1})), \\
\Xi_{\omega 1}(t_k) &= \beta \mathcal{S}(t_{k+1}) + (1-\beta)\mathcal{R}(t_{k+1}) \\
&\quad -(1-\beta)\mathrm{T}_k D^T(t_k)\mathcal{R}(t_{k+1}) B(t_k) \Theta^{-1}(t_k) B^T(t_k) \mathcal{R}(t_{k+1}), \\
P_{u0}(t_k) &= -\Pi_{u0}^{-1}(t_k) B^T(t_k) \mathcal{R}(t_{k+1}) \Xi_{u0}(t_k)(\mathrm{T}_k A(t_k) + I), \\
P_{\omega 0}(t_k) &= -\Pi_{\omega 0}^{-1}(t_k) D^T(t_k) \Xi_{\omega 0}(t_k)(\mathrm{T}_k A(t_k) + I), \\
P_{u1}(t_k) &= -\Pi_{u1}^{-1}(t_k) B^T(t_k) \mathcal{R}(t_{k+1}) \Xi_{u1}(t_k)(\mathrm{T}_k A(t_k) + I), \\
P_{\omega 1}(t_k) &= -\Pi_{\omega 1}^{-1}(t_k) D^T(t_k) \Xi_{\omega 1}(t_k)(\mathrm{T}_k A(t_k) + I), \\
\mathcal{X}_{\mathcal{S}_0}(t_k) &= A(t_k) + D(t_k) P_{\omega 0}(t_k),\ \mathcal{X}_{\mathcal{R}_0}(t_k) \\
&= A(t_k) + B(t_k) P_{u0}(t_k) + D(t_k) P_{\omega 0}(t_k), \\
\mathcal{X}_{\mathcal{S}_1}(t_k) &= A(t_k) + D(t_k) P_{\omega 1}(t_k),\ \mathcal{X}_{\mathcal{R}_1}(t_k) \\
&= A(t_k) + B(t_k) P_{u1}(t_k) + D(t_k) P_{\omega 1}(t_k)
\end{aligned}
$$

Lemma 5.7. *[162] For any time function $x(t_k)$ and $y(t_k)$, there exists the following property of delta operator*

$$\delta(x(t_k)y(t_k)) = y(t_k)\delta x(t_k) + x(t_k)\delta y(t_k) + T_k \delta x(t_k)\delta y(t_k),$$

where T_k is the sampling period.

Theorem 5.8. *With specific transmit and DoS attack strategies, information set $\mathcal{I}(t_k)$, a fixed $\gamma > 0$, and finite time-level K, the following constraint and conclusions are presented for the physical-layer game $\mathcal{G}_2$*

1. *The game $\mathcal{G}_2$ exists unique saddle-point solution if and only if*

$$\Theta(t_k) > 0,\ \Lambda_0(t_k) < 0,\ \Lambda_1(t_k) < 0, \forall k \in \boldsymbol{K}, \tag{5.27}$$

where $\mathcal{R}(t_k)$ and $\mathcal{S}(t_k)$ satisfy the Riccati recursive equations (5.25) and (5.26), respectively. Furthermore, set that $\mathcal{R}(t_K) = \mathcal{S}(t_K) = Q^K$.

2. *If matrices $\Pi_{u0}(t_k)$, $\Pi_{\omega 0}(t_k)$, $\Pi_{u1}(t_k)$, and $\Pi_{\omega 1}(t_k)$ are invertible, under Condition 1, the feedback saddle-equilibrium $(\mu^*(\mathcal{I}(t_k)), \nu^*(\mathcal{I}(t_k)))$ is given by*

- $v(t_{k-1}) = 0$

$$\begin{cases} u_0(t_k) = \mu^*(\mathcal{I}(t_k)) = P_{u0}(t_k)x(t_k), \\ \omega_0(t_k) = \nu^*(\mathcal{I}(t_k)) = P_{\omega 0}(t_k)x(t_k), \end{cases} \tag{5.28}$$

- $v(t_{k-1}) = 1$

$$\begin{cases} u_1(t_k) = \mu^*(\mathcal{I}(t_k)) = P_{u1}(t_k)x(t_k), \\ \omega_1(t_k) = \nu^*(\mathcal{I}(t_k)) = P_{\omega 1}(t_k)x(t_k), \end{cases} \tag{5.29}$$

3. The corresponding system performance is

$$J_K^* = \frac{\Sigma}{\alpha+\beta}(\beta\mathcal{S}(t_0) + \alpha\mathcal{R}(t_0)) \tag{5.30}$$

Proof. Let us construct the quadratic cost function $V(x(t_k))$ as

$$V(x(t_k)) = \begin{cases} \mathbb{E}\{x^T(t_k)\mathcal{S}(t_k)x(t_k)\},\ v(t_{k-1}) = 0 \\ \mathbb{E}\{x^T(t_k)\mathcal{R}(t_k)x(t_k)\},\ v(t_{k-1}) = 1 \end{cases}$$

For step t_{k+1}, we have

$$V(x(t_{k+1})) = \begin{cases} \mathbb{E}\{\mathrm{T}_k\delta(x^T(t_k)\mathcal{S}(t_k)x(t_k)) + x^T(t_k)\mathcal{S}(t_k)x(t_k)\},\ v(t_k) = 0 \\ \mathbb{E}\{\mathrm{T}_k\delta(x^T(t_k)\mathcal{R}(t_k)x(t_k)) + x^T(t_k)\mathcal{R}(t_k)x(t_k)\},\ v(t_k) = 1 \end{cases}$$

Furthermore,

$$V(x(t_{k+1})) = \begin{cases} \mathrm{T}_k^2\delta x_{\mathcal{S}}^{\mathrm{T}}(t_k)\mathcal{S}(t_{k+1})\delta x_{\mathcal{S}}(t_k) + \mathrm{T}_k\delta x_{\mathcal{S}}^{\mathrm{T}}(t_k)\mathcal{S}(t_{k+1})x(t_k) \\ +\mathrm{T}_k x^T(t_k)\mathcal{S}(t_{k+1})\delta x_{\mathcal{S}}(t_k) + x^T(t_k)\mathcal{S}(t_{k+1})x(t_k),\ v(t_k) = 0 \\ \mathrm{T}_k^2\delta x_{\mathcal{R}}^{\mathrm{T}}(t_k)\mathcal{R}(t_{k+1})\delta x_{\mathcal{R}}(t_k) + \mathrm{T}_k\delta x_{\mathcal{R}}^{\mathrm{T}}(t_k)\mathcal{R}(t_{k+1})x(t_k) \\ +\mathrm{T}_k x^T(t_k)\mathcal{R}(t_{k+1})\delta x_{\mathcal{R}}(t_k) + x^T(t_k)\mathcal{R}(t_{k+1})x(t_k),\ v(t_k) = 1 \end{cases}$$

where

$$\delta x_{\mathcal{S}}(t_k) = A(t_k)x(t_k) + D(t_k)\omega(t_k)$$
$$\delta x_{\mathcal{R}}(t_k) = A(t_k)x(t_k) + B(t_k)u(t_k) + D(t_k)\omega(t_k)$$

By using dynamic programming, when $v(t_{k-1}) = 0$ the cost function at t_k is

$$
\begin{aligned}
&V(x(t_k)) \\
&= \mathbb{E}\{x^T(t_k)\mathcal{S}(t_k)x(t_k)\} \\
&= \min_{u_0(t_k)} \max_{\omega_0(t_k)} \mathbb{E}\left\{\mathtt{T}_k x^T(t_k)Qx(t_k)\right. \\
&\quad \left. +\mathtt{T}_k v(t_k)u_0^T(t_k)Ru_0(t_k) - \mathtt{T}_k\gamma^2\omega_0^2(t_k) + V(x(t_{k+1}))\right\} \\
&= \min_{u_0(t_k)} \max_{\omega_0(t_k)} \mathbb{E}\left\{\mathtt{T}_k x^T(t_k)Qx(t_k) + \mathtt{T}_k v(t_k)u_0^T(t_k)Ru_0(t_k) - \mathtt{T}_k\gamma^2\omega_0^2(t_k)\right. \\
&\quad +\mathbb{P}(v(t_k)=0|v(t_{k-1})=0)(\mathtt{T}_k^2\delta x_{\mathcal{S}}^{\mathtt{T}}(t_k)\mathcal{S}(t_{k+1})\delta x_{\mathcal{S}}(t_k) \\
&\quad +\mathtt{T}_k\delta x_{\mathcal{S}}^{\mathtt{T}}(t_k)\mathcal{S}(t_{k+1})x(t_k) + \mathtt{T}_k x^T(t_k)\mathcal{S}(t_{k+1})\delta x_{\mathcal{S}}(t_k) \\
&\quad +x^T(t_k)\mathcal{S}(t_{k+1})x(t_k)) \\
&\quad +\mathbb{P}(v(t_k)=1|v(t_{k-1})=0)(\mathtt{T}_k^2\delta x_{\mathcal{R}}^{\mathtt{T}}(t_k)\mathcal{R}(t_{k+1})\delta x_{\mathcal{R}}(t_k) \\
&\quad \left. +\mathtt{T}_k\delta x_{\mathcal{R}}^{\mathtt{T}}(t_k)\mathcal{R}(t_{k+1})x(t_k) + \mathtt{T}_k x^T(t_k)\mathcal{R}(t_{k+1})\delta x_{\mathcal{R}}(t_k)\right\} \\
&\quad +x^T(t_k)\mathcal{R}(t_{k+1})x(t_k)) \\
&= \min_{u_0(t_k)} \max_{\omega_0(t_k)} \left\{\mathtt{T}_k x^T(t_k)Qx(t_k) + \mathtt{T}_k\alpha u_0^T(t_k)Ru_0(t_k) - \mathtt{T}_k\gamma^2\omega_0^2(t_k)\right. \\
&\quad +(1-\alpha)((\mathtt{T}_k A(t_k)+I)x(t_k) + \mathtt{T}_k D(t_k)\omega_0(t_k))^T\mathcal{S}(t_{k+1})\left((\mathtt{T}_k A(t_k)+I)x(t_k)\right. \\
&\quad +\mathtt{T}_k D(t_k)\omega_0(t_k)) + \alpha\left((\mathtt{T}_k A(t_k)+I)x(t_k) + \mathtt{T}_k B(t_k)u_0(t_k)\right. \\
&\quad +\mathtt{T}_k D(t_k)\omega_0(t_k))^T\mathcal{R}(t_{k+1})\left((\mathtt{T}_k A(t_k)+I)x(t_k)\right. \\
&\quad \left. +\mathtt{T}_k B(t_k)u_0(t_k) + \mathtt{T}_k D(t_k)\omega_0(t_k))\right\}
\end{aligned} \tag{5.31}
$$

Under $\Theta(t_k) > 0$ and $\Lambda_0(t_k) < 0$ in Condition 1 of Theorem 5.8, one has that $V(x(t_k))$ is convex with respect to $u_0(t_k)$ yet it is concave in $\omega_0(t_k)$. The first order derivation of $V(x(t_k))$ in $u_0(t_k)$ and $\omega_0(t_k)$ are given as

$$
\begin{aligned}
\frac{\partial V(x(t_k))}{\partial u_0(t_k)} &= 2\alpha\mathtt{T}_k Ru_0(t_k) + 2\alpha\mathtt{T}_k^2 B^T(t_k)\mathcal{R}(t_{k+1})B(t_k)u_0(t_k) \\
&\quad +2\alpha\mathtt{T}_k B^T(t_k)\mathcal{R}(t_{k+1})(\mathtt{T}_k A(t_k)+I)x(t_k) \\
&\quad +2\alpha\mathtt{T}_k^2 B^T(t_k)\mathcal{R}(t_{k+1})D(t_k)\omega_0(t_k)
\end{aligned}
$$

$$
\begin{aligned}
\frac{\partial V(x(t_k))}{\partial \omega_0(t_k)} &= -2\gamma^2\mathtt{T}_k\omega_0(t_k) \\
&\quad +2(1-\alpha)\mathtt{T}_k D^T(t_k)\mathcal{S}(t_{k+1})(\mathtt{T}_k A(t_k)+I)x(t_k) \\
&\quad +2(1-\alpha)\mathtt{T}_k^2 D^T(t_k)\mathcal{S}(t_{k+1})D(t_k)\omega_0(t_k) \\
&\quad +2\alpha\mathtt{T}_k^2 D^T(t_k)\mathcal{R}(t_{k+1})D(t_k)\omega_0(t_k) \\
&\quad +2\alpha\mathtt{T}_k D^T(t_k)\mathcal{R}(t_{k+1})(\mathtt{T}_k A(t_k)+I)x(t_k) \\
&\quad +2\alpha\mathtt{T}_k^2 D^T(t_k)\mathcal{R}(t_{k+1})B(t_k)u_0(t_k)
\end{aligned}
$$

Thus, the first order sufficient and necessary condition for the convexity and concavity are

$$\begin{aligned}&(R+\mathtt{T}_kB^T(t_k)\mathcal{R}(t_{k+1})B(t_k))u_0(t_k)\\&=-(B^T(t_k)\mathcal{R}(t_{k+1})(\mathtt{T}_kA(t_k)+I)x(t_k)\\&\quad+\mathtt{T}_kB^T(t_k)\mathcal{R}(t_{k+1})D(t_k)\omega_0(t_k))\end{aligned}\tag{5.32}$$

and

$$\begin{aligned}&(-\gamma^2+(1-\alpha)\mathtt{T}_kD^T(t_k)\mathcal{S}(t_{k+1})D(t_k)\\&+\alpha\mathtt{T}_kD^T(t_k)\mathcal{R}(t_{k+1})D(t_k))\omega_0(t_k)\\&=-((1-\alpha)D^T(t_k)\mathcal{S}(t_{k+1})(\mathtt{T}_kA(t_k)+I)x(t_k)\\&+\alpha D^T(t_k)\mathcal{R}(t_{k+1})(\mathtt{T}_kA(t_k)+I)x(t_k)\\&+\alpha D(t_k)^T\mathcal{R}(t_{k+1})B(t_k)u_0(t_k))\end{aligned}\tag{5.33}$$

Combining (5.32) and (5.33), the saddle-point equilibrium $\mu^*(\mathcal{I}(t_k))$ and $\nu^*(\mathcal{I}(t_k))$ is obtained as (5.28). Similarly, substituting (5.32) and (5.33) into (5.31), the Riccati recursive equation is obtained as (5.25).

With the same technique, when $v(t_{k-1})=1$ sets up, the saddle-point equilibrium (5.29) and Riccati recursive equation (5.26) are concluded.

The optimal value J_p^* for finite-time horizon system performance optimization problem is

$$\begin{aligned}J_p^*&=\mathbb{E}\{V(x(t_0))\}\\&=\mathbb{P}(v(t_{-1})=0)(\mathbb{E}\{x^T(t_0)\mathcal{S}(t_0)x(t_0)\})\\&\quad+\mathbb{P}(v(t_{-1})=1)(\mathbb{E}\{x^T(t_0)\mathcal{R}(t_0)x(t_0)\})\end{aligned}$$

Since the Markov packets dropout process is assumed to be irreducible and stationary, it is concluded that $\mathbb{P}(v(t_{-1})=0)=\beta/(\alpha+\beta)$, $\mathbb{P}(v(t_{-1})=1)=\alpha/(\alpha+\beta)$ and then

$$J_p^*=\frac{1}{\alpha+\beta}\mathrm{tr}(\beta\Sigma\mathcal{S}(t_0)+\alpha\Sigma\mathcal{R}(t_0))\tag{5.34}$$

This completes the proof.

The optimal disturbance attenuation level γ^* is introduced as

$$\gamma^*=\inf\{\gamma\mid\Lambda_0(t_k)<0\text{ and }\Lambda_1(t_k)<0,\ \forall k\in\mathbf{K}\}\tag{5.35}$$

which means that Problem 5.2 admits a unique saddle point if $\gamma>\gamma^*$.

Remark 5.9. Note that the physical-layer is in a small time scale relative to the cyber-layer. Thus, the delta operator approach is used in physical-layer to overcome the numerical defects for high frequency sampling [16, 160]. In a control system, all signals are sampled by a sampler, and an appropriate sampling period interval should be chosen based on network load conditions. The sampling period in delta operator systems is an explicit parameter, thus it is easy to observe and analyze the control effect with different network load conditions. Therefore, in this chapter time varying sampling periods are applied for control system in delta domain which can present many advantages when network load conditions are changing.

5.3.3 Coupled Design for the WNCS

By using the value iteration or Q-learning method, the mixed-saddle point equilibrium strategies are obtained as $\mathbf{f}^*(s)$ and $\mathbf{g}^*(s)$. Then, the average transmit power of transmitter and attacker is given as

$$\bar{p}(s,n) = \mathbf{f}^*(s)\mathbf{p}^T, \ \bar{w}(s,n) = \mathbf{g}^*(s)\mathbf{w}^T, \ s \in \Pi \tag{5.36}$$

According to (5.3) and (5.9), one has that

$$\alpha(s,n) = 1 - 2\mathbf{Q}\left(\sqrt{\kappa\frac{\zeta(s)\bar{p}(s,n)}{\eta(s)\bar{w}(s,n)+\sigma^2}}\right) \tag{5.37}$$

That is, the cyber layer impacts physical system performance through parameter α. On the other hand, the players in cyber-layer can detect the running state of the physical system and take the system performance into consideration, i.e., $J_p(s)$ is a part of the cost function of the transmitter and attacker as in equation (5.4).

Algorithm 5 and 6 are proposed for the coupled design. The algorithms involve iteration for computing the stationary mixed saddle-point equilibrium of the hierarchical game. Specially, since the game here is zero-sum and finite, the value iteration method of Algorithm 5 converges to the stationary saddle-point equilibrium [120], [112]. The convergence of the Q-learning method is shown in Theorem 5.6. In what follows, the detailed implementation steps are provided in Algorithm 5 and 6.

5.4 Simulation Example

To demonstrate the validity of the proposed method, the design scheme is applied into a load frequency control problem of a two-area interconnected power system. In the power system model, the control signal is transmitted by wireless network assuming that an DoS attacker exists, which is shown as Figure 5.2. The system equation of a two-area interconnected power system is

$$\dot{x}(t) = Ax(t) + Bu(t) + FP_d(t), \tag{5.38}$$

Algorithm 5 Value iteration algorithm for known transition probability

Require: $p_m \in \mathbf{p}$, $w_l \in \mathbf{w}$.

1: Initialize ρ, $J_c(s,0), \forall s \in \Pi$ and give a small enough scalar ε.
2: **while** $\|\mathbf{J}_c(n+1) - \mathbf{J}_c(n)\| > \varepsilon$ **do**
3: **for** $s = 1, 2, \cdots, S$ **do**
4: Calculate recursive Riccati equations and optimal control strategies according to equations (5.25), (5.26), (5.28), and (5.29). Obtain the optimal system performance $J_p^*(s,n)$, and then compute cost $r_n(s, p_m, w_l)$ using (5.4).
5: Calculate the cost matrix $\mathbf{Q}(s)$ using (5.16).
6: Find $J_c^*(s, n+1)$ by solving the following linear programming problem (LP1).

$$
\begin{aligned}
\text{(LP1)} \quad & 1/J_c(s,n+1) = \max_{\tilde{y}(s)} \tilde{y}^T(s)\mathbf{1}_m \\
\text{s.t.} \quad & \mathbf{Q}^T(s)\tilde{y}(s) \leq \mathbf{1}_n \\
& \tilde{y}(s) \geq 0.
\end{aligned}
$$

7: **end for**
8: $n = n + 1$.
9: **end while**
10: The optimal mixed strategy $\mathbf{F}^*$ is obtained using $\mathbf{f}^*(s) = \tilde{y}(s)J_c^*(s,n)$, $s \in \Pi$. For attacker, the optimal mixed strategy $\mathbf{G}^*$ is obtained by solving the dual problem of (LP1).
11: Output the optimal transmit and control strategies under different dynamic network environment induced system mode.

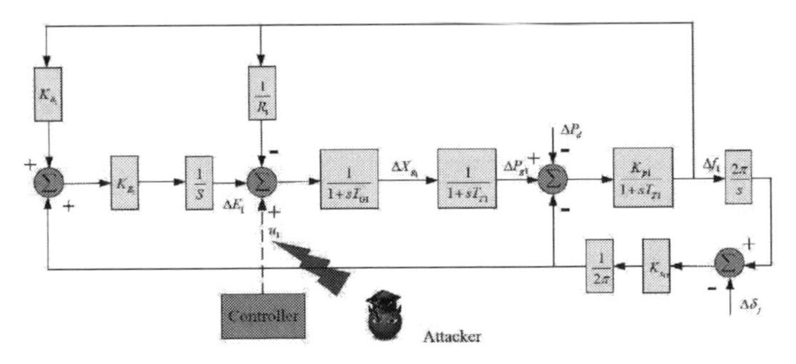

Figure 5.2 Structure of load frequency control system under DoS attack.

where

Algorithm 6 Q-learning algorithm for unknown transition probability

---

Require: $p_m \in \mathbf{p}$, $w_l \in \mathbf{w}$.

1: Initialize ρ, N, $\mathcal{Q}(s, p_m, w_l)$, $\forall s \in \Pi, m \in \mathcal{M}, l \in \mathcal{L}$.
2: **while** $n \leq N$ **do**
3: Choose action p_m. Observe the action of attacker w_l and the next Markov mode s'.
4: Calculate recursive Riccati equations and optimal control strategies with equations (5.25), (5.26), (5.28), and (5.29). Obtain the optimal system performance $J_p^*(s, n)$. Calculate cost $r_n(s, p_m, w_l)$ by using (5.4).
5: Calculate $J_c^*(s')$ and mixed strategy $\mathbf{f}^*(s')$, $\mathbf{g}^*(s')$ by using (5.19).
6: Update $\mathbf{Q}_{n+1}(s)$ with (5.18).
7: $n = n + 1$.
8: **end while**
9: Output the optimal transmit and control strategies $\mathbf{f}^*(s)$, $\mathbf{g}^*(s)$, $\forall s \in \Pi$, $\mu^*(\mathcal{I}(t_k))$, and $\nu^*(\mathcal{I}(t_k))$.

$$
x(t) = \left[x^{1T}(t)\; x^{2T}(t)\right]^T, \; u(t) = \left[u^{1T}(t)\; u^{2T}(t)\right]^T,
$$

$$
A = \begin{bmatrix} A^{11} & A^{12} \\ A^{21} & A^{22} \end{bmatrix}, \; B = \mathrm{diag}\left[B^1\; B^2\right], \; F = \left[F^1\; F^2\right]^T,
$$

$$
A^{ii} = \begin{bmatrix} -\frac{1}{\mathrm{T}_{p_i}} & \frac{K_{p_i}}{\mathrm{T}_{p_i}} & 0 & 0 & -\frac{K_{p_i}}{2\pi \mathrm{T}_{p_i}} \sum_{j \in \mathbf{S}, j \neq i} K_{s_{ij}} \\ 0 & -\frac{1}{\mathrm{T}_{\mathrm{T}_i}} & \frac{1}{\mathrm{T}_{\mathrm{T}_i}} & 0 & 0 \\ -\frac{1}{R_i \mathrm{T}_{G_i}} & 0 & -\frac{1}{\mathrm{T}_{G_i}} & \frac{1}{\mathrm{T}_{G_i}} & 0 \\ K_{E_i} K_{B_i} & 0 & 0 & 0 & \frac{K_{E_i}}{2\pi} \sum_{j \in \mathbf{S}, j \neq i} K_{s_{ij}} \\ 2\pi & 0 & 0 & 0 & 0 \end{bmatrix},
$$

$$
B^i = \left[0\; 0\; \frac{1}{\mathrm{T}_{G_i}}\; 0\; 0\right]^T, \; F^i = \left[\frac{K_{p_i}}{\mathrm{T}_{p_i}}\; 0\; 0\; 0\; 0\right],
$$

$$
A^{ij} = \begin{bmatrix} 0\,0\,0\,0 & -\frac{K_{p_i}}{2\pi \mathrm{T}_{p_i}} K_{s_{ij}} \\ 0\,0\,0\,0 & 0 \\ 0\,0\,0\,0 & 0 \\ 0\,0\,0\,0 & \frac{K_{E_i}}{2\pi} K_{s_{ij}} \\ 0\,0\,0\,0 & 0 \end{bmatrix}, \; x^i(t) = \begin{bmatrix} \Delta f_i(t) \\ \Delta P_{g_i}(t) \\ \Delta X_{g_i}(t) \\ \Delta E_i(t) \\ \Delta \delta_i(t) \end{bmatrix}, \; i, j \in \{1, 2\}.
$$

Variables $\Delta f_i(t)$, $\Delta P_{g_i}(t)$, $\Delta X_{g_i}(t)$, $\Delta E_i(t)$, and $\Delta \delta_i(t)$ are the changes of frequency, power output, governor valve position, integral control, and rotor angle deviation, respectively. $\Delta P_{d_i}(t) \in \mathbb{R}^k$ is the vector of load disturbance. Parameters T_{p_i}, $\mathrm{T}_{\mathrm{T}_i}$, and T_{G_i} are time constants of power system, turbine, and governor, respectively. Constants K_{p_i}, K_{E_i}, K_{B_i} are power system gain, integral control gain, and frequency bias factor, and $K_{s_{ij}}$ is the interconnection gain between area i and j ($i \neq j$). Parameter R_i is the speed regulation coefficient. Some basic parameters of the system are shown in Table I of [96].

In the attack model, two network modes are considered, i.e., $\Pi = \{1, 2\}$. The sets of channel gains are $\Xi = [0.5, 0.2]$, and $\Gamma = [0.3, 0.1]$. The transition probability from mode 1 to mode 2 is assumed to be 0.5, and from mode 2 to mode 1 is 0.3. The strategy sets of transmitter and attacker are $\mathbf{p} = [1, 3]$, $\mathbf{w} = [0.5, 2]$. Set network parameter $\kappa(1) = 0.8$, $\kappa(2) = 0.6$ and background noise $\sigma^2 = 0.05$. Background packets dropout rate in two network modes are given as $\beta(1) = 0.4$ and $\beta(2) = 0.6$. For the control system model, the sampling period is uniformly distributed in [0.04, 0.06]. The weighting matrices are $Q^K = Q = I_{10\times10}$, $R = I_{2\times2}$ and $\mathcal{S}(t_K) = \mathcal{R}(t_K) = Q^K$. Set the attenuation level as $\gamma = 35$, the variance of initial state as $\Sigma = 1$, and the finite time-level as $K = 300$.

Set the initial value as $J_c(1, 0) = 0$, $J_c(2, 0) = 0.2$, and the optimization accuracy $\varepsilon = 0.001$. By using Algorithm 5, the evolution of optimal values for game $\mathcal{G}_1$ with $\rho = 0.5$ and $\rho = 0.05$ is shown in Figure 5.3. Set that $N = 20$, initial value $Q(s, p_m, w_l) = 0, \forall s \in \Pi, m \in \mathcal{M}, l \in \mathcal{L}$. The evolution of optimal values by using the Q-learning method is shown in Figure 5.4. The optimal values of Algorithm 5 and Algorithm 6 are compared in Table 5.1 and Table 5.2. It is concluded that using the Q-learning method, the close results to value iteration method are obtained, which illustrates the Q-learning method is effective when the transition probability is unknown. Under the optimal mixed strategies, the evolution of control input and disturbance are shown as Figure 5.5 and Figure 5.6.

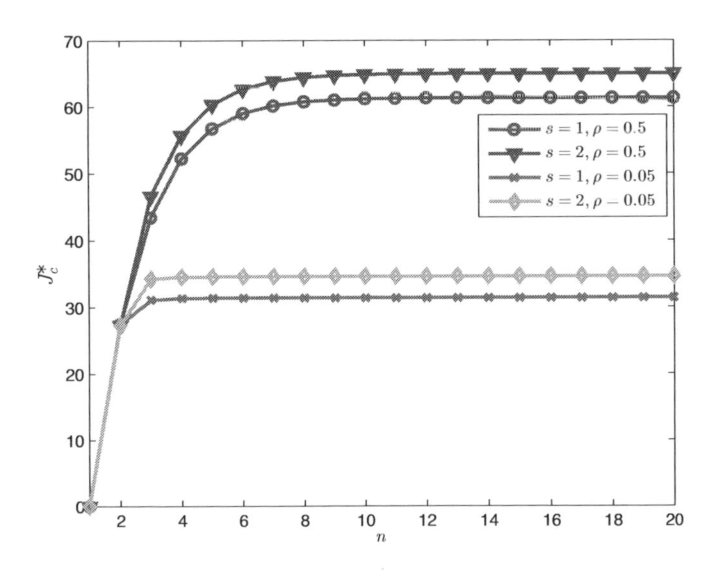

Figure 5.3 Evolution of cost value J_c^* by Algorithm 5.

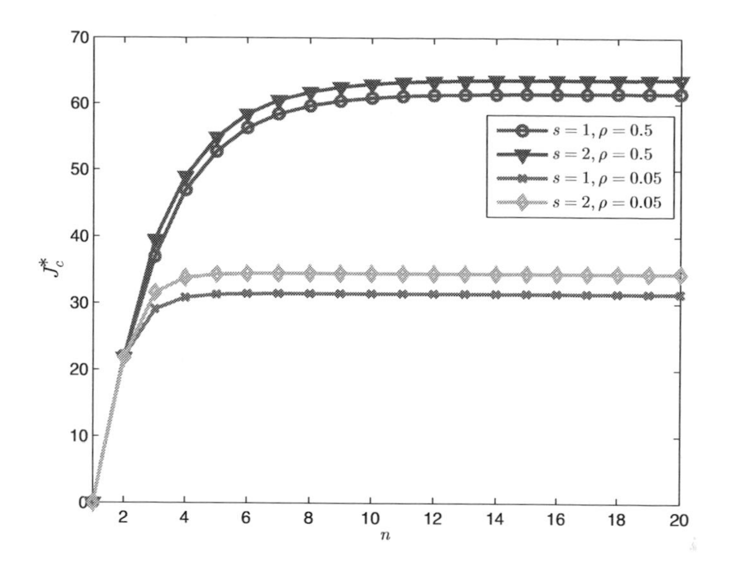

Figure 5.4 Evolution of cost value J_c^* by Algorithm 6.

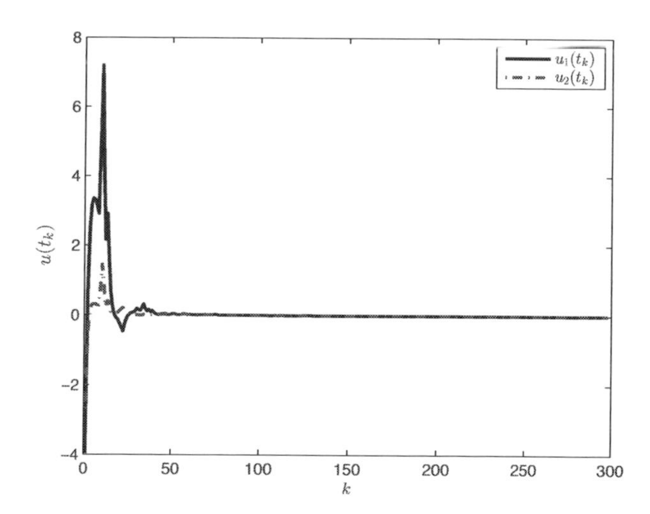

Figure 5.5 The evolution of control input $u(t_k)$.

Table 5.1 Comparison of optimal values between Algorithm 5 and Algorithm 6

	$[J_c^*(1)\ J_c^*(2)]$	$[J_p^*(1)\ J_p^*(2)]$	$[\alpha(1)\ \alpha(2)]$
Value Iteration $\rho = 0.5$	[62.0645 66.5901]	[586.7915 672.3531]	[0.6970 0.6546]
Value Iteration $\rho = 0.05$	[31.3810 34.6398]	[583.1298 651.7269]	[0.7013 0.6716]
Q-learning $\rho = 0.5$	[61.4967 63.6107]	[582.8587 650.3495]	[0.7016 0.6728]
Q-learning $\rho = 0.05$	[31.4410 34.4610]	[582.8587 650.3194]	[0.7016 0.6732]

Table 5.2 Comparison of mixed strategies between Algorithm 5 and Algorithm 6

	$\mathbf{f}(1)$	$\mathbf{f}(2)$	$\mathbf{g}(1)$	$\mathbf{g}(2)$
Value Iteration $\rho = 0.5$	[0.6930 0.3070]	[0.8818 0.1182]	[0.2427 0.7573]	[0.2709 0.7291]
Value Iteration $\rho = 0.05$	[0.6879 0.3121]	[0.8588 0.1412]	[0.2495 0.7505]	[0.2847 0.7153]
Q-learning $\rho = 0.5$	[0.6875 0.3125]	[0.8571 0.1429]	[0.2500 0.7500]	[0.2857 0.7143]
Q-learning $\rho = 0.05$	[0.6875 0.3125]	[0.8563 0.1437]	[0.2500 0.7500]	[0.2854 0.7146]

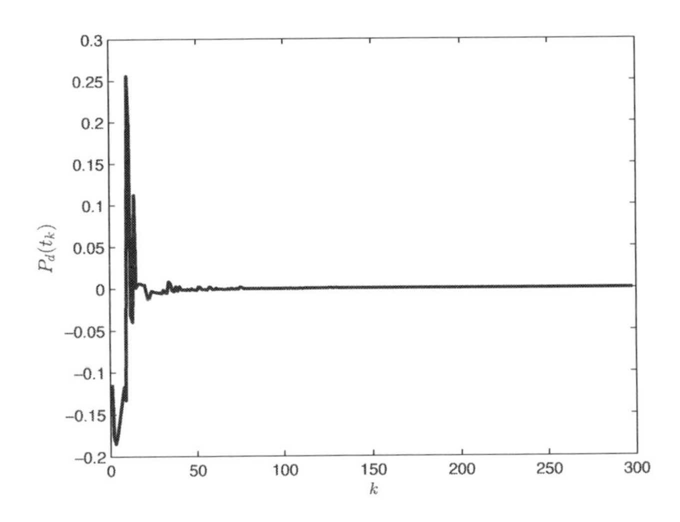

Figure 5.6 The evolution of disturbance $P_d(t_k)$.

Figure 5.7 depicts the relation curves between physical-layer system performance J_p^* versus per power cost c_1, which shows that when the unit power cost of the transmitter increases, the system performance of physical-layer J_p^* increases. This illustrates that when the cost of per power consumption increases, the transmitter is not willing to transmit information with large power, so that system performance is deteriorated in physical layer. The infimum of disturbance attenuation level versus parameter α is shown in Figure

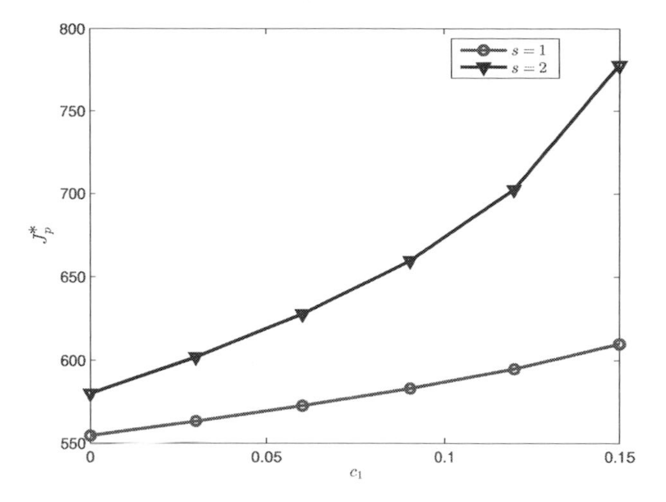

Figure 5.7 System performance J_p^* versus c_1.

5.8, which means that the disturbance attenuation level beyond the curve can be achieved by using the proposed zero-sum game approach.

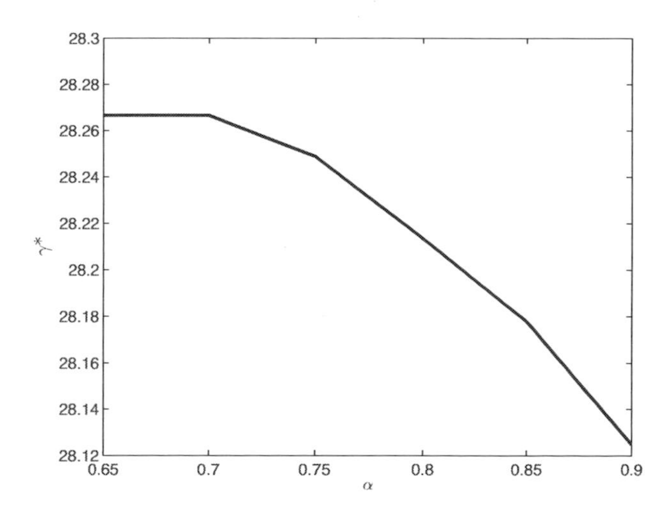

Figure 5.8 Disturbance attenuation level γ^* versus α.

5.5 Conclusion

In this chapter, a hierarchical game approach has been investigated for the resilient control problem of the WNCS under DoS attack. A layered model integrating the cyber security and physical control process has been established. The zero-sum Markov game has been exploited to model the interaction between transmitter and DoS attacker in the cyber-layer. Under the attack-induced packets dropout, control system with disturbance has been modeled in delta domain to overcome the numerical stiffness for high frequency sampling in the physical layer. The optimal power transmit and H_∞ control strategies have been derived in the form of saddle-point equilibriums by using value iteration and Q-learning methods. Finally, a simulation example has been provided to verify the effectiveness of the proposed methodology.

Chapter 6
A Bayesian Game Approach to Secure WNCSs

6.1 Introduction

Wireless networks are susceptible to interference from outside, which is a serious problem for WNCSs. To analyze jamming attacks on WNCSs, game theory that acts as a powerful tool has been employed to model interactions between legitimate users and malicious jammers [22]. Stochastic game framework for anti-jamming defense design is proposed with time-varying spectrum environment in a cognitive radio network [135]. In [117], a Bayesian jamming game between a legitimate transmitter and a smart jammer is discussed when there exists incomplete information for every network user. The Stackelberg game is a well-developed and appropriate method to cope with hierarchical interactions among players in the anti-jamming field [155]. Furthermore, an anti-jamming Bayesian Stackelberg game with incomplete information is proposed in [66]. In the cyber layer of a WNCS, the interaction between a malicious jammer and a legitimate user is reasonable to take as a Bayesian Stackelberg game in incomplete information environment. As mentioned in [171, 170, 173], resilient control strategies that focus on maintaining the operational normalcy expand the scope of classical control methods. Then to make NCSs achieve optimal performances, H_∞ minimax theory is employed while a plant controller is designed under the worst case [16].

In this chapter, a resilient control problem is investigated when WNCSs suffer malicious jamming attacks in cyber-layers. In the presence of incomplete information, a game framework between a malicious jammer and a legitimate user is established as a Bayesian Stackelberg game. The legitimate user is shown as the leader who takes actions first while the malicious jammer is the follower. Moreover, an H_∞ minimax control problem is studied to guarantee an optimal WNCS performance under disturbances. A coupled design between cyber- and physical-layers in WNCSs is presented to eliminate observation errors on transmitting power by an iterative algorithm.

The remainder of this chapter is organized as follows. In Section 6.2, a WNCS framework and a Bayesian Stackelberg game equilibrium definition in wireless communication channels are prescribed. Section 6.3 discusses a game process in the cyber-layer and an H_∞ minimax controller in the physical-layer. Meanwhile, a coupled design between cyber- and physical-layers is shown with a dynamical algorithm in this section. In Section 6.4, a numerical example is provided to illustrate the effectiveness of theoretical results. Conclusions of this chapter are drawn in Section 6.5.

6.2 Problem Formulation

6.2.1 WNCS Framework

A WNCS is pictured as suffering attacks from a smart jammer in Figure 6.1, where the two wireless communication channels are susceptible to malicious jamming attacks. Particularly, there exists a certain probability of being lost

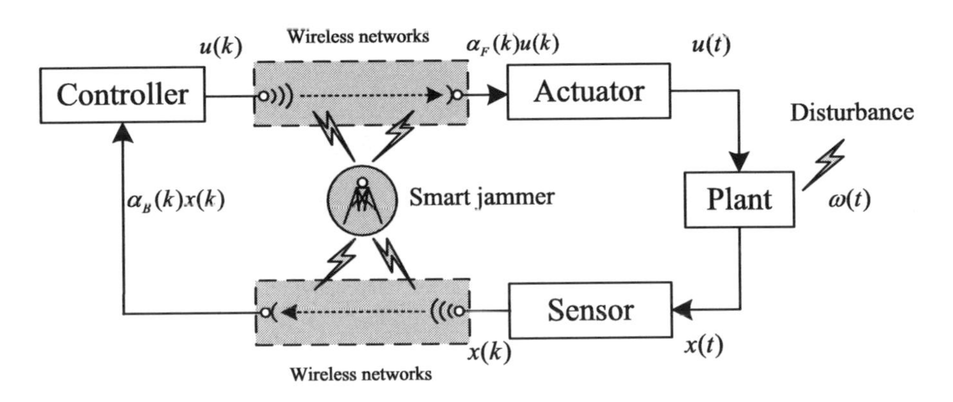

Figure 6.1 A WNCS suffering jamming attacks and physical disturbances.

for data packets in the wireless channels. Meanwhile, the plant is also interfered by outside disturbances in the physical-layer of the WNCS. According to Figure 6.1, a control system subject to jamming attacks is described as follows:

$$x(k+1) = Ax(k) + \alpha(k)Bu(k) + D\omega(k), \tag{6.1}$$

where $x(k) \in \mathbb{R}^n$ and $u(k) \in \mathbb{R}^m$ are the state variable and control signal, respectively, and $\omega(k) \in \mathbb{R}^q$ is the outside disturbance. Matrices A, B and D are known with appropriate dimensions. Let stochastic variables

$\alpha_F(k), \alpha_B(k) \in \mathbb{I}[0,1]$ represent conditions of packets dropout in forward and backward channels, respectively. Referring to [173], the round-trip packets dropout indicator is denoted by variable $\alpha(k) = \alpha_F(k)\alpha_B(k)$ which subjects to the Bernoulli distribution. There exists

$$\Pr\{\alpha(k) = 1\} = \mathbb{E}\{\alpha(k)\} \triangleq \bar{\alpha},$$
$$\Pr\{\alpha(k) = 0\} = \mathbb{E}\{1 - \alpha(k)\} \triangleq 1 - \bar{\alpha}.$$

Thus $\bar{\alpha} \in [0,1]$ represents the Obtained Rate of Data-packets (ORD). In the finite horizon case, a system quadratic performance index is introduced as a cost-to-go function as follows

$$J(\mathbf{u}, \mathbf{w}) = \mathbb{E}_{\alpha(i)}\Bigg\{x(N)^T Q(N) x(N) + \sum_{i=0}^{N-1} [x(i)^T Q(i) x(i) + \alpha(i) u(i)^T R(i) u(i) - \gamma^2 \omega(i)^T \omega(i)]\Bigg\}, \tag{6.2}$$

where $Q(N) \geq 0$, $Q(i) \geq 0$, and $R(i) > 0$, $i \in \mathbb{I}[0, N-1]$, vectors $\mathbf{u} = \left[u(0)^T \ \cdots \ u(N-1)^T\right]^T$ and $\mathbf{w} = \left[\omega(0)^T \ \cdots \ \omega(N-1)^T\right]^T$, scalar γ represents the disturbance attenuation level. In this chapter, $\mathbb{E}_{\alpha(i)}\{\cdot\}$ is simplified as $\mathbb{E}\{\cdot\}$. Then both the control objective and the disturbance aim are expressed as

$$\min_{\mathbf{u}} \max_{\mathbf{w}} J(\mathbf{u}, \mathbf{w}).$$

in this chapter. Moreover, a Nash equilibrium policy in control systems is defined in the following.

Definition 6.1. [84] For given strategy pair $(\mathbf{u}^*, \mathbf{w}^*)$, if the following inequality

$$J(\mathbf{u}^*, \mathbf{w}) \leq J(\mathbf{u}^*, \mathbf{w}^*) \leq J(\mathbf{u}, \mathbf{w}^*)$$

holds, then the strategy pair $(\mathbf{u}^*, \mathbf{w}^*)$ is called a Nash equilibrium strategy.

6.2.2 Wireless Communication Channel

Note that wireless communication channels are commonly used to deliver data packets. Because of power fading and channel interference, random data-drops always occur in transmission. To model this situation, a feasible approach has been presented based on SINR in [81]. Using Quadrature Amplitude Modulation (QAM), the communication between the sensor and receiver is assumed to over an Additive White Gaussian Noise (AWGN) network. Similar to [82] and [180], the relationship between the Symbol Error Rate (SER) and Signal-to-Noise Ratio (SNR) is revealed by digital communication theory as follows

$$\mathrm{SER} = 2Q\left(\sqrt{\xi \mathrm{SNR}}\right), \ Q(x) \triangleq \frac{1}{\sqrt{2\pi}} \int_x^{\infty} \exp(-\eta^2/2)\mathrm{d}\eta,$$

where $\xi > 0$ is a parameter. In the presence of a malicious jammer at time k, the SINR for communication networks is rewritten as

$$\mathrm{SINR} = \frac{\rho^{uc} p(k)}{\rho^{jc} w(k) + \delta^2},$$

where $p(k)$ is the transmission power used by the sensor, δ^2 is the additive white Gaussian noise power, $w(k)$ is the interference power from attacker [109], ρ^{uc} is the channel gain between legitimate users and receivers, ρ^{jc} is the channel gain between the malicious jammers and receivers. Note that the legitimate users are controllers and sensors. Then transmission of input $u(k)$ between the controller and actuator is characterized by a binary random process $\{\alpha_F(k)\}$, i.e.,

$$\alpha_F(k) = \begin{cases} 1, & \text{if } u(k) \text{ arrives error-free at time } k, \\ 0, & \text{otherwise (regarded as dropout).} \end{cases}$$

Then at time k, ORD $\bar{\alpha}_F(k)$ is given from [81] that

$$\bar{\alpha}_F(k) = 1 - 2Q\left(\sqrt{\frac{\xi \rho^{uc} p(k)}{\rho^{jc} w(k) + \delta^2}}\right), \tag{6.3}$$

Usually, the forward and backward channels share the same links in WNCSs. Furthermore, ORD $\bar{\alpha}(k)$ in control systems is obtained from equation (6.3) as

$$\bar{\alpha}(k) = \bar{\alpha}_F(k)\bar{\alpha}_B(k) = \left[1 - 2Q\left(\sqrt{\frac{\xi \rho^{uc} p(k)}{\rho^{jc} w(k) + \delta^2}}\right)\right]^2, \tag{6.4}$$

Note that ORD $\bar{\alpha}(k)$ is not only dependent on transmission powers used by sensors and controllers, but it is also affected by interference powers from malicious jammers. Furthermore, concrete relationships between legitimate users and malicious jammers are discussed with transmission powers in the future.

6.2.3 Beyasian Stackelberg Game Equilibrium

Consider the case that a legitimate user transmits a signal at time k, whereas a malicious jammer transmits a jamming signal after sensing the signal from user. The decision-making process between the user and jammer is seen as a Stackelberg game. Let C_u and C_j denote the transmission costs per unit power for the user and jammer, respectively. Note that both opposite channel state information and transmission cost information are hardly achieved by the user or jammer. Moreover, joint probability distributions of the channel information are entirely obtained by both sides.

Assumption 6.1 *[66] For the legitimate user, the channel gain ρ^{jc} between the malicious jammer and receiver has R positive states, i.e., $\rho_1^{jc}, \cdots, \rho_R^{jc}$. The transmission cost C_u has H positive states, i.e., $C_{u1}, C_{u2}, \cdots, C_{uH}$. Joint probability distributions of the two-dimensional discrete random variable (ρ^{jc}, C_u) is $\eta_u(\rho_r^{jc}, C_{uh})$ with $\sum_{r=1}^{R}\sum_{h=1}^{H}\eta_u(\rho_r^{jc}, C_{uh}) = 1$.*

Assumption 6.2 *[66] For the malicious jammer, the channel gain ρ^{uc} between the network user and receiver has M positive states, i.e., $\rho_1^{uc}, \cdots, \rho_M^{uc}$. The transmission cost C_j has N positive states, i.e., $C_{j1}, C_{j2}, \cdots, C_{jN}$. Joint probability distributions of the two-dimensional discrete random variable (ρ^{uc}, C_j) is $\eta_j(\rho_m^{uc}, C_{jn})$ with $\sum_{m=1}^{M}\sum_{n=1}^{N}\eta_j(\rho_m^{uc}, C_{jn}) = 1$.*

By Assumption 6.1 and 6.2, the utility function of a user based on SINR is shown as

$$\mu_u\left(p(k), w(k)\right) = \sum_{r=1}^{R}\sum_{h=1}^{H}\eta_u(\rho_r^{jc}, C_{uh})\left(\frac{\rho^{uc}p(k)}{\rho_r^{jc}w(k)+\delta^2} - C_{uh}p(k)\right). \quad (6.5)$$

Similarly, the utility function of a jammer is given as

$$\mu_j\left(p(k), w(k)\right) = \sum_{m=1}^{M}\sum_{n=1}^{N}\eta_j(\rho_m^{uc}, C_{jn})\left(\frac{-\rho_m^{uc}p(k)}{\rho^{jc}w(k)+\delta^2} - C_{jn}w(k)\right). \quad (6.6)$$

Furthermore, a dynamic Bayesian Stackelberg game between the user and the jammer is formulated with decision pair $(p(k), w(k))$ updating in each time step. The legitimate user is a leader, whereas the malicious jammer acts as a follower. The strategies for both the user and jammer is the selections of their transmission power $(p(k), w(k))$.

Note that Stackelberg equilibrium is an equilibrium solution concept for the proposed dynamic Bayesian Stackelberg game. Similar to [147], for the smart jammer, an observation error on the transmission power of the user is denoted by

$$\varepsilon(k) = \frac{|\tilde{p}(k) - p(k)|}{p(k)},$$

where $\tilde{p}(k)$ is an observation value of $p(k)$ at time k. Given the transmission power of the legitimate user, an optimization problem on the jammer under uncertainties of the channel gain ρ^{uc} and the transmission cost C_j is formulated as

$$\begin{aligned} \max_{w(k)\geq 0} \ & \mu_j(\tilde{p}(k), w(k)), \\ \text{s.t.} \ & w(k) \leq w_M, \end{aligned} \tag{6.7}$$

where w_M denotes the maximum transmission power of the malicious jammer. Furthermore, an optimal transmission power of the jammer is shown as

$$w(k)^* = \arg \max_{0\leq w(k)\leq w_M} \mu_j(\tilde{p}(k), w(k)).$$

For the legitimate user, uncertainties of the channel gain ρ^{jc} and the transmission cost C_u are considered in the same way. Then the optimal strategy is obtained by solving the following optimization problem as

$$\begin{aligned} \max_{p(k)\geq 0} \ & \mu_u(p(k), w\,[p(k)]), \\ \text{s.t.} \ & p(k) \leq p_M, \end{aligned} \tag{6.8}$$

where p_M denotes the maximum transmission power of the legitimate user. Moreover, an optimal transmission power of the user is given as

$$p(k)^* = \arg \max_{0\leq p(k)\leq p_M} \mu_u(p(k), w\,[p(k)]).$$

Then a definition of the Stackelberg equilibrium solution for the proposed game is given in the following.

Definition 6.2. [52] An optimal strategy pair $(p(k)^*, w(k)^*)$ constitutes a Stackelberg equilibrium for the proposed dynamic game. For $p(k) \geq 0$ and $w(k) \geq 0$, the following conditions are satisfied with

$$\begin{aligned} \mu_u(p(k)^*, w(k)^*) &\geq \mu_u(p(k), w(k)^*), \\ \mu_j(p(k)^*, w(k)^*) &\geq \mu_j(p(k)^*, w(k)). \end{aligned}$$

Additionally, the existence and uniqueness of the Stackelberg equilibrium solution for the proposed dynamic game are absolutely necessary in this chapter.

6.3 Main Results

6.3.1 Best Responses for Cyber-Layer Game

Note that the malicious jammer acts as a follower compared to the legitimate user in the proposed dynamic Bayesian Stackelberg game. For the jammer, the random variable ρ^{uc} is independent on variable C_j. The channel gain ρ^{uc} has M positive states, i.e., $\rho_1^{uc}, \cdots, \rho_M^{uc}$ with probabilities $\phi_1, \cdots, \phi_M$ and $\sum_{m=1}^{M} \phi_m = 1$. Similarly, the transmission cost C_j has N positive states, i.e., $C_{j1}, \cdots, C_{jN}$ with probabilities $\kappa_1, \cdots, \kappa_N$ and $\sum_{n=1}^{N} \kappa_n = 1$. By solving optimization problem (6.7), the best response of the malicious jammer is derived in the following theorem.

Theorem 6.3. *Consider the control system (6.1) with wireless communications. The best response of the malicious jammer by solving optimization problem (6.7) is shown as*

$$w\left[\tilde{p}(k)\right] = \frac{1}{\rho^{jc}} \left(\sqrt{\frac{\rho^{jc} \sum_{m=1}^{M} \rho_m^{uc} \phi_m \tilde{p}(k)}{\sum_{n=1}^{N} C_{jn} \kappa_n + \lambda}} - \delta^2 \right)^+, \tag{6.9}$$

where $(\cdot)^+ \triangleq \max(\cdot, 0)$. *The malicious jammer will stop transmitting only if the transmission power of the legitimate user satisfies*

$$\tilde{p}(k) \leq \frac{\left(\sum_{n=1}^{N} C_{jn} \kappa_n + \lambda\right) (\delta^2)^2}{\rho^{jc} \sum_{m=1}^{M} \rho_m^{uc} \phi_m},$$

where λ *is a nonnegative dual variable introduced by the following Lagrange function.*

$$L(p(k), w(k), \lambda) = -\frac{\sum_{m=1}^{M} \rho_m^{uc} \phi_m \tilde{p}(k)}{\delta^2 + \rho^{jc} w(k)} - \sum_{n=1}^{N} C_{jn} \kappa_n + \lambda(w_M - w(k)).$$

Proof. Main proof of this theorem is similar with Lemma 1 in [66]. Note that Lemma 1 in [66] has made a proof on a static game, while conclusions in this chapter are drawn on the dynamic game. Variables $(\tilde{p}(k), w(k))$ are dependent on states of system (6.1), and a detailed updating algorithm is presented in the following subsection. So the proof of Theorem 6.3 can refer to Lemma 1 in [66].

For the legitimate user, the random variable ρ^{jc} is independent on variable C_u. The channel gain ρ^{jc} has R positive states, i.e., $\rho_1^{jc}, \cdots, \rho_R^{jc}$ with probabilities $\varphi_1, \cdots, \varphi_R$ and $\sum_{r=1}^{R} \varphi_r = 1$. The transmission cost C_u has H positive states, i.e., $C_{u1}, \cdots, C_{uH}$ with probabilities $\tau_1, \cdots, \tau_H$ and $\sum_{h=1}^{H} \tau_h = 1$. An optimal strategy is shown from solving optimization problem (6.8) in the following theorem.

Theorem 6.4. *As for the control system (6.1) with wireless communica-tions, the optimal strategy $p(k)^*$ of the legitimate user by solving optimization problem (6.8) is shown as*

$$p(k)^* = \begin{cases} P_1, & \sum_{h=1}^{H} C_{uh}\tau_h \leq \Gamma \\ P_2, & \Gamma < \sum_{h=1}^{H} C_{uh}\tau_h \leq \frac{\rho}{\delta^2} \\ 0, & \frac{\rho}{\delta^2} < \sum_{h=1}^{H} C_{uh}\tau_h, \end{cases} \tag{6.10}$$

where

$$P_1 = \frac{\left(\sum_{n=1}^{N} C_{jn}\kappa_n + \lambda\right)\rho^2}{4\left(\sum_{h=1}^{H} C_{uh}\tau_h + \varpi\right)^2\left(\sum_{m=1}^{M}\rho_m\phi_m\right)}\left(\sum_{r=1}^{R}\varphi_r\sqrt{\frac{1}{\rho_r^{jc}}}\right)^2,$$

$$P_2 = \sum_{r=1}^{R}\varphi_r\frac{\left(\sum_{n=1}^{N} C_{jn}\kappa_n + \lambda\right)(\delta^2)^2}{\rho_r^{jc}\sum_{m=1}^{M}\rho_m\phi_m},$$

$$\Gamma = \frac{\rho\sum_{r=1}^{R}\varphi_r\sqrt{\frac{1}{\rho_r^{jc}}}}{2\delta^2\sqrt{\sum_{r=1}^{R}\frac{\varphi_r}{\rho_r^{jc}}}} - \varpi,$$

and ϖ is a nonnegative dual variable introduced by the following Lagrange function.

$$L(p(k), w[p(k)], \varpi) = \sum_{r=1}^{R}\varphi_r\rho^{uc}\sqrt{\frac{p(k)\left(\sum_{n=1}^{N} C_{jn}\kappa_n + \lambda\right)}{\rho_r^{jc}\sum_{m=1}^{M}\rho_m\phi_m}} - \sum_{h=1}^{H} C_{uh}\tau_h p(k) + \varpi(p_M - p(k)).$$

Proof. In the dynamic Stackelberg game, the user acts as a leader to decide a transmission power firstly. Note that Lemma 2 in [66] has given an optimal transmission power of the user in a static Stackelberg game. When $w(k)$, λ and ϖ change as observation error $\varepsilon(k)$ diminishing, $p(k)^*$ is adjusted in the dynamic game. Furthermore, the main proof is similar with Lemma 2 in [66] and is omitted here.

According to the dual optimization theory, optimization problems (6.7) and (6.8) are transformed into

$$\max_{k\geq 0} \; p(k, \varpi)^*,$$
$$\text{s.t.} \quad \varpi \geq 0,$$

and

$$\max_{k\geq 0} \ w(\tilde{p}(k)^*, \lambda)^*,$$
$$\text{s.t.} \ \ \lambda \geq 0,$$

respectively. By sub-gradient update method in [33], the dual variables λ and ϖ are derived to

$$\lambda^{l+1} = \left[\lambda^l - \Omega_\lambda^l(w_M - w(k))\right]^+,$$
$$\varpi^{l+1} = \left[\varpi^l - \Omega_\varpi^l(p_M - p(k))\right]^+,$$

where l denotes the iteration number, and Ω_λ^l and Ω_ϖ^l are the iteration steps. Note that the time complexity of $w[\tilde{p}(k)]$ and $p(k)$ is $O(1)$, and an overall time complexity of the proposed game is $O(l)$. In this subsection, the best responses of the malicious jammer and the legitimate user have been derived with incomplete channel information for both sides of the controller.

6.3.2 Optimal Controller Design

In this subsection, an optimal controller for the discrete control system (6.1) is designed based on H_∞ optimal control theory [193]. If the controller is affected by the cyber-layer game with an ORD, then the malicious jammer not only aims to jam communication channels but also to reduce control system performances. Note that the ORD is derived by the SINR from equation (6.4). Thereby, the utility functions (6.5)-(6.6) are also designed based on the SINR. That is, the SINR influences the control system performances.

Considering jamming attacks in this chapter, the existence and uniqueness of a Stackelberg equilibrium in the Bayesian Stackelberg game have been presented by Lemmas 3 and 4 in [66]. Then the ORD is also unique when the dynamic Bayesian Stackelberg game arrives at the Stackelberg equilibrium. Because the observation error $\varepsilon(k)$ exists for jammers and is adjusted in the dynamic game, ORD $\bar{\alpha}(k)$ changes in each step. Furthermore, the optimal controllers of system (6.1) are studied based on changeable ORDs in the following theorem.

Theorem 6.5. *Consider the control system (6.1) with wireless communications. For a linear-quadratic dynamic game between the controller and the disturbance in the physical-layer, ORD $\bar{\alpha}(k)$ is obtained from the cyber-layer game by N and γ. Since the control strategies are independent on current values of disturbances, the following conclusions are presented as*

1). There exists a unique Nash equilibrium if

$$\gamma^2 I - D^T Z(k+1)D > 0, \ \ k \in \mathbb{I}[1, N-1], \tag{6.11}$$

where the non-increasing sequence $Z(k) > 0$ is generated by Riccati recursion.

2). The Riccati recursion is $Z(N) = Q(N)$. For all $k \in \mathbb{I}[1, N-1]$, there exists

$$Z(k) = Q(k) + \bar{\alpha}(k)P_{u(k)}^T B^T R(k) B P_{u(k)} - \gamma^2 P_{\omega(k)}^T P_{\omega(k)} + \\ + \left[A - \bar{\alpha}(k)BP_{u(k)} + DP_{\omega(k)}\right]^T Z(k+1) \left[A - \bar{\alpha}(k)BP_{u(k)} + DP_{\omega(k)}\right],$$

where

$$\begin{aligned} P_{u(k)} &= \left[R(k) + B^T Z(k+1)B + \bar{\alpha}(k)B^T Z(k+1)D(\gamma^2 I \right. \\ &\quad \left. - D^T Z(k+1)D)^{-1} D^T Z(k+1)B\right]^{-1} \\ &\quad \times \left[B^T Z(k+1)(A + D(\gamma^2 I - D^T Z(k+1)D)^{-1} D^T Z(k+1)A)\right], \\ P_{\omega(k)} &= (\gamma^2 I - D^T Z(k+1)D)^{-1} D^T Z(k+1)(A - \bar{\alpha}(k)BP_{u(k)}). \end{aligned}$$

3). Under conditions 1) and 2), the optimal controller and disturbance are designed as

$$u(k)^* = -P_{u(k)}x(k), \;\; \omega(k)^* = P_{\omega(k)}x(k).$$

4). Under condition 1), the Nash equilibrium value of the game is given by

$$J(\boldsymbol{u}^*, \boldsymbol{w}^*) = x(0)^T Z(0) x(0),$$

where $x(0)$ is the initial state value.

Proof. According to [14], the following proof is obtained by employing the dynamic programming and induction method. Let $k = N$, it follows the cost-to-go function that

$$V(x(N)) = x(N)^T Q(N) x(N). \tag{6.12}$$

With the dynamic programming, the cost at time $k = N-1$ is obtained from (6.2) as

$$\begin{aligned} V(x(N-1)) = & \min_{u(N-1)} \max_{w(N-1)} \mathbb{E}\left\{x(N-1)^T Q(N-1)x(N-1)\right. \\ & + \alpha(N-1)u(N-1)^T R(N-1)u(N-1) \\ & \left. - \gamma^2 \omega(k)^T \omega(k) + V(N)\right\} \\ = & \min_{u(N-1)} \max_{w(N-1)} \mathbb{E}\left\{x(N-1)^T Q(N-1)x(N-1)\right. \\ & + \alpha(N-1)u(N-1)^T R(N-1)u(N-1) - \gamma^2 \omega(k)^T \omega(k) \\ & + (Ax(N-1) + \alpha(N-1)Bu(N-1) + D\omega(N-1)^T)^T Q(N) \\ & \left. \times (Ax(N-1) + \alpha(N-1)Bu(N-1) + D\omega(N-1)\right\}. \end{aligned} \tag{6.13}$$

It is easy to obtain that function $V(x(N-1))$ is strictly convex for $u(N-1)$ for the second derivative of

$$\frac{\partial^2 V(N-1)}{\partial u(N-1)^2} = 2\left[R(N-1) + \alpha(N-1)B^T Q(N-1)B\right] > 0,$$

and function $V(x(N-1))$ is strictly concave for $\omega(N-1)$ when

$$\frac{\partial^2 V(N-1)}{\partial \omega(N-1)^2} = 2\left[-\gamma^2 I + D^T Q(N)D\right] < 0.$$

Moreover, the first derivative of (6.13) for $u(N-1)$ and $\omega(N-1)$ are used to obtain the minimizer and maximizer for $V(x(N-1))$, respectively.

$$\begin{aligned}\frac{\partial V(N-1)}{\partial u(N-1)} &= (R(N-1) + \bar{\alpha}(N-1)B^T Q(N)B)u(N-1)\\ &\quad + u(N-1)^T(R(N-1) + \bar{\alpha}(N-1)B^T Q(N)B)\\ &\quad + B^T Q(N)(Ax(N-1) + D\omega(N-1))\\ &\quad + (Ax(N-1) + D\omega(N-1))^T Q(N)B\\ &= 0, \end{aligned} \tag{6.14}$$

$$\begin{aligned}\frac{\partial V(N-1)}{\partial \omega(N-1)} &= (-\gamma^2 I + D^T Q(N)D)\omega(N-1)\\ &\quad + \omega(N-1)^T(-\gamma^2 I + D^T Q(N)D)\\ &\quad + D^T Q(N)(Ax(N-1) + \bar{\alpha}(N-1)Bu(N-1))\\ &\quad + (Ax(N-1) + \bar{\alpha}(N-1)Bu(N-1))^T Q(N)D\\ &= 0. \end{aligned} \tag{6.15}$$

From equations (6.14) and (6.15), an optimal strategy pair $(u(N-1)^*, \omega(N-1)^*)$ is obtained as

$$\begin{aligned} u(N-1)^* &= -(R(N-1) + B^T Q(N)B))^{-1}B^T Q(N)(Ax(N-1)\\ &\quad + D\omega(N-1)^*), \end{aligned} \tag{6.16}$$

$$\begin{aligned} \omega(N-1)^* &= (\gamma^2 I - D^T Q(N)D)^{-1}D^T Q(N)(Ax(N-1)\\ &\quad + \bar{\alpha}(N-1)Bu(N-1)^*). \end{aligned} \tag{6.17}$$

Note that both $u(N-1)^*$ and $\omega(N-1)^*$ are the two functions of $x(N-1)$, i.e.,

$$u(N-1)^* = -P_{u(N-1)}x(N-1),\ \omega(N-1)^* = P_{u(N-1)}x(N-1). \tag{6.18}$$

By substituting (6.18) into (6.16) and (6.17), it is obtained that

$$
\begin{aligned}
P_{u(N-1)} = & \left[R(N-1) + B^T Q(N) B + \bar{\alpha}(N-1) B^T Q(N) D \right.\\
& \times (\gamma^2 I - D^T Q(N) D)^{-1} D^T Q(N) B\left.\right]^{-1} \\
& \times \left[B^T Q(N)(A + D(\gamma^2 I - D^T Q(N) D)^{-1} D^T Q(N) A)\right], \\
P_{\omega(N-1)} = & (\gamma^2 I - D^T Q(N) D)^{-1} D^T Q(N)(A - \bar{\alpha}(N-1) B P_{u(N-1)}). \quad (6.19)
\end{aligned}
$$

Substituting (6.18) into (6.13), $V(x(N-1))$ is also obtained as

$$
\begin{aligned}
V(x(N-1)) = & \, x(N-1)^T Q(N-1) x(N-1) \\
& + \bar{\alpha}(N-1)[B P_{u(N-1)} x(N-1)]^T R(N-1) B P_{u(N-1)} x(N-1) \\
& - \gamma^2 [P_{\omega(N-1)} x(N-1)]^T P_{\omega(N-1)} x(N-1) \\
& + x(N-1)^T [A - \bar{\alpha}(N-1) B P_{u(N-1)} + D P_{\omega(N-1)}]^T \\
& \times Q(N)[A - \bar{\alpha}(N-1) B P_{u(N-1)} + D P_{\omega(N-1)}] x(N-1) \\
= & \, x(N-1)^T \left\{ Q(N-1) + \bar{\alpha}(N-1) P_{u(N-1)}^T B^T R(N-1) B P_{u(N-1)} \right. \\
& - \gamma^2 P_{\omega(N-1)}^T P_{\omega(N-1)} + \left[A - \bar{\alpha}(N-1) B P_{u(N-1)} + D P_{\omega(N-1)}\right]^T \\
& \left. \times Q(N) \left[A - \bar{\alpha}(N-1) B P_{u(N-1)} + D P_{\omega(N-1)}\right] \right\} x(N-1) \\
\triangleq & \, x(N-1)^T Z(N-1) x(N-1). \quad (6.20)
\end{aligned}
$$

Moreover, equation (6.20) implies that the cost-to-go function $V(x(N-1))$ has a linear quadratic structure. By the same way, both the Nash equilibrium solution $(u(N-2)^*, \omega(N-2)^*)$ and value function $V(x(N-2))$ are obtained. The value of game $x(0)^T Z(0) x(0)$ is obtained by carrying out the procedure from (6.12) to (6.20) for all $k \leq N-2$ recursively. Furthermore, sequence $\{Z(k)\}, k \in \mathbb{I}[1, N-1]$ is non-negative definite and non-increasing by

$$
\begin{aligned}
& V(x(N-1)) \\
= & \, x(N-1)^T Z(N-1) x(N-1) \\
= & \max_{\omega(N-1)} \min_{u(N-1)} \mathbb{E}\left\{ x(N-1)^T Q(N-1) x(N-1) \right. \\
& + \alpha(N-1) u(N-1)^T R(N-1) u(N-1) \\
& \left. - \gamma^2 \omega(N-1)^T \omega(N-1) + x(N)^T Z(N) x(N) \right\} \\
= & \max_{\omega(N-1)} \mathbb{E}\left\{ x(N-1)^T Q(N-1) x(N-1) \right. \\
& + \alpha(N-1) x(N-1)^T P_{u(N-1)}^T B^T R(N-1) \\
& \left. \times P_{u(N-1)} B x(N-1) - \gamma^2 \omega(N-1)^T \omega(N-1) + x(N)^T Z(N) x(N) \right\} \\
\geq & \, \mathbb{E}\left\{ x(N-1)^T Q(N-1) x(N-1) \right. \\
& + \alpha(N-1) x(N-1)^T P_{u(N-1)}^T B^T R(N-1) \\
& \left. \times P_{u(N-1)} B x(N-1) + x(N)^T Z(N) x(N) | \omega(N-1) = 0 \right\} \\
> & \, 0. \quad (6.21)
\end{aligned}
$$

From inequality (6.21), it is easy to obtain that $Z(N-1) > 0$ and $Z(N-1) - Z(N) \geq 0$. Then the same procedure is repeated for all $k \geq N-2$ to obtain general results as shown in this theorem.

The optimal disturbance attenuation level γ^* is introduced by $\gamma^* \triangleq \inf\{\gamma : \gamma \in \Sigma\}$, where set Σ is given as

$$\Sigma \triangleq \left\{\gamma > 0 : \gamma^2 I - D^T Z(k+1) D > 0, k \in \mathbb{I}[1, N-1]\right\}.$$

6.3.3 Coupled Design

In this subsection, an online updating algorithm for a jammer is included with incomplete information. Note that a Stackelberg game is considered between a smart jammer and a legitimate attacker in the cyber-layer of a WNCS. An optimal controller in the physical-layer is designed based on a Nash equilibrium game between the controller and disturbances. Then system performance information is regarded as an indicator for the smart jammer to adjust its jamming strategy, because ORD $\bar{\alpha}(k)$ affects the choices of optimal controller and disturbance. A concrete relationship between $\bar{\alpha}(k)$ and $u(k)^*$ is presented in the following.

Theorem 6.6. *For given $Q(N)$, $Q(k)$ and $R(k), k \in \mathbb{I}[1, N-1]$, the relationship between optimal controller $u(k)^*$ and $\bar{\alpha}(k) \in [0,1]$ obeys the following properties*

$$\begin{cases} \text{If } \Theta(k) > 0, \text{ then } u(k)^* < 0 \text{ and } \frac{\partial u(k)^*}{\partial \bar{\alpha}(k)} > 0, \\ \text{If } \Theta(k) \leq 0, \text{ then } u(k)^* \geq 0 \text{ and } \frac{\partial u(k)^*}{\partial \bar{\alpha}(k)} \leq 0, \end{cases} \tag{6.22}$$

where

$$\Theta(k) = \left[B^T Z(k+1)(A + D(\gamma^2 I - D^T Z(k+1)D)^{-1} D^T Z(k+1)A)\right] x(k).$$

Proof. From equations (6.18) and (6.19), it is easy to obtain that

$$\begin{aligned} &\frac{\partial u(k)}{\partial \bar{\alpha}(k)} \\ &= \left[R(k) + B^T Z(k+1)B + \bar{\alpha}(k) B^T Z(k+1) D \right. \\ &\quad \left. \times (\gamma^2 I - D^T Z(k+1)D)^{-1} D^T Z(k+1) B\right]^{-2} \\ &\quad \times \left[B^T Z(k+1) D(\gamma^2 I - D^T Z(k+1)D)^{-1} D^T Z(k+1) B\right] \\ &\quad \times \left[B^T Z(k+1)(A + D(\gamma^2 I - D^T Z(k+1)D)^{-1}\right. \\ &\quad \left. \times D^T Z(k+1)A)\right] x(k). \end{aligned} \tag{6.23}$$

Since inequalities

$$B^T Z(k+1)D(\gamma^2 I - D^T Z(k+1)D)^{-1} D^T Z(k+1)B > 0,$$
$$R(k) + B^T Z(k+1)B + \bar{\alpha}(k)B^T Z(k+1)$$
$$+\bar{\alpha}(k)B^T Z(k+1)D(\gamma^2 I - D^T Z(k+1)D)^{-1} D^T Z(k+1)B > 0$$

hold, the properties in (6.22) is completed by combining equations (6.18) and (6.19).

From Theorem 6.6, it is obtained that $\|u(k)^*\|$ is small when the ORD $\bar{\alpha}(k)$ becomes large. Considering the Stackelberg game in the cyber-layer, there exists an observation error $\varepsilon(k)$ on user power for malicious jammer. Then the strategy of the jammer is not at a Stackelberg equilibrium point, and utility $\mu_j(\tilde{p}(k), w(k))$ is not maximal actually. From Theorem 6.6, state $x(k)$ of the control system (6.1) with controller $u(k)^*$ is observed by the smart jammer. Then state $x(k)$ is exploited as an indicator to adjust attack strategies in the cyber-layer game. The detailed adjusting algorithm of jammer is presented in the following.

Algorithm 7 Online updating algorithm for smart jammer

Require: Given $Q(N)$, $Q(i)$ and $R(i)$, $i \in \mathbb{I}[1, N-1]$, N, γ, Δ, system matrices A, B and D, set $r = 0$.

1: A legitimate user sends packets via wireless networks with power $p(0)^*$ from equations (6.10) in Theorem 6.4.
2: A smart jammer observes user power as $\tilde{p}(0)$ and launch jamming attacks with power $w(0)$ from equation (6.9) in Theorem 6.3.
3: **for** $k = 1 : N$ **do**
4: Design and send optimal $u(k)^*$ by controller according to network conditions.
5: Estimates and observes system states as $\tilde{x}(k+1)$ and $x(k+1)$ by jammer, respectively.
6: **if** $\|\tilde{x}(k+1) - x(k+1)\| > \|D\omega(k)\|$ **then**
7: **if** $\|\tilde{x}(k+1)\| < \|x(k+1)\|$ **then**
8: Set $w(k) = w(k-1) - \frac{\Delta}{2^r}$.
9: **else if** $\|\tilde{x}(k+1)\| > \|x(k+1)\|$ **then**
10: Set $w(k) = w(k-1) + \frac{\Delta}{2^r}$.
11: Set $r = r + 1$.
12: **end if**
13: **else if** $\|\tilde{x}(k+1) - x(k+1)\| \leq \|D\omega(k)\|$ **then**
14: Break.
15: **end if**
16: **end for**

6.4 Numerical Example

In this example, a resilient control problem which is associated with the Uninterrupted Power System (UPS) is investigated with presented methods. When wireless networks are interfered by a malicious jammer in a networked UPS, a main control objective is to make use of a PWM inverter to keep the ouBacsarB95ut AC voltage at a desired setting and undistorted. The discrete-time system model (6.1) with sampling time 10 ms at the half-load operating point as in [156] is obtained with the following parameters

$$A = \begin{bmatrix} 0.9226 & -0.6330 & 0 \\ 1.0 & 0 & 0 \\ 0 & 1.0 & 0 \end{bmatrix}, \; B = \begin{bmatrix} 1 \\ 0 \\ 0 \end{bmatrix}, \; D = \begin{bmatrix} 0.5 \\ 0 \\ 0.2 \end{bmatrix}. \qquad (6.24)$$

Thereby, the initial conditions are chosen as $x(0) = [0.5\ 0.5\ 0.5]$, the disturbance attenuation level $\gamma = 1$, parameters $Q(N) = Q(i) = I_{3\times3}$, $N = 100$ and $R(i) = I_{1\times1}$, $i \in \mathbb{I}[1, N-1]$.

For the cyber-layer, incomplete network conditions are managed by the Bayesian transformation for both game players. The transmission costs C_u and C_j are the same probability distribution which has two positive states $[0.1,\ 0.3]$ with probability distribution $[0.5,\ 0.5]$. The parameters are given as $\delta^2 = 0.1$, $p_M = w_M = 5$. For the average channel gain $\rho^{uc} = 0.55$ and $\rho^{jc} = 0.6$, the incomplete channel gain with two states are given as $[0.35,\ 0.75]$ and $[0.4,\ 0.8]$, respectively, with the same probability distribution $[0.5,\ 0.5]$. Then the optimal decision of network users is $p = 1.1703$. When the observation error $\varepsilon(0) = 0.2$ exists in calculations of jammer, decisions of jammer will be always suboptimal without updating algorithms as $w = 1.9049$. The estimated controller that is considered by jammer to be applied in UPSs is designed, and the desired system state curves are drawn in Figure 6.2. However, if decisions of network users are performed referring to the optimal jamming strategy, then the real state responses of UPS are displayed in Figure 6.3. Comparing Figures 6.2 and 6.3, the control performance of UPS is influenced by estimated error ε. For eliminating observation error ε, a coupled updating algorithm is applied in the cyber-layer game. From observing real-time states of the UPS are displayed in Figure 6.4 and Figure 6.5, respectively. Comparing Figures 6.4 and 6.5, the influence of the observation error ε on the UPS is limited gradually. In Figures 6.5 and 6.6, it is found that the observation error ε diminishes and tends to zero with the application of the updating algorithm. The coupled design in this chapter is shown to be errorless in this scenario.

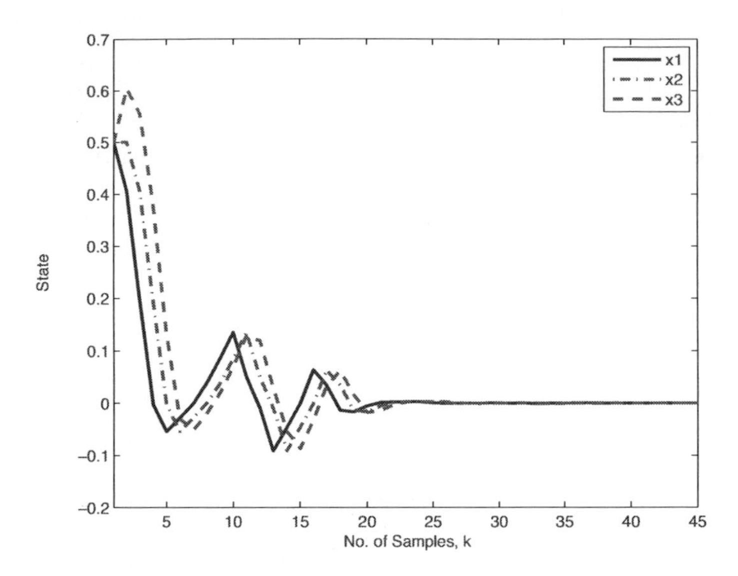

Figure 6.2 Estimated state responses of UPS by jammer under suboptimal jamming attacks.

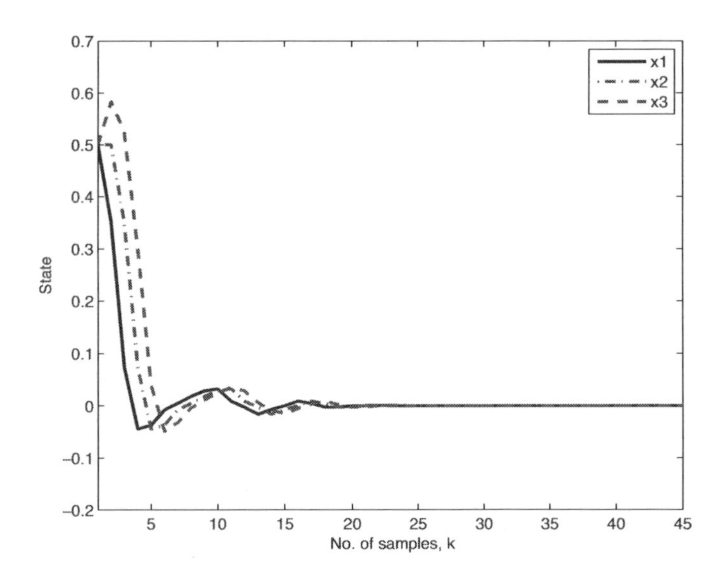

Figure 6.3 Real state responses of UPS under suboptimal jamming attacks.

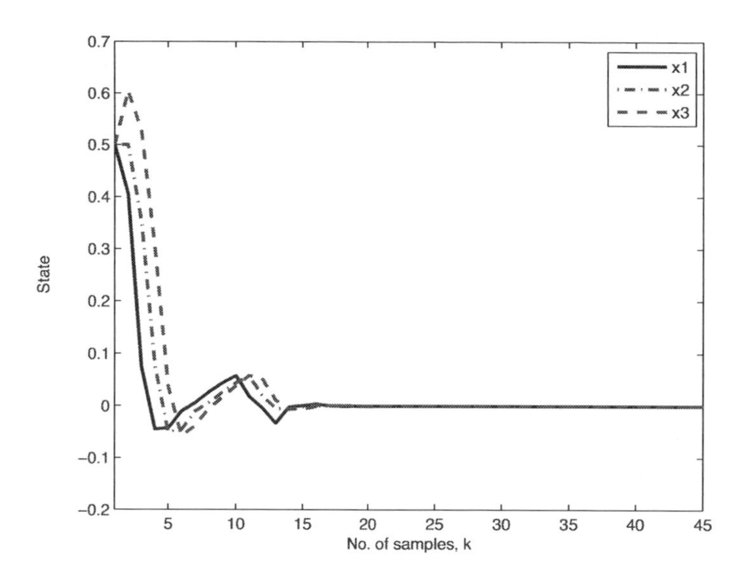

Figure 6.4 Estimated state responses of UPS under updating jamming attack.

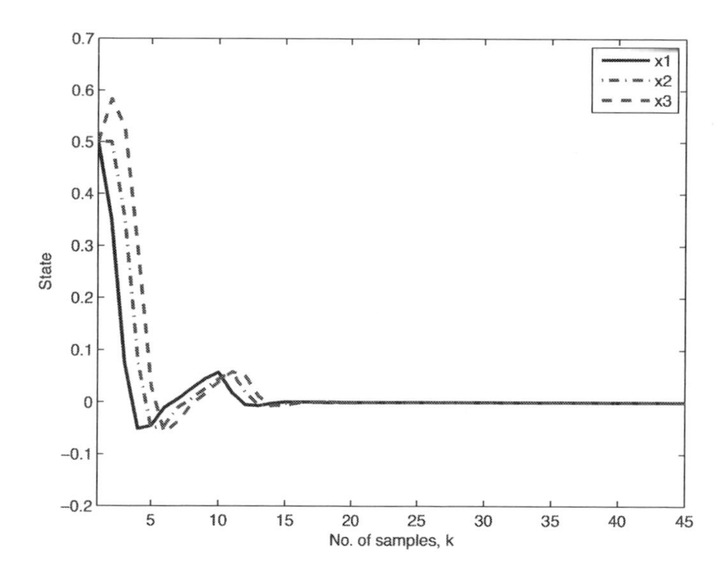

Figure 6.5 Real state responses of UPS under updating jamming attack.

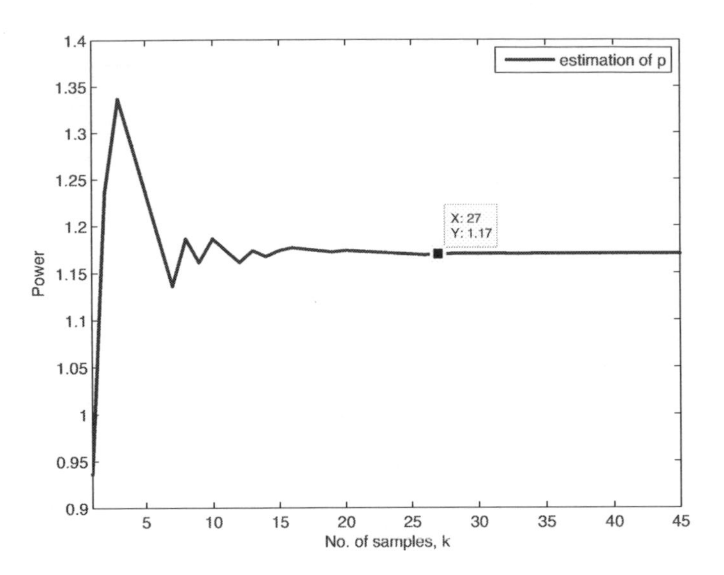

Figure 6.6 Power adjustment of jammer using updating algorithm.

6.5 Conclusion

In this chapter, the resilient control problem has been studied when the cyber layer suffers malicious jamming attacks in WNCSs. With incomplete information existing on both the smart jammer and the legitimate user, a Bayesian Stackelberg game approach has been exploited to analyze the cyber layer game. Moreover, an H_∞ minimax controller has been designed in the physical-layer with the jam-induced packets dropout. Then a coupled design between cyber- and physical-layers has been presented to eliminate observation error ε on transmission powers. The cyber layer game has arrived at Bayesian Stackelberg equilibrium with the provided updating algorithm. Finally, simulation results have demonstrated the feasibility of designs presented in this chapter.

Part III
Application of Resilient Control to Power System

Chapter 7
Quantifying the Impact of Attacks on NCSs

7.1 Introduction

In recent years, security of NCSs has received considerable attention since the networks of NCSs render the control system a potential target to a variety of attacks [32]. Since control systems can be regarded as the connection of the information world and physical world, any successful attacks on NCSs might lead to significant loss of property or even human lives. Owing to the rapid development of sensing techniques, sampling intervals of modern industrial control systems are normally quite small, and a sampled-data problem becomes very critical in system design. The delta operator approach has been well recognized in addressing sampling issues for NCSs [160]. Numerical-stiffness problems resulting from the fast sampling protocol can be circumvented by using the delta operator approach [162]. Actually, it has already been reported in [81] and [103] that systems in a number of critical infrastructures were compromised by a series of attacks. Up to now, cyber attacks compromise NCSs mainly by corrupting the integrity or availability of the measurement and actuator data. It is noted that the physical attack is also addressed, which is often launched along with the cyber attacks. Therefore, a natural idea is to develop a more comprehensive attack model for NCSs that simultaneously includes integrity attacks, availability attacks, and physical attacks. On the other hand, how to assess the security level before adversarial incidents occur is also significant, because one can verify whether the system remains within the safety region with the applied security or control strategies.

In this chapter, a comprehensive attack model is proposed covering randomly occurring DoS attacks, deception attacks, and physical attacks for the NCSs. In the presence of the considered attacks, the so-called ϵ-NE is put forward to describe the control performance. Then, in virtue of the stochastic analysis techniques, an upper bound for the possible attack-induced performance-loss is provided explicitly, and the corresponding convex optimization algorithm is given to compute such an upper bound.

The rest of this chapter is organized as follows: Section 7.2 presents the problem formulation. In Section 7.3.1, we derive the optimal control strategies in the delta-domain. Section 7.3.2 shows the estimation for ϵ-level of NE under randomly occurring DoS attacks, deception attacks, and physical attacks. A numerical example on power grid is conducted in Section 7.4. Finally, the conclusion is drawn in Section 7.5.

7.2 Problem Formulation

In this chapter, the NCSs in the all-around adversary environment is taken into account as is shown in Figure 7.1. The considered NCS is composed of

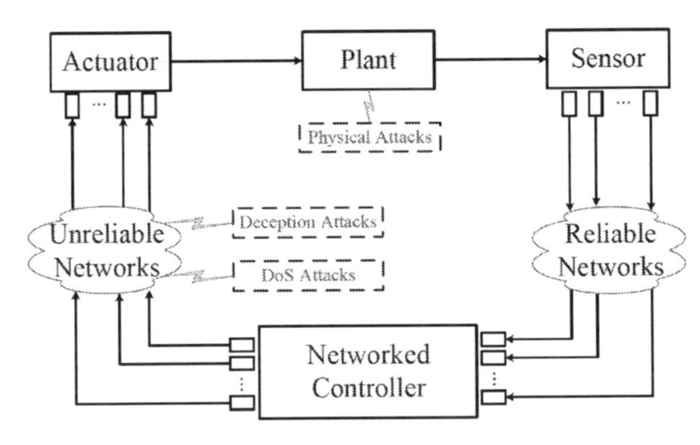

Figure 7.1 The structure of NCS in the all-around adversary environment.

actuator, plant, sensor, and controller, where the controller and actuator are connected over the communication network. A comprehensive attacking scenario is considered where the attackers are capable of launching both cyber and physical attacks. The cyber attack to be exploited includes both DoS and deception-like attacks. It can be seen from Figure 7.1 that a DoS attack prevents the actuator receiving the control command by jamming the communication channel, and the deception attack corrupts the control command by modifying the signal. Furthermore, it is worth mentioning that the physical attack is also addressed, which is often launched along with the cyber attack. For example, in order to render the deception attack undetectable, the attackers can pump water out of the irrigation system while corrupting the measurement of the water level [10]. Next, we are in a position to present the delta-domain system model on the finite horizon ($k \in \mathbf{K} := [0, K-1]$) under all-around attacks:

$$\delta x(t_k) = Ax(t_k) + Bu(t_k) + \eta(t_k)f(t_k), \tag{7.1}$$

where $x(t_k) \in \mathbb{R}^{n_x}$ is the system state, $u(t_k) \in \mathbb{R}^{n_u}$ is the control input, $f(t_k) \in \mathbb{R}^{n_f}$ is physical attack signal injected by the anomalies, normally referred to fault signal or nonlinearities [127]. A and B are matrices in the delta-domain with appropriate dimensions. In general, these matrices have the time-varying property, but the time subscript on them are omitted for ease of notation. The matrices can be partitioned as $B = [B^1 \quad B^2 \quad \cdots \quad B^r]$ and the control input received by the actuator yields

$$u(t_k) = \Xi(t_k)[u^{1T}(t_k)\ u^{2T}(t_k)\ \cdots\ u^{rT}(t_k)]^T, \tag{7.2}$$

where $\Xi(t_k)$ describes the occurrence of the DoS attack as

$$\Xi(t_k) = \text{diag}\{\beta_1(t_k)\ \beta_2(t_k)\ \cdots\ \beta_r(t_k)\},$$

with the indicator $\beta_i(t_k)$, $i \in \mathbf{R} := \{1, \cdots, r\}$ being the Bernoulli distributed white sequence. Suppose that the adversaries are resource-limited as in [177], that is, the attacks are not arbitrarily launched. Therefore, the physical attack $f(t_k)$ satisfies $\|f(t_k)\|^2 < \gamma_1$, where γ_1 is a constant. In the cyber layer, the control input $u(t_k)$ is corrupted by both DoS attacks and deception attacks, and hence the control input $u^i(t_k)$ received by the actuator yields

$$u^i(t_k) = P^i(t_k)x(t_k) + \alpha_i(t_k)\Delta u^i(t_k), \tag{7.3}$$

where $\Delta u^i(t_k) \in \mathbb{R}^{u_i}$ represents the deception attack signal and $\Delta u(t_k) := [\Delta u^{1T}(t_k)\ \Delta u^{2T}(t_k)\ \cdots\ \Delta u^{rT}(t_k)]^T$ satisfies $\|\Delta u(t_k)\|^2 < \gamma_2$ with γ_2 being a constant. The indicator $\alpha_i(t_k)$ is a Bernoulli distributed white sequence. Furthermore, the indicators $\eta(t_k)$, $\alpha_i(t_k)$ and $\beta_i(t_k)$, $i \in \mathbf{R}$ are uncorrelated with each other and possess the following stochastic properties

$$\begin{aligned}
&\mathbb{P}\{\eta(t_k) = 1\} = \bar{\eta},\ \mathbb{P}\{\eta(t_k) = 0\} = \eta,\\
&\mathbb{P}\{\alpha_i(t_k) = 1\} = \bar{\alpha}_i,\ \mathbb{P}\{\alpha_i(t_k) = 0\} = \alpha_i,\\
&\mathbb{P}\{\beta_i(t_k) = 1\} = \bar{\beta}_i,\ \mathbb{P}\{\beta_i(t_k) = 0\} = \beta_i,
\end{aligned}$$

where $\bar{\eta} \in [0,1]$, $\bar{\alpha}_i \in [0,1]$ and $\bar{\beta}_i \in [0,1]$ are priori known constants. Substituting (7.2) and (7.3) into system (7.1), the following closed-loop system in the all-around adversary environment is obtained as

$$\begin{aligned}
(\Sigma_e) : \delta x(t_k) = Ax(t_k) + \sum_{i=1}^{r} \beta_i(t_k)B^i(P^i(t_k)x(t_k)\\
+\alpha_i(t_k)\Delta u^i(t_k)) + \eta(t_k)f(t_k).
\end{aligned} \tag{7.4}$$

For the purpose of describing the impacts from the attacks, system 7.1 is said to be a nominal/ideal system (Σ_n) if no attacks are launched, i.e., $\eta = 1$, $\alpha_i = 1$, and $\bar{\beta}_i = 1$. The fact that any one of these conditions is absent can lead to the abnormality of the considered system and possible performance degradation. The associated cost function of system (Σ_n) is given as:

$$J^i(u^i, u^{-i}) = x^T(t_K)Q_\delta^{iK}x(t_K) + \mathtt{T}_s \sum_{k=0}^{K-1} (\mathcal{W}_1(t_k) + \mathcal{W}_2(t_k)), i \in \mathbf{R}, \quad (7.5)$$

where

$$\mathcal{W}_1(t_k) = x^T(t_k)Q_\delta^i x(t_k) + u^{iT}(t_k)R_\delta^{ii}u^i(t_k)$$

and

$$\mathcal{W}_2(t_k) = \sum_{j=1, j\neq i}^{r} \varepsilon_j u^{jT}(t_k)R_\delta^{ij}u^j(t_k).$$

The scaler ε_j describes the 'fairness' of the game, which is a small positive scalar indicating that the ith player needn't be punished too much by others' faults. For the nominal system (Σ_n), pure NE is employed to describe the final outcome of the noncooperative game, i.e., $J^{i*} := J^i(u^{i*}, u^{-i*}) \leq J^i(u^i, u^{-i*})$. However, a deviation from the pure NE is inevitably caused due to the aggregate impacts on (Σ_e) from the all-around attacks. To describe such attack-induced impacts, the following concept of ϵ-NE is introduced as

$$\hat{J}^i(u^*) \leq J^i(u^*) + \epsilon_K^i, \ i \in \mathbf{R},$$

where $\hat{J}^i$ is the cost function of system (Σ_e) with the same structure as J^i; the scalar ϵ_K^i characterizes the deviation from the pure NE. We are now in a position to present the problem to be dealt with in the chapter as follows: We aim at developing the multitasking optimal control strategy in the delta-domain for system (Σ_n) such that the individual cost function (7.5) is minimized. In addition, an explicit upper bound of the ϵ_K^i is to be provided such that the maximum attack-induced impacts can be quantified. Before ending this section, the following lemmas are provided to develop our main results.

Lemma 7.1. *[162] For any time function $x(t_k)$ and $y(t_k)$, the following property of delta operator holds*

$$\delta(x(t_k)y(t_k)) = y(t_k)\delta x(t_k) + x(t_k)\delta y(t_k) + T_s\delta x(t_k)\delta y(t_k)$$

where T_s is the sampling period.

Lemma 7.2. *[68] Let x and y be any n-dimensional real vectors, and Λ be an $n \times n$ symmetric positive semi-definite matrix. Then the following inequality holds.*

$$x^T y + y^T x \leq x^T \Lambda x + y^T \Lambda^{-1} y.$$

Lemma 7.3. *[26] Suppose that X is a random variable and function $g(\cdot)$ is concave and finite. Then the following inequality holds.*

$$\mathbb{E}\{g(X)\} \leq g(\mathbb{E}\{X\}).$$

Lemma 7.4. *[57] Let $C = [C_{ij}]_{p\times p}$ be a real value matrix and $D = diag\{d_1\ d_2\ \cdots\ d_p\}$ be a diagonal random matrix. Then one has that*

$$\mathbb{E}\{DCD^T\} = \begin{bmatrix} \mathbb{E}\{d_1^2\} & \mathbb{E}\{d_1d_2\} & \cdots & \mathbb{E}\{d_1d_p\} \\ \mathbb{E}\{d_2d_1\} & \mathbb{E}\{d_2^2\} & \cdots & \mathbb{E}\{d_2d_p\} \\ \vdots & \vdots & \ddots & \vdots \\ \mathbb{E}\{d_pd_1\} & \mathbb{E}\{d_pd_2\} & \cdots & \mathbb{E}\{d_p^2\} \end{bmatrix} \circ C$$

where notation $\circ$ is the Hadamard product.

7.3 Main Results

In this section, multitasking optimal control strategies in delta-domain are first derived. Then performance degradation caused by the attacks are estimated and an upper bound of the performance degradation is provided.

7.3.1 Multitasking Optimal Control Strategy

In this subsection, the nominal system (Σ_n) is considered and the multitasking optimal control strategies are provided in the finite- and infinite-time horizon, respectively.

Theorem 7.5. *Considering the ideal system (Σ_n) with cost function (7.5), the following conclusions are presented.*

1) There exists a unique feedback NE solution if the inequality $R_\delta^{ii} + T_sB^{iT}Z^i(t_{k+1})B^i > 0$ holds and the matrix $\Theta(k) = (\Theta_{ij}(k))$ is invertible with $\Theta_{ii}(t_k) = R_\delta^{ii} + T_sB^{iT}Z^i(t_{k+1})B^is$ and $\Theta_{ij}(t_k) = T_sB^{iT}Z^i(t_{k+1})B^j$, where $Z^i(t_k)$ satisfies the following backward recursions

$$\begin{aligned} -\delta Z^i(t_k) = &\ \mathcal{Q}_0(t_k) + \mathcal{Q}_1(t_k) + \mathcal{Q}_2^T(t_k)Z^i(t_{k+1}) \\ &+ Z^i(t_{k+1})\mathcal{Q}_2(t_k) + T_s\mathcal{Q}_2^T(t_k)Z^i(t_{k+1})\mathcal{Q}_2(t_k) \end{aligned} \quad (7.6a)$$

$$Z^i(t_k) = Z^i(t_{k+1}) - T_s\delta Z^i(t_k),\ Z^i(t_K) = Q_\delta^{iK} \quad (7.6b)$$

with

$$\mathcal{Q}_0(t_k) = Q^i_\delta + P^{iT}(t_k)R^{ii}_\delta P^i(t_k),$$
$$\mathcal{Q}_1(t_k) = \sum_{j\neq i}^{r} \varepsilon_j P^{jT}(t_k)R^{ij}_\delta P^j(t_k)$$
$$\mathcal{Q}_2(t_k) = A + \sum_{j=1}^{r} B^j P^j(t_k).$$

2) Under condition 1), the optimal control strategy of system (Σ_n) *is given by* $u^{i*}(t_k) = P^i(t_k)x(t_k)$*, where* $P^i(t_k)$ *satisfies*

$$(R^{ii}_\delta + \mathrm{T}_s B^{iT} Z^i(t_{k+1})B^i)P^i(t_k) + \mathrm{T}_s B^{iT} Z^i(t_{k+1}) \times \sum_{j\neq i}^{r} B^j P^j(t_k) = -B^{iT} Z^i(t_{k+1})(\mathrm{T}_s A + I). \tag{7.7}$$

Proof: The delta-domain cost function at time t_k is constructed as $V^i(x(t_{k+1})) = x^T(t_{k+1})Z^i(t_{k+1})x(t_{k+1})$ with $Z^i(t_{k+1}) > 0$. According to Lemma 7.1, one has that

$$\begin{aligned} V^i(x(t_{k+1})) &= \mathrm{T}_s\delta(x^T(t_k)Z^i(t_k)x(t_k)) + x^T(t_k)Z^i(t_k)x(t_k) \\ &= \mathrm{T}_s\delta x^T(t_k)Z^i(t_{k+1})x(t_k) + \mathrm{T}_s x^T(t_k)Z^i(t_{k+1})\delta x(t_k) \\ &\quad + \mathrm{T}_s^2\delta x^T(t_k)Z^i(t_{k+1})\delta x(t_k) + x^T(t_k)Z^i(t_{k+1})x(t_k). \end{aligned}$$

It follows from the dynamic programming algorithm [16] that

$$\begin{aligned} V^i(x(t_k)) &= \min_{u^i(t_k)} \{\mathrm{T}_s\mathcal{W}_1(t_k) + \mathcal{Q}_1(t_k) + V^i(x(t_{k+1}))\} \\ &= \min_{u^i(t_k)} \{\mathrm{T}_s\mathcal{W}_1(t_k) + \mathcal{Q}_1(t_k) + \mathrm{T}_s\mathcal{Q}_3^T(t_k)Z^i(t_{k+1})x(t_k) \\ &\quad + \mathrm{T}_s x^T(t_k)Z^i(t_{k+1})\mathcal{Q}_3(t_k) + \mathrm{T}_s^2\mathcal{Q}_3^T(t_k)Z^i(t_{k+1})\mathcal{Q}_3(t_k) \\ &\quad + x^T(t_k)Z^i(t_{k+1})x(t_k)\} \end{aligned} \tag{7.8}$$

where $\mathcal{Q}_3(t_k) = Ax(t_k) + \sum_{j=1}^{r} B^j u^j(t_k)$. Since the second-derivative of (7.8) yields $R^{ii}_\delta + \mathrm{T}_s B^T Z^i(t_{k+1})B > 0$, then the cost function $V^i(t_k)$ is strictly convex with respect to $u^i(t_k)$. Taking the first-derivative of equation (7.8) with respect to $u^i(t_k)$ and letting it equal to zero, one can obtain the optimal control strategy $u^{i*}(t_k) = P^i(t_k)x(t_k)$ with $P^i(t_k)$ given by (7.7). Furthermore, there exists a unique NE if $\Theta(t_k)$ satisfying $\Theta(t_k)P(t_k) = \Upsilon(t_k)$ is invertible, where

$$\begin{cases} \Theta_{ii}(t_k) := R^{ii}_\delta + \mathrm{T}_s B^{iT} Z^i(t_{k+1})B^i, \\ \Theta_{ij}(t_k) := \mathrm{T}_s B^{iT} Z^i(t_{k+1})B^j, \\ \Upsilon_i(t_k) := -B^{iT} Z^i(t_{k+1})(\mathrm{T}_s A + I), \\ P(t_k) := [P^{1T}(t_k)\ P^{2T}(t_k)\ \cdots\ P^{rT}(t_k)]^T. \end{cases}$$

Substituting $u^i(t_k) = P^i(t_k)x(t_k)$ back into (7.8), we have (7.6). The proof is completed.

Remark 7.6. The equivalent discrete-time results of Theorem 7.5 are obtained by substituting $A_z = \mathtt{T}_s A + I$, $B_z = \mathtt{T}_s B$, $Z(t_k) = Z(k)$, $P(t_k) = P(k)$ into recursions (7.6) and (7.7). On the other hand, the equivalent continuous-time system results of Theorem 7.5 are verified by the following arguments. With definition (2.2), one has that

$$\begin{cases} \lim_{\mathtt{T}_s \to 0} \delta Z^i(t_k) = \dot{Z}^i(t), \\ P^i(t) = -(R_\delta^{ii})^{-1} B^{iT} Z^i(t). \end{cases} \tag{7.9}$$

Substituting (7.9) into recursion (7.6) results in

$$\begin{aligned} & Q_\delta^i + Z^{iT}(t) B^i (R_\delta^{ii})^{-1} R_\delta^{ii} (R_\delta^{ii})^{-1} B^{iT} Z^i(t) + \sum_{j \neq i}^{r} \varepsilon_j Z^{jT}(t) B^j (R_\delta^{jj})^{-1} \\ & \times R_\delta^{ij} (R_\delta^{jj})^{-1} B^{jT} Z^j(t) + \tilde{F}^T(t) Z^i(t) + Z^{iT}(t) \tilde{F}(t) + \dot{Z}(t) = 0 \end{aligned} \tag{7.10}$$

where $\tilde{F}(t) = A - \sum_{j=1}^{r} B^j (R_\delta^{jj})^{-1} B^{jT} Z^j(t)$. Note that (7.9) and (7.10) are consistent with the results of the continuous-time systems in [16]. Therefore, the fact that Theorem 7.5 unifies the results in the discrete- and continuous-domain is verified.

In the following, we extend the finite-horizon multitasking optimal control problem to the infinite-horizon case. It is assumed that $Q_\delta^i = Q_\delta^{iK}$. First, sufficient conditions are provided such that the recursion (7.6) becomes convergent, and then the stability problem of system (Σ_n) is verified. Without loss of generality, the time index of (7.6b) is reversed by using the time reverse notation $\tilde{Z}(t_k) := Z(t_K - t_k)$. Before the statement of Theorem 7.7, the following notations are introduced with hope to simplify the presentation.

$$\mathcal{F}^i := \mathtt{T}_s A + I + \mathtt{T}_s \sum_{j \neq i}^{r} B^j P_\infty^j, \quad \mathcal{E}^i := \mathtt{T}_s Q_\delta^i + \mathtt{T}_s \sum_{j \neq i}^{r} \varepsilon_j P_\infty^{jT} R_\delta^{ij} P_\infty^j$$

Theorem 7.7. *Consider system (Σ_n) and cost function (7.5) with $K \to \infty$. Let $\{P_\infty^j\}_{j \neq i}^r$ be given. If $(\mathcal{F}^i, \sqrt{\mathcal{E}^i})$ is observable, we have $\lim_{k \to \infty} \tilde{Z}^i(t_k) = Z_\infty^i$. Furthermore, the following system*

$$\delta x(t_k) = (A + \sum_{j \neq i}^{r} B^j P_\infty^j + B^i P_\infty^i) x(t_k)$$

can be stabilized by $P_\infty^i = -(R_\delta^{ii} + T_s B^{iT} Z_\infty^i B^i)^{-1} B^{iT} Z_\infty^i \mathcal{F}^i$.

Proof: The recursion (7.6) can be rearranged as

$$\tilde{Z}^i(t_{k+1}) = \mathcal{F}^{iT}\tilde{Z}^i(t_k)\mathcal{F}^i + \mathcal{E}^i - \mathtt{T}_s\mathcal{F}^{iT}(t_k)\tilde{Z}^i(t_k)\hat{B}^i \times(\mathtt{T}_s R_\delta^{ii} + \hat{B}^{iT}Z^i(t_k)\hat{B}^i)^{-1}\hat{B}^{iT}\tilde{Z}^i(t_k)\mathcal{F}_i. \tag{7.11}$$

where $\hat{B}^i = \mathtt{T}_s B^i$. Then, it follows from [64] that $\lim_{k\to\infty}\tilde{Z}^i(t_k) = Z_\infty^i$. The steady state Riccati equation yields

$$Z_\infty^i = \mathcal{F}^{iT}Z_\infty^i\mathcal{F}^i + \mathtt{T}_s Q_\delta^i + \mathtt{T}_s\sum_{j\neq i}^{r}\varepsilon_j P_\infty^{jT}R_\delta^{ij}P_\infty^j + \mathtt{T}_s\mathcal{F}^{iT}Z_\infty^i B^i P_\infty^i + \mathtt{T}_s P_\infty^{iT}B^{iT}Z_\infty^i\mathcal{F}^i + \mathtt{T}_s P_\infty^{iT}R_\delta^{ii}P_\infty^i + \mathtt{T}_s^2 P_\infty^{iT}B^{iT}Z_\infty^i B^i P_\infty^i. \tag{7.12}$$

It is observed that

$$\begin{aligned} & x^T(t_{k+1})Z_\infty^i x(t_{k+1}) - x^T(t_k)Z_\infty^i x(t_k) \\ &= (\mathtt{T}_s\delta x(t_k) + x(t_k))^T Z_\infty^i(\mathtt{T}_s\delta x(t_k) + x(t_k)) - x^T(t_k)Z_\infty^i x(t_k) \\ &= x^T(t_k)(\mathcal{F}^{iT}Z_\infty^i\mathcal{F}^i + \mathtt{T}_s\mathcal{F}^{iT}Z_\infty^i B^i P_\infty^i + T_s P_\infty^{iT}B^{iT}Z_\infty^i\mathcal{F}^i \\ &\quad + \mathtt{T}_s^2 P_\infty^{iT}B^{iT}Z_\infty^i B^i P_\infty^i - Z_\infty^i)x(t_k). \end{aligned} \tag{7.13}$$

By substituting (7.12) into (7.13), one has

$$x^T(t_{k+1})Z_\infty^i x(t_{k+1}) = x^T(t_0)Z_\infty^i x(t_0) - \sum_{l=0}^{k}\mathbb{E}\{x^T(t_l)(T_s Q_\delta^i + \mathtt{T}_s P_\infty^{iT}R_\delta^{ii}P_\infty^i + \mathtt{T}_s\sum_{j\neq i}^{r}\varepsilon_j P_\infty^{jT}R_\delta^{ij}P_\infty^j)x(t_k)\}, \tag{7.14}$$

which further implies that $\lim_{k\to\infty}\mathbb{E}\{x^T(t_k)(\mathtt{T}_s P_\infty^{iT}R_\delta^{ii}P_\infty^i + \mathcal{E}^i)x(t_k)\} = 0$. Since $(\mathcal{F}^i, \sqrt{\mathcal{E}^i})$ is observable and $R_\delta^{ij} > 0$, we have $\mathbb{E}\{\|x(t_k)\|\} \to 0$ as $t_k \to \infty$. The proof is completed.

7.3.2 Robustness Analysis of ϵ-NE

Up to now, the feedback gain $P(t_k)$ (respectively, P_∞) has been obtained for system (Σ_n) in a finite-time horizon (respectively, infinite-time horizon) case. Due to the aggregate effects of DoS attacks, deception attacks, and physical attacks in system (Σ_e), a certain deviation from the pure NE is caused. In this subsection, the estimations for epsilon level of ϵ-NE are provided in finite- and infinite-time horizons, respectively. Before providing the main results, the following denotations and illustrations are presented: let $x^*(t_k)$ (respectively, $x(t_k)$) be the state vector of system (Σ_n) (respectively, (Σ_e)) if the optimal feedback control strategy (7.7) is adopted, and

$$
\begin{aligned}
&e(t_k) = x(t_k) - x^*(t_k),\ V_1(t_k) = e^T(t_k)S_1(t_k)e(t_k),\ \bar{Q}^i_\delta(t_K) = \mathtt{T}_s Q^{iK}_\delta,\\
&\bar{Q}^i_\delta(t_k) = \mathtt{T}_s Q^i_\delta + \mathtt{T}_s P^{iT}(t_k) R^{ii}_\delta P^i(t_k) + \mathtt{T}_s \sum_{j\neq i}^{r} \varepsilon_j P^{jT}(t_k) R^{ij}_\delta P^j(t_k),\\
&B = [B^1\ B^2\ \cdots\ B^r],\ P(t_k) = [P^{1T}(t_k)\ P^{2T}(t_k)\ \cdots\ P^{rT}(t_k)]^T,\\
&\Delta u(t_k) = [\Delta u^{1T}(t_k)\ \Delta u^{2T}(t_k)\ \cdots\ \Delta u^{rT}(t_k)]^T,\\
&u^*(t_k) = [u^{1*T}(t_k)\ u^{2*T}(t_k)\ \cdots\ u^{r*T}(t_k)]^T,
\end{aligned}
$$

$$
\begin{aligned}
\lambda_{x^*}(t_k) = \lambda_{\max}\{&P^T(t_k)(\mathcal{D}_9 \circ (B^T S_1(t_{k+1})(\mathtt{T}_s \Lambda_1^{-1} + \mathtt{T}_s^2 \bar{\eta} \Lambda_8) S_1(t_{k+1}) B))\\
&\times P(t_k) + \mathtt{T}_s^2 P^T(t_k)(\mathcal{D}_2 \circ (B^T S_1(t_{k+1}) B))^T \Lambda_4^{-1} (\mathcal{D}_2 \circ (B^T\\
&\times S_1(t_{k+1}) B)) P(t_k) + \mathtt{T}_s^2 P^T(t_k)(\mathcal{D}_6 \circ (B^T S_1(t_{k+1}) B))^T \Lambda_7\\
&\times (\mathcal{D}_6 \circ (B^T S_1(t_{k+1}) B)) P(t_k) + \mathtt{T}_s^2 P^T(t_k)(\mathcal{D}_3 \circ (B^T S_1(t_{k+1})\\
&\times B)) P(t_k)\},\\
\lambda_f(t_k) = \lambda_{\max}\big\{&\mathtt{T}_s \bar{\eta} S_1(t_{k+1}) \Lambda_3^{-1} S_1(t_{k+1}) + \mathtt{T}_s^2 \bar{\eta} S_1(t_{k+1})\\
&+ \mathtt{T}_s^2 \bar{\eta} S_1(t_{k+1}) \Lambda_6^{-1} S_1(t_{k+1}) + \mathtt{T}_s^2 \bar{\eta} \Lambda_8^{-1}\\
&+ \mathtt{T}_s^2 \bar{\eta} S_1(t_{k+1}) \Lambda_9^{-1} S_1(t_{k+1})\big\},\\
\lambda_{\Delta u}(t_k) = \lambda_{\max}\{&\mathtt{T}_s^2 \bar{\eta} (\mathcal{D}_7 \circ (B^T \Lambda_9 B)) + \mathtt{T}_s^2 (\mathcal{D}_5 \circ (B^T S_1(t_{k+1}) B))^T \Lambda_5^{-1}\\
&\times (\mathcal{D}_5 \circ (B^T S_1(t_{k+1}) B)) + \mathtt{T}_s^2 \Lambda_7^{-1} + \mathtt{T}_s^2 (\mathcal{D}_4 \circ (B^T S_1(t_{k+1}) B))\\
&+ \mathtt{T}_s (\mathcal{D}_7 \circ (B^T S_1(t_{k+1}) \Lambda_2^{-1} S_1(t_{k+1}) B))\},
\end{aligned}
$$

$$\mathcal{G}_1 = \mathrm{diag}\{\bar{\beta}_1\ \bar{\beta}_2\ \cdots\ \bar{\beta}_r\},\ \mathcal{G}_2 = \mathrm{diag}\{\bar{\beta}_1 - 1\ \bar{\beta}_2 - 1\ \cdots\ \bar{\beta}_r - 1\},$$
$$\mathcal{G}_3 = \mathrm{diag}\{\bar{\beta}_1\bar{\alpha}_1\ \bar{\beta}_2\bar{\alpha}_2\ \cdots\ \bar{\beta}_r\bar{\alpha}_r\},$$
$$\mathcal{D}_1 = \begin{bmatrix} \bar{\beta}_1 & \bar{\beta}_1\bar{\beta}_2 & \cdots & \bar{\beta}_1\bar{\beta}_r \\ \bar{\beta}_2\bar{\beta}_1 & \bar{\beta}_2 & \cdots & \bar{\beta}_2\bar{\beta}_r \\ \vdots & \vdots & \ddots & \vdots \\ \bar{\beta}_r\bar{\beta}_1 & \bar{\beta}_r\bar{\beta}_2 & \cdots & \bar{\beta}_r \end{bmatrix},$$
$$\mathcal{D}_2 = \begin{bmatrix} 0 & \bar{\beta}_1(\bar{\beta}_2 - 1) & \cdots & \bar{\beta}_1(\bar{\beta}_r - 1) \\ \bar{\beta}_2(\bar{\beta}_1 - 1) & 0 & \cdots & \bar{\beta}_2(\bar{\beta}_r - 1) \\ \vdots & \vdots & \ddots & \vdots \\ \bar{\beta}_r(\bar{\beta}_1 - 1) & \bar{\beta}_r(\bar{\beta}_2 - 1) & \cdots & 0 \end{bmatrix},$$
$$\mathcal{D}_3 = \begin{bmatrix} 1 - \bar{\beta}_1 & (\bar{\beta}_1 - 1)(\bar{\beta}_2 - 1) & \cdots & (\bar{\beta}_1 - 1)(\bar{\beta}_r - 1) \\ (\bar{\beta}_2 - 1)(\bar{\beta}_1 - 1) & 1 - \bar{\beta}_2 & \cdots & (\bar{\beta}_2 - 1)(\bar{\beta}_r - 1) \\ \vdots & \vdots & \ddots & \vdots \\ (\bar{\beta}_r - 1)(\bar{\beta}_1 - 1) & (\bar{\beta}_r - 1)(\bar{\beta}_2 - 1) & \cdots & 1 - \bar{\beta}_r \end{bmatrix},$$
$$\mathcal{D}_4 = \begin{bmatrix} \bar{\beta}_1\bar{\alpha}_1 & \bar{\beta}_1\bar{\alpha}_1\bar{\beta}_2\bar{\alpha}_2 & \cdots & \bar{\beta}_1\bar{\alpha}_1\bar{\beta}_r\bar{\alpha}_r \\ \bar{\beta}_2\bar{\alpha}_2\bar{\beta}_1\bar{\alpha}_1 & \bar{\beta}_2\bar{\alpha}_2 & \cdots & \bar{\beta}_2\bar{\alpha}_2\bar{\beta}_r\bar{\alpha}_r \\ \vdots & \vdots & \ddots & \vdots \\ \bar{\beta}_r\bar{\alpha}_r\bar{\beta}_1\bar{\alpha}_1 & \bar{\beta}_r\bar{\alpha}_r\bar{\beta}_2\bar{\alpha}_2 & \cdots & \bar{\beta}_r\bar{\alpha}_r \end{bmatrix},$$

$$\mathcal{D}_5 = \begin{bmatrix} \bar{\beta}_1\bar{\alpha}_1 & \bar{\beta}_1\bar{\beta}_2\bar{\alpha}_2 & \cdots & \bar{\beta}_1\bar{\beta}_r\bar{\alpha}_r \\ \bar{\beta}_2\bar{\beta}_1\bar{\alpha}_1 & \bar{\beta}_2\bar{\alpha}_2 & \cdots & \bar{\beta}_2\bar{\beta}_r\bar{\alpha}_r \\ \vdots & \vdots & \ddots & \vdots \\ \bar{\beta}_r\bar{\beta}_1\bar{\alpha}_1 & \bar{\beta}_r\bar{\beta}_2\bar{\alpha}_2 & \cdots & \bar{\beta}_r\bar{\alpha}_r \end{bmatrix}.$$

$$\mathcal{D}_6 = \begin{bmatrix} 0 & (\bar{\beta}_1 - 1)\bar{\beta}_2\bar{\alpha}_2 & \cdots & (\bar{\beta}_1 - 1)\bar{\beta}_r\bar{\alpha}_r \\ (\bar{\beta}_2 - 1)\bar{\beta}_1\bar{\alpha}_1 & 0 & \cdots & (\bar{\beta}_2 - 1)\bar{\beta}_r\bar{\alpha}_r \\ \vdots & \vdots & \ddots & \vdots \\ (\bar{\beta}_r - 1)\bar{\beta}_1\bar{\alpha}_1 & (\bar{\beta}_r - 1)\bar{\beta}_2\bar{\alpha}_2 & \cdots & 0 \end{bmatrix},$$

$$\mathcal{D}_7 = \begin{bmatrix} \bar{\beta}_1^2\bar{\alpha}_1^2 & \bar{\beta}_1\bar{\alpha}_1\bar{\beta}_2\bar{\alpha}_2 & \cdots & \bar{\beta}_1\bar{\alpha}_1\bar{\beta}_r\bar{\alpha}_r \\ \bar{\beta}_2\bar{\alpha}_2\bar{\beta}_1\bar{\alpha}_1 & \bar{\beta}_2^2\bar{\alpha}_2^2 & \cdots & \bar{\beta}_2\bar{\alpha}_2\bar{\beta}_r\bar{\alpha}_r \\ \vdots & \vdots & \ddots & \vdots \\ \bar{\beta}_r\bar{\alpha}_r\bar{\beta}_1\bar{\alpha}_1 & \bar{\beta}_r\bar{\alpha}_r\bar{\beta}_2\bar{\alpha}_2 & \cdots & \bar{\beta}_r^2\bar{\alpha}_r^2 \end{bmatrix},$$

$$\mathcal{D}_8 = \begin{bmatrix} \bar{\beta}_1^2 & \bar{\beta}_1\bar{\beta}_2 & \cdots & \bar{\beta}_1\bar{\beta}_r \\ \bar{\beta}_2\bar{\beta}_1 & \bar{\beta}_2^2 & \cdots & \bar{\beta}_2\bar{\beta}_r \\ \vdots & \vdots & \ddots & \vdots \\ \bar{\beta}_r\bar{\beta}_1 & \bar{\beta}_r\bar{\beta}_2 & \cdots & \bar{\beta}_r^2 \end{bmatrix},$$

$$\mathcal{D}_9 = \begin{bmatrix} (\bar{\beta}_1 - 1)^2 & (\bar{\beta}_1 - 1)(\bar{\beta}_2 - 1) & \cdots & (\bar{\beta}_1 - 1)(\bar{\beta}_r - 1) \\ (\bar{\beta}_2 - 1)(\bar{\beta}_1 - 1) & (\bar{\beta}_2 - 1)^2 & \cdots & (\bar{\beta}_2 - 1)(\bar{\beta}_r - 1) \\ \vdots & \vdots & \ddots & \vdots \\ (\bar{\beta}_r - 1)(\bar{\beta}_1 - 1) & (\bar{\beta}_r - 1)(\bar{\beta}_2 - 1) & \cdots & (\bar{\beta}_r - 1)^2 \end{bmatrix}$$

In what follows, we are ready to provide the estimation for the epsilon level of ϵ-NE over the finite horizon.

Theorem 7.8. *Consider system* (Σ_e) *with cost function* $\hat{J}^i(u^*)$. *For given optimal control strategies* u^{i*} *obtained from Theorem 7.5 and positive define matrices*$\{\Lambda_i\}_{i=1}^9$ *and* L_1, *if there exists a positive definite matrix* $S_1(t_k)$, $\forall k \in \boldsymbol{K}$ *such that the following recursions hold:*

$$\begin{aligned} \mathcal{H}(S_1(t_k)) := {} & T_s A^T S_1(t_{k+1})A + A^T S_1(t_{k+1}) + S_1(t_{k+1})A + (T_s A + I)^T \\ & \times(\Lambda_1 + \Lambda_2 + \bar{\eta}\Lambda_3)(T_s A + I) + (T_s A + I)^T S_1(t_{k+1})B\mathcal{G}_1 P(t_k) \\ & + P^T(t_k)\mathcal{G}_1 B^T S_1(t_{k+1})(T_s A + I) + T_s P^T(t_k)\Lambda_4 P(t_k) \\ & + T_s\bar{\eta}P^T(t_k)(\mathcal{D}_8 \circ (B^T\Lambda_6 B))P(t_k) \\ & + T_s P^T(t_k)(\mathcal{D}_1 \circ (B^T S_1(t_{k+1})B))P(t_k) + T_s P^T(t_k)\Lambda_5 P(t_k) \\ & + \delta S_1(t_k) + \frac{1}{T_s}L_1 = 0 \end{aligned}$$

$$S_1(t_k) = S_1(t_{k+1}) - T_s\delta S_1(t_k), \;\; S_1(t_K) = I,$$

then the ϵ*-NE yields* $\hat{J}^i(u^*) \leq J^i(u^*) + \epsilon_K^i$, *where*

$$\epsilon_K^i \leq \max_{\forall k \in \boldsymbol{K}} \lambda_{\max}\{\bar{Q}_\delta^i(t_k)\} \Bigg(\Psi + \lambda_{\min}^{-1}\{L_1\}V_1(t_0) + 2\sqrt{\sum_{k=0}^{K} \|x^*(t_k)\|^2(\Psi + \lambda_{\min}^{-1}\{L_1\}V_1(t_0))} \Bigg) \tag{7.15}$$

with

$$\Psi = \lambda_{\min}^{-1}\{L_1\} \sum_{k=0}^{K} (\lambda_f(t_k)\gamma_1 + \lambda_{\Delta u}(t_k)\gamma_2 + \lambda_{x^*}(t_k)\|x^*(t_k)\|^2).$$

Proof: The cost functions for system (Σ_n) and (Σ_e) are written as

$$\begin{cases} J^i(u^*) = \sum_{k=0}^{K} x^{*T}(t_k)\bar{Q}_\delta^i(t_k)x^*(t_k), \\ \hat{J}^i(u^*) = \mathbb{E}\left\{ \sum_{k=0}^{K} x^T(t_k)\bar{Q}_\delta^i(t_k)x(t_k) \right\}. \end{cases}$$

It follows from the definition of ϵ_K^i that

$$\begin{aligned} \epsilon_K^i &= \mathbb{E}\{\hat{J}^i(u^*) - J^i(u^*)\} \\ &= \mathbb{E}\left\{ \sum_{k=0}^{K} x^T(t_k)\bar{Q}_\delta^i(t_k)x(t_k) - \sum_{k=0}^{K} x^{*T}(t_k)\bar{Q}_\delta^i(t_k)x^*(t_k) \right\} \\ &= \mathbb{E}\left\{ \sum_{k=0}^{K} (x(t_k) - x^*(t_k))^T \bar{Q}_\delta^i(t_k)(x(t_k) - x^*(t_k) + 2x^*(t_k)) \right\} \\ &\leq \mathbb{E}\left\{ \sum_{k=0}^{K} \|x(t_k) - x^*(t_k)\|^2_{\bar{Q}_\delta^i(t_k)} + 2\|x(t_k) - x^*(t_k)\|\|\bar{Q}_\delta^i(t_k)x^*(t_k)\| \right\} \\ &\leq \mathbb{E}\left\{ \max_{\forall k \in \mathbf{K}} \lambda_{\max}\{\bar{Q}_\delta^i(t_k)\} \left(\sum_{k=0}^{K} \|x(t_k) - x^*(t_k)\|^2 \right.\right. \\ &\quad \left.\left. +2\sum_{k=0}^{K} \|x(t_k) - x^*(t_k)\|\|x^*(t_k)\| \right) \right\} \\ &\leq \max_{\forall k \in \mathbf{K}} \lambda_{\max}\{\bar{Q}_\delta^i(t_k)\} \left(\mathbb{E}\left\{ \sum_{k=0}^{K} \|e(t_k)\|^2 \right\} \right. \\ &\quad \left. +2\sqrt{\mathbb{E}\left\{ \sum_{k=0}^{K} \|e(t_k)\|^2 \right\}}\sqrt{\sum_{k=0}^{K} \|x^*(t_k)\|^2} \right) \end{aligned}$$

The last inequality holds from Cauchy inequality and Lemma 7.3. Consider the following error dynamics

$$\delta e(t_k) = \bar{A}e(t_k) + \sum_{j=1}^{r}(\beta_j(t_k) - 1)B^j P^j(t_k)x^*(t_k)$$
$$+\eta(t_k)f(t_k) + \sum_{j=1}^{r}\beta_j(t_k)\alpha_j(t_k)B^j \Delta u^j(t_k) \tag{7.16}$$

where $\bar{A} = A + \sum_{j=1}^{r}\beta_j(t_k)B^j P^j(t_k)$. According to the definition of $V_1(t_k)$ and (7.16), one has that

$$\begin{aligned}
\mathtt{T}_s\mathbb{E}\{\delta V_1(t_k)\} &= \mathbb{E}\{e^T(t_{k+1})S_1(t_{k+1})e(t_{k+1}) - e^T(t_k)S_1(t_k)e(t_k)\} \\
&= \mathbb{E}\{((\mathtt{T}_s A + I + \mathtt{T}_s\sum_{j=1}^{r}\beta_j(t_k)B^j P^j(t_k))e(t_k) \\
&\quad +\mathtt{T}_s\sum_{j=1}^{r}(\beta_j(t_k) - 1)B^j P^j(t_k)x^*(t_k) + \mathtt{T}_s\eta(t_k)f(t_k) \\
&\quad +\mathtt{T}_s\sum_{j=1}^{r}\beta_j(t_k)\alpha_j(t_k)B^j \Delta u^j(t_k))^T S_1(t_{k+1}) \\
&\quad \times((\mathtt{T}_s A + I + \mathtt{T}_s\sum_{j=1}^{r}\beta_j(t_k)B^j P^j(t_k))e(t_k) \\
&\quad +\mathtt{T}_s\sum_{j=1}^{r}(\beta_j(t_k) - 1)B^j P^j(t_k)x^*(t_k) + \mathtt{T}_s\eta(t_k)f(t_k) \\
&\quad +\mathtt{T}_s\sum_{j=1}^{r}\beta_j(t_k)\alpha_j(t_k)B^j \Delta u^j(t_k)) - e^T(t_k)S_1(t_k)e(t_k)\}
\end{aligned}$$

It then follows from Lemma 7.4 that

$$\begin{aligned}
&\mathtt{T}_s\mathbb{E}\{\delta V_1(t_k)\} \\
&= \mathbb{E}\{\mathtt{T}_s e^T(t_k)(\mathtt{T}_s A + I)^T S_1(t_{k+1})B\mathcal{G}_1 P(t_k)e(t_k) + \mathtt{T}_s\bar{\eta}e^T(t_k)(\mathtt{T}_s A + I)^T \\
&\quad \times S_1(t_{k+1})f(t_k) + \mathtt{T}_s e^T(t_k)P^T(t_k)\mathcal{G}_1 B^T S_1(t_{k+1})(\mathtt{T}_s A + I)e(t_k) \\
&\quad +\mathtt{T}_s\bar{\eta}f^T(t_k)S_1(t_{k+1})(\mathtt{T}_s A + I)e(t_k) + \mathtt{T}_s e^T(t_k)(\mathtt{T}_s A + I)^T S_1(t_{k+1}) \\
&\quad \times B\mathcal{G}_3\Delta u(t_k) + \mathtt{T}_s e^T(t_k)(\mathtt{T}_s A + I)^T S_1(t_{k+1})B\mathcal{G}_2 P(t_k)x^*(t_k) \\
&\quad +\mathtt{T}_s\Delta u^T(t_k)\mathcal{G}_3 B^T S_1(t_{k+1})(\mathtt{T}_s A + I)e(t_k) + \mathtt{T}_s x^{*T}(t_k)P^T(t_k)\mathcal{G}_2 B^T \\
&\quad \times S_1(t_{k+1})(\mathtt{T}_s A + I)e(t_k) + \mathtt{T}_s^2 e^T(t_k)P^T(t_k)(\mathcal{D}_2 \circ (B^T S_1(t_{k+1})B)) \\
&\quad \times P(t_k)x^*(t_k) + \mathtt{T}_s^2\bar{\eta}x^{*T}(t_k)P^T(t_k)\mathcal{G}_2 B^T S_1(t_{k+1})f(t_k) + \mathtt{T}_s^2 x^{*T}(t_k)
\end{aligned}$$

$$\begin{aligned}
&\times P^T(t_k)(\mathcal{D}_2^T \circ (B^T S_1(t_{k+1})B))P(t_k)e(t_k) + \mathtt{T}_s^2\bar{\eta} f^T(t_k)S_1(t_{k+1})B \\
&\times \mathcal{G}_2 P(t_k)x^*(t_k) + \mathtt{T}_s^2\bar{\eta} e^T(t_k)P^T(t_k)\mathcal{G}_1 B^T S_1(t_{k+1})f(t_k) + \mathtt{T}_s^2 x^{*T}(t_k) \\
&\times P^T(t_k)(\mathcal{D}_6 \circ (B^T S_1(t_{k+1})B))\Delta u(t_k) + \mathtt{T}_s^2\bar{\eta} f^T(t_k)S_1(t_{k+1})B\mathcal{G}_1 \\
&\times P(t_k)e(t_k) + \mathtt{T}_s^2 \Delta u^T(t_k)(\mathcal{D}_6^T \circ (B^T S_1(t_{k+1})B))P(t_k)x^*(t_k) \\
&+\mathtt{T}_s^2 e^T(t_k)P^T(t_k)(\mathcal{D}_5 \circ (B^T S_1(t_{k+1})B))\Delta u(t_k) + \mathtt{T}_s^2\bar{\eta}\Delta u^T(t_k)\mathcal{G}_3 B^T \\
&\times S_1(t_{k+1})f(t_k) + \mathtt{T}_s^2 \Delta u^T(t_k)(\mathcal{D}_5^T \circ (B^T S_1(t_{k+1})B))P(t_k)e(t_k) \\
&+\mathtt{T}_s^2\bar{\eta} f^T(t_k)S_1(t_{k+1})B\mathcal{G}_3\Delta u(t_k) + e^T(t_k)(\mathtt{T}_s A + I)^T S_1(t_{k+1})(\mathtt{T}_s A \\
&+I)e(t_k) + \mathtt{T}_s^2 e^T(t_k)P^T(t_k)(\mathcal{D}_1 \circ (B^T S_1(t_{k+1})B))P(t_k)e(t_k) \\
&+\mathtt{T}_s^2 x^{*T}(t_k)P^T(t_k)(\mathcal{D}_3 \circ (B^T S_1(t_{k+1})B))P(t_k)x^*(t_k) + \mathtt{T}_s^2\bar{\eta} f^T(t_k) \\
&\times S_1(t_{k+1})f(t_k) + \mathtt{T}_s^2 \Delta u^T(t_k)(\mathcal{D}_4 \circ (B^T S_1(t_{k+1})B))\Delta u(t_k) \\
&-e^T(t_k)S_1(t_k)e(t_k)\}.
\end{aligned}$$

Noting Lemma 7.2 and the upper bound of parameters $\Delta u(t_k)$ and $f(t_k)$, one has that

$$\begin{aligned}
\mathtt{T}_s\mathbb{E}\{\delta V_1(t_k)\} \leq \mathbb{E}\{&e^T(t_k)((\mathtt{T}_s A + I)^T(S_1(t_{k+1}) + \mathtt{T}_s\Lambda_1 + \mathtt{T}_s\Lambda_2 + \mathtt{T}_s\bar{\eta}\Lambda_3) \\
&\times(\mathtt{T}_s A + I) + \mathtt{T}_s^2 P^T(t_k)\Lambda_4 P(t_k) + \mathtt{T}_s(\mathtt{T}_s A + I)^T S_1(t_{k+1}) \\
&\times B\mathcal{G}_1 P(t_k) + \mathtt{T}_s P^T(t_k)\mathcal{G}_1 B^T S_1(t_{k+1})(\mathtt{T}_s A + I) \\
&+\mathtt{T}_s^2 P^T(t_k)\Lambda_5 P(t_k) + \mathtt{T}_s^2\bar{\eta} P^T(t_k)(\mathcal{D}_8 \circ (B^T\Lambda_6 B))P(t_k) \\
&+\mathtt{T}_s^2 P^T(t_k)(\mathcal{D}_1 \circ (B^T S_1(t_{k+1})B))P(t_k) - S_1(t_k) \\
&++L_1)e(t_k)\lambda_f(t_k)\gamma_1 + \lambda_{\Delta u}(t_k)\gamma_2 + \lambda_{x^*}(t_k)\|x^*(t_k)\|^2 \\
&-e^T(t_k)L_1 e(t_k)\}. \qquad (7.17)
\end{aligned}$$

Selcct a positive definite matrix $S_1(t_k), \forall k \in \mathbf{K}$ such that

$$\begin{aligned}
&\mathtt{T}_s A^T S_1(t_{k+1})A + A^T S_1(t_{k+1}) + S_1(t_{k+1})A + (\mathtt{T}_s A + I)^T \\
&\times(\Lambda_1 + \Lambda_2 + \bar{\eta}\Lambda_3)(\mathtt{T}_s A + I) + \mathtt{T}_s P^T(t_k)\Lambda_4 P(t_k) \\
&+(\mathtt{T}_s A + I)^T S_1(t_{k+1})B\mathcal{G}_1 P(t_k) + P^T(t_k)\mathcal{G}_1 B^T S_1(t_{k+1}) \\
&\times(\mathtt{T}_s A + I) + \mathtt{T}_s\bar{\eta} P^T(t_k)(\mathcal{D}_8 \circ (B^T\Lambda_6 B))P(t_k) \\
&+\mathtt{T}_s P^T(t_k)\Lambda_5 \times P(t_k) + \mathtt{T}_s P^T(t_k)(\mathcal{D}_1 \circ (B^T S_1(t_{k+1})B))P(t_k) \\
&+\delta S_1(t_k) + \frac{1}{\mathtt{T}_s}L_1 = 0. \qquad (7.18)
\end{aligned}$$

Then

$$\begin{aligned}
\mathtt{T}_s\mathbb{E}\left\{\sum_{k=0}^{K}\delta V_1(t_k)\right\} \leq &\sum_{k=0}^{K}(\lambda_f(t_k)\gamma_1 + \lambda_{\Delta u}(t_k)\gamma_2 + \lambda_{x^*}(t_k)\|x^*(t_k)\|^2) \\
&-\mathbb{E}\left\{\sum_{k=0}^{K}e^T(t_k)L_1 e(t_k)\right\}. \qquad (7.19)
\end{aligned}$$

Furthermore, one has that

$$\lambda_{\min}\{L_1\}\mathbb{E}\left\{\sum_{k=0}^{K}\|e(t_k)\|^2\right\} \le \sum_{k=0}^{K}(\lambda_f(t_k)\gamma_1 + \lambda_{\Delta u}(t_k)\gamma_2 + \lambda_{x^*}(t_k)\|x^*(t_k)\|^2) \\ +\mathbb{E}\{V_1(t_0) - V_1(t_k)\},$$

which leads to

$$\mathbb{E}\left\{\sum_{k=0}^{K}\|e(t_k)\|^2\right\} \le \lambda_{\min}^{-1}\{L_1\}\sum_{k=0}^{K}(\lambda_f(t_k)\gamma_1 + \lambda_{\Delta u}(t_k)\gamma_2 \\ +\lambda_{x^*}(t_k)\|x^*(t_k)\|^2) + \lambda_{\min}^{-1}\{L_1\}V_1(t_0). \quad (7.20)$$

Finally, substituting (7.20) into (7.16) results in (7.15). The proof is completed.

Note that, if $\mathcal{H}(S_1(t_k))$ satisfies the inequality $\mathcal{H}(S_1(t_k)) < 0$, the inequality (7.19) also holds. Therefore, an iterative LMI method which is less conservative than Theorem 7.8 is proposed in Algorithm 7.

Remark 7.9. It follows from Theorem 7.8 that the following argument can be easily verified: if $\Delta u(t_k) = f(t_k) \equiv 0$ and $\bar{\beta}_i \equiv 1$, the so-called ϵ-NE will reduce to the pure NE with $\epsilon_K^i \equiv 0$.

Up to this end, an upper bound of ϵ-NE for the finite-horizon multitasking optimal control has been provided in the delta-domain. In this part, an infinite-horizon case is dealt with. Suppose that $f(t_k)$, $\Delta u^i(t_k)$ and $x^*(t_k)$ belong to $l_2[0,\infty)$, i.e.,

$$\|f(t_k)\|_2^2 \le \gamma_1^\infty,\ \|\Delta u^i(t_k)\|_2^2 \le \gamma_2^\infty,\ \|x^*(t_k)\|_2^2 \le \gamma_3^\infty.$$

Some notations are presented as follows:

$$\begin{aligned}
&\bar{Q}_\delta^{i,\infty} = \mathrm{T}_s Q_\delta^i + \mathrm{T}_s P_\infty^{iT} R_\delta^{ii} P_\infty^i + \mathrm{T}_s \sum_{j\neq i}^{r} \varepsilon_j P_\infty^{jT} R_\delta^{ij} P_\infty^j, \\
&e(t_k) = x(t_k) - x^*(t_k), V_1^\infty(t_k) = e^T(t_k) S_1^\infty e(t_k), \\
&P_\infty = [P_\infty^{1T}\ P_\infty^{2T}\ \cdots\ P_\infty^{rT}]^T, \\
&\lambda_{x^*}^\infty = \lambda_{\max}\{P_\infty^T(\mathcal{D}_9 \circ (B^T S_1^\infty(\mathrm{T}_s\bar{\Lambda}_1^{-1} + \mathrm{T}_s^2\bar{\eta}\bar{\Lambda}_8)S_1^\infty B))P_\infty \\
&\qquad +\mathrm{T}_s^2 P_\infty^T(\mathcal{D}_3 \circ (B^T S_1^\infty B))P_\infty + \mathrm{T}_s^2 P_\infty^T(\mathcal{D}_2 \circ (B^T S_1^\infty B))^T \bar{\Lambda}_4^{-1} \\
&\qquad \times(\mathcal{D}_2 \circ (B^T S_1^\infty B))P_\infty + \mathrm{T}_s^2 P_\infty^T(\mathcal{D}_6 \circ (B^T S_1^\infty B))^T \bar{\Lambda}_7 \\
&\qquad \times(\mathcal{D}_6 \circ (B^T S_1^\infty B))P_\infty\}, \\
&\lambda_f^\infty = \lambda_{\max}\{\mathrm{T}_s\bar{\eta}S_1^\infty\bar{\Lambda}_3^{-1}S_1^\infty + \mathrm{T}_s^2\bar{\eta}S_1^\infty + \mathrm{T}_s^2\bar{\eta}S_1^\infty\bar{\Lambda}_6^{-1}S_1^\infty + \mathrm{T}_s^2\bar{\eta}\bar{\Lambda}_8^{-1} \\
&\qquad +\mathrm{T}_s^2\bar{\eta}S_1^\infty\bar{\Lambda}_9^{-1}S_1^\infty\}, \\
&\lambda_{\Delta u}^\infty = \lambda_{\max}\{\mathrm{T}_s^2\bar{\eta}(\mathcal{D}_7 \circ (B^T\bar{\Lambda}_9 B)) + \mathrm{T}_s^2(\mathcal{D}_5 \circ (B^T S_1^\infty B))^T \bar{\Lambda}_5^{-1}(\mathcal{D}_5 \circ (B^T S_1^\infty \\
&\qquad \times B)) + \mathrm{T}_s^2(\mathcal{D}_4 \circ (B^T S_1^\infty B)) + \mathrm{T}_s^2\bar{\Lambda}_7^{-1} + \mathrm{T}_s(\mathcal{D}_7 \circ (B^T S_1^\infty \bar{\Lambda}_2^{-1} S_1^\infty B))\}.
\end{aligned}$$

In the following theorem, an estimation for the epsilon level of ϵ-NE is given for the infinite-horizon case.

Theorem 7.10. *Consider system (Σ_e) and the cost function $\hat{J}^i(u^*)$ with $K \to \infty$. For given optimal control strategies u^{i*} obtained from Theorem 2 and positive definite matrices $\{\bar{\Lambda}_i\}_{i=1}^{9}$ and $\bar{L}_1$, if there exists a positive definite matrix S_1^∞ such that the following Riccati equation holds*

$$\begin{aligned}\mathcal{H}_\infty(S_1^\infty) := {} & T_s A^T S_1^\infty A + A^T S_1^\infty + S_1^\infty A + (T_s A + I)^T(\bar{\Lambda}_1 + \bar{\Lambda}_2 + \bar{\eta}\bar{\Lambda}_3) \\ & \times (T_s A + I) + (T_s A + I)^T S_1^\infty B \mathcal{G}_1 P_\infty + P_\infty^T \mathcal{G}_1 B^T S_1^\infty (T_s A + I) \\ & + T_s \bar{\eta} P_\infty^T (\mathcal{D}_8 \circ (B^T \bar{\Lambda}_6 B)) P_\infty + T_s P_\infty^T \bar{\Lambda}_4 P_\infty + T_s P_\infty^T \bar{\Lambda}_5 P_\infty \\ & + T_s P_\infty^T (\mathcal{D}_1 \circ (B^T S_1^\infty B)) P_\infty + \frac{1}{T_s}\bar{L}_1 = 0, \end{aligned} \tag{7.21}$$

then the ϵ-NE yields

$$\hat{J}^i(u^*) \leq J^i(u^*) + \epsilon_\infty^i,$$

where

$$\begin{aligned}\epsilon_\infty^i \leq {} & \lambda_{\max}\{\bar{Q}_\delta^{i,\infty}\}(\Psi^\infty + \lambda_{\min}^{-1}\{\bar{L}_1\}V_1^\infty(t_0) \\ & + 2\sqrt{\gamma_3^\infty(\Psi^\infty + \lambda_{\min}^{-1}\{\bar{L}_1\}V_1^\infty(t_0)))} \end{aligned} \tag{7.22}$$

with

$$\Psi^\infty = \lambda_{\min}^{-1}\{\bar{L}_1\}\{\lambda_f^\infty \gamma_1^\infty + \lambda_{\Delta u}^\infty \gamma_2^\infty + \lambda_{x^*}^\infty \gamma_3^\infty\}.$$

Proof: The proof is similar to that of Theorem 7.8 and is omitted here.

The following convex optimization algorithm, which is less conservative than Theorem 7.10, is provided to compute the epsilon level of ϵ-NE for the infinite-horizon case.

Algorithm 8 Calculation of the epsilon level ϵ_∞^i

1: Initialize system matrix parameters in (7.1). Derive the optimal feedback gains $P_\infty^i, i \in \mathbf{R}$ in Theorem 2.
2: **repeat**
3: Select positive definite matrices $\{\bar{\Lambda}_i\}_{i=1}^{9}$ and $\bar{L}_1$.
4: Solve LMI problems $\mathcal{H}_\infty(S_1^\infty) < 0$ to obtain S_1^∞.
5: **until** $S_1^\infty > 0$
6: Output the feasible solutions S_1^∞.
7: Compute $\lambda_{\Delta u}^\infty, \lambda_f^\infty, \lambda_{x^*}^\infty$ and $\lambda_{\min}^{-1}\{\bar{L}_1\}$. The scalar ϵ_∞^i is obtained by substituting these parameters into 7.22.

7.4 Numerical Examples

In order to demonstrate the validity and applicability, we apply the proposed design scheme to the load frequency control problem of a two-area interconnected power system. Consider the two-area interconnected power system model in the all-around adversary environment as follows:

$$\dot{x}(t) = Ax(t) + B(u(t) + \Delta u(t)) + f(t), \tag{7.23}$$

where

$$\begin{aligned}
&x(t) = [x^{1T}(t)\ x^{2T}(t)]^T, u(t) = [u^{1T}(t)\ u^{2T}(t)]^T,\\
&\Delta u(t) = [\Delta u^{1T}(t)\ \Delta u^{2T}(t)]^T,\\
&A = \begin{bmatrix} A^{11} & A^{12} \\ A^{21} & A^{22} \end{bmatrix}, \quad B = \mathrm{diag}\{B^1\ B^2\},\\
&A^{ii} = \begin{bmatrix} -\frac{1}{T_{p_i}} & \frac{K_{p_i}}{T_{p_i}} & 0 & 0 & -\frac{K_{p_i}}{2\pi T_{p_i}} \sum\limits_{j \neq i} K_{s_{ij}} \\ 0 & -\frac{1}{T_{T_i}} & \frac{1}{T_{T_i}} & 0 & 0 \\ -\frac{1}{R_i T_{G_i}} & 0 & -\frac{1}{T_{G_i}} & \frac{1}{T_{G_i}} & 0 \\ K_{E_i} H_{B_i} & 0 & 0 & 0 & \frac{K_{E_i}}{2\pi} \sum\limits_{j \neq i} K_{s_{ij}} \\ 2\pi & 0 & 0 & 0 & 0 \end{bmatrix},\\
&A^{ij} = \begin{bmatrix} 0\,0\,0\,0 & -\frac{K_{p_i}}{2\pi T_{p_i}} K_{s_{ij}} \\ 0\,0\,0\,0 & 0 \\ 0\,0\,0\,0 & 0 \\ 0\,0\,0\,0 & \frac{K_{E_i}}{2\pi} K_{s_{ij}} \\ 0\,0\,0\,0 & 0 \end{bmatrix}, \quad B^i = \begin{bmatrix} 0\ 0\ \frac{1}{T_{G_i}}\ 0\ 0 \end{bmatrix}^T,\\
&x^i(t) = [\Delta f_i(t)\ \Delta P_{g_i}(t)\ \Delta X_{g_i}(t)\ \Delta E_i(t)\ \Delta \delta_i(t)]^T,
\end{aligned}$$

$\Delta u(t)$, $\Delta f_i(t)$, $\Delta P_{g_i}(t)$, $\Delta X_{g_i}(t)$, $\Delta E_i(t)$, and $\Delta\delta_i(t)$ are the deception attack, change of frequency, power output, governor valve position, integral control, and rotor angle deviation, respectively. Parameters T_{G_i}, T_{t_i}, and T_{p_i} are time constants of governor, turbine, and power system, respectively. Parameters K_{p_i}, R_i, K_{E_i} and K_{B_i} are the power system gain, speed regulation coefficient, integral control gain and frequency bias factor, respectively. Parameter $K_{s_{ij}}$ is an interconnection gain between area i and j $(i \neq j)$. The specific values of the these parameters can be found in [96].

Let us choose the initial value as $x(t_0) = [0\ 500\ 0\ 0\ 0\ 0\ 0\ 0\ 0\ 500]^T$. The deception attack is given by $\Delta u(t_k) = 0.001[\cos(t_k)\ \sin(t_k)]^T$, and the physical attack is given by $f(t_k) = 0.1/(10 + t_k)$. The parameters $\bar{\eta}$, $\bar{\alpha}_j$, and $\bar{\beta}_j$, $j \in \mathbf{R}$ are set by $\bar{\eta} = 0.1$, $\bar{\alpha}_j = 0.995$ and $\bar{\beta}_j = 0.1$, and other parameters can be found in Table 7.1. According to inequality (7.15) in Theorem 7.8, one has that $\epsilon_K^1 = 4.1582 * 10^{10}$ and $\epsilon_K^2 = 4.2119 * 10^8$. The values of the cost

Table 7.1 System parameters

Parameter	K	T_s	Q_δ^1	Q_δ^2	Λ_1	Λ_2	Λ_3
Value	80	0.05	$500000I$	$5000I$	$0.5I$	$0.001I$	$0.01I$
Parameter	Λ_4	Λ_5	Λ_6	Λ_7	Λ_8	Λ_9	L_1
Value	I	$0.0001I$	$0.01I$	$0.001I$	$0.01I$	$0.01I$	$1.1I$

functions are computed as $J^1 = 2.5349 * 10^{11}$ and $J^2 = 2.5350 * 10^9$. Let us define $\phi^i := \frac{\epsilon_K^i}{J^i} * 100\%$ and we have $\phi^1 = 16.40\%$ and $\phi^2 = 16.62\%$. Figure 7.2 depicts the relation curves of ϕ^i versus delivery rate $\bar{\beta}$. It is observed that the larger packet delivery rate $\bar{\alpha}$ leads to smaller ϕ^i. Figure 7.3 depicts the

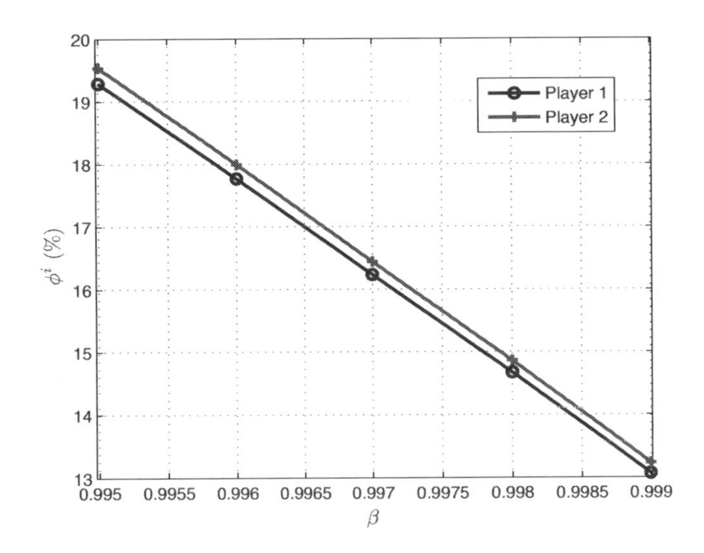

Figure 7.2 ϕ^i versus packet delivery rates $\bar{\beta}$.

relation curves between ϵ_K^i and the upper bound of deception attack γ_2, which shows that larger γ_2 leads to larger ϕ^i.

Let us reselect the initial condition $x(t_0) = [0\ 5\ 0\ 0\ 0\ 0\ 0\ 0\ 0\ 0\ 5]^T$, sampling interval $\mathrm{T}_s = 0.1$, $\bar{\alpha}_i = 0.9$, and $\Delta u(t_k) = 200 * [\cos(t_k)\ \ \sin(t_k)]/t_k^2$. Other parameters are the same as the setup above. Figure 7.4 and Figure 7.5 plot the state evolutions of frequency changes $\Delta f_1(t_k)$ and $\Delta f_2(t_k)$ in four attacking scenarios, where we can see that the DoS attack, deception attack, and physical attack all contribute to the performance degradation.

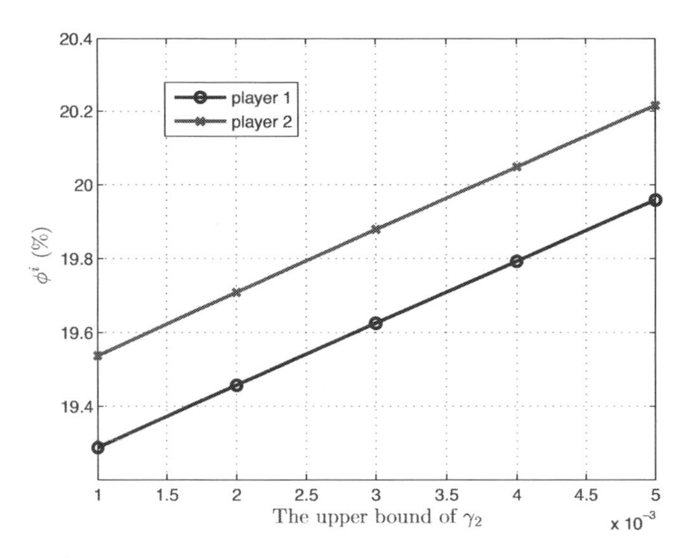

Figure 7.3 ϕ^i versus the upper bound γ_2.

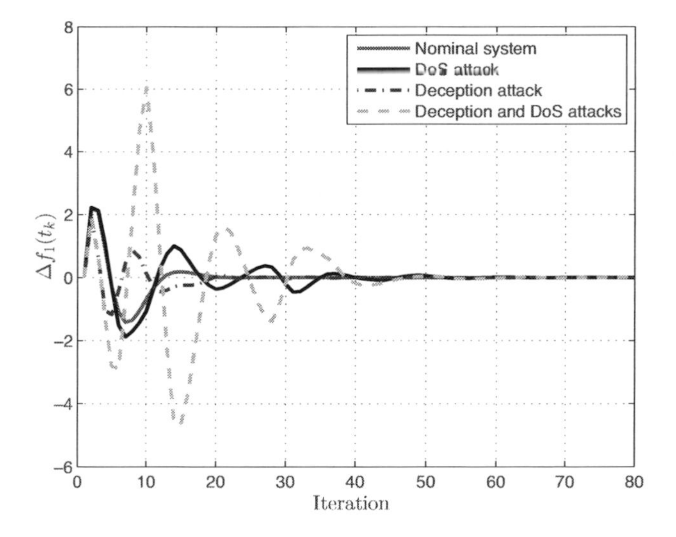

Figure 7.4 The state evolution of $\Delta f_1(t_k)$.

7.5 Conclusion

In this chapter, the multitasking optimal control problem has been investigated for a class of NCSs in simultaneous presence of DoS attacks, deception attacks, and physical attacks. The NE strategies for the exploited system have

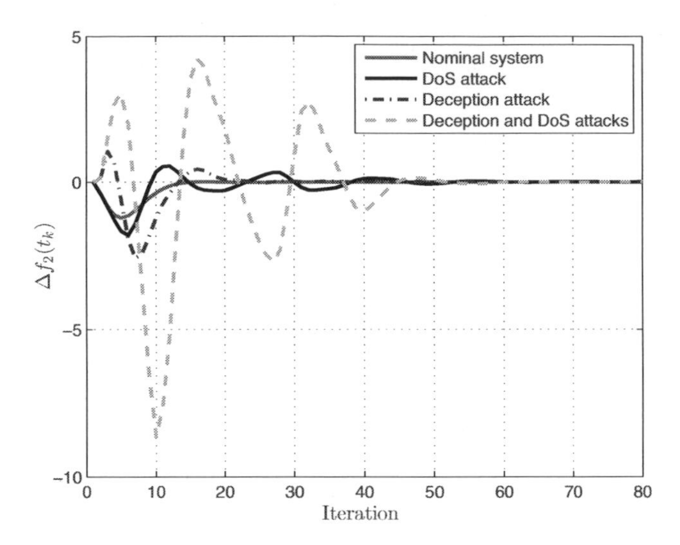

Figure 7.5 The state evolution of $\Delta f_2(t_k)$.

been obtained in the delta-domain in terms of finite- and infinite-time horizons, respectively. In order to quantify the impacts of the considered attacks, an upper bound for the epsilon level of the ϵ-NE has been provided explicitly, and the corresponding convex optimization algorithms have been given to compute such an upper bound. Finally, a numerical example has been given to demonstrate the validity and advantage of the proposed methodology.

Chapter 8
Resilient Control of CPS against Intelligent Attacker

8.1 Introduction

In this chapter, we consider the RCS compromised by the intelligent DoS attacker, which can adjust its strategy according to the knowledge of the defender's security profile. In the cyber layer, some IDSs are deployed to raise alarms once an anomaly behavior is detected such that it can be removed automatically. Thus, the DoS attackers have to go through the IDSs first before it can compromise the control system. When it comes to the underlying physical layer, H_∞ minimax control is used to extenuate the external disturbance. The architecture of the RCS is shown in Figure 8.1, where the adversaries from both the cyber layer and physical layer are considered.

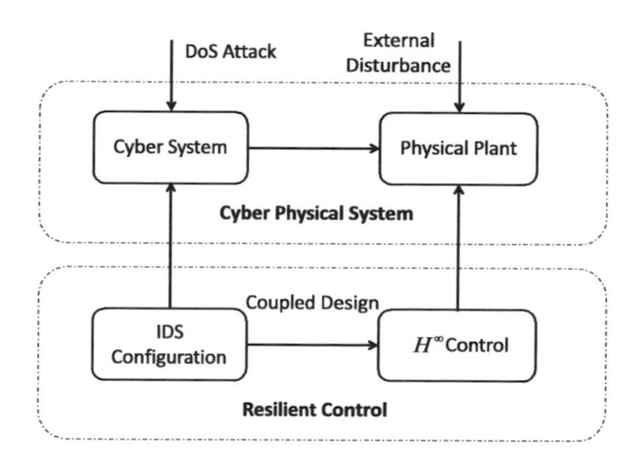

Figure 8.1 Layered Design of RCS: H_∞ minimax control is developed to provide robustness to the physical plant, while IDSs are optimally configured to secure the system. The coupled design between the cyber and physical part leads to a system-level resilient design of the CPS.

Note that IDSs have to be configured appropriately, since we need to find a tradeoff between system performance and security enforcement levels [190]. Here we propose a hierarchical decision structure corresponding to IDS configuration and H_∞ minimax control. Since the DoS attacker is intelligent and doesn't act simultaneously with the defender, the interaction between IDS configuration and DoS attacks is modeled as a static infinite Stackelberg game. In the underlying physical layer, we consider the Full-Information (FI) case of the H_∞ minimize control, which is also modeled as a difference Stackelberg game [14]. The coupled design of IDS and H_∞ control are formulated within the Stackelberg-in-Stackelberg framework. It is worth mentioning that, in practice, the cost function of the malicious attacker is unavailable. Hence, we use Generic Algorithm (GA) to find the Stackelberg equilibrium solution.

The rest of the chapter is organized as follows. In Section 8.2, the design objectives of RCS are provided in terms of joint optimal defense strategies. In Section 8.3, we provide analytical solutions for both cyber layer static game and physical layer difference game. A coupled design method is then proposed to find the joint optimal defense strategies. In Section 8.4, our method is applied to the voltage regulation of UPS. Finally, the conclusion is drawn in Section 8.5.

8.2 Problem Setting

8.2.1 Hierarchical Model for RCS

We first make the following assumption which is used throughout this chapter.

Assumption 8.1 *The package drops that are caused by the background traffic under the normal status have been compensated, and the C-A package drops are all caused by DoS attacks.*

A Stackelberg-in-Stackelberg structure is proposed for RCS, where the interaction between the IDS configuration and attacker is formulated as the inner game $\mathcal{G}_1$, and FI H_∞ minimax control is modeled as the outer game $\mathcal{G}_2$. The solution of $\mathcal{G}_1$ is propagated to $\mathcal{G}_2$ and affects its outcomes. This structure is consistent with the actual situation since the cyber attacker has to go through the cyber defense facilities first to actually exert influence on the underlying control system. Specifically, let us illustrate the hierarchical game stage by stage. The players of the inner game $\mathcal{G}_1$ are denoted as $\mathbf{P}_{\mathbf{a}_i}$, $i \in \{1, 2\}$, and the players of the outer game $\mathcal{G}_2$ are $\mathbf{P}_{\mathbf{b}_j}$, $j \in \{1, 2\}$. For $\mathcal{G}_1$, $\mathbf{P}_{\mathbf{a}_1}$ and $\mathbf{P}_{\mathbf{a}_2}$ represent the IDSs and intelligent DoS attacker, respectively. The strategy vector of $\mathbf{P}_{\mathbf{a}_1}$ is defined as $\mathbf{u}_a^1 := \left[u_{a1}^1 \; u_{a2}^1 \; \cdots \; u_{am}^1 \right]^T$, while the strategy vector of $\mathbf{P}_{\mathbf{a}_2}$ is $\mathbf{u}_a^2 := \left[u_{a1}^2 \; u_{a2}^2 \; \ldots \; u_{an}^2 \right]^T$. The players choose strategies from the strategy spaces defined as $U_a^1 := \{\mathbf{u}_a^1 \subset \mathbb{R}^m : u_{ai}^1 \geq 0, \;\; i = 1, \ldots, m\}$ and

$U_a^2 := \{\mathbf{u}_a^2 \subset \mathbb{R}^n : u_{aj}^2 \geq 0, \quad j = 1, \ldots, n\}$. The intrusion detection model in [3] is employed, where the cost functions of the IDSs and intelligent attacker are denoted as

$$J_a^1(\mathbf{u}_a^1, \mathbf{u}_a^2) := r(\mathbf{u}_a^2)^T \mathcal{P} \bar{\mathcal{Q}} \mathbf{u}_a^1 + (\mathbf{u}_a^1)^T diag(\varphi) \mathbf{u}_a^1 + c^1(\mathcal{Q}\mathbf{u}_a^2 - \bar{\mathcal{Q}}\mathbf{u}_a^1), \quad (8.1)$$
$$J_a^2(\mathbf{u}_a^1, \mathbf{u}_a^2) := -r(\mathbf{u}_a^2)^T \mathcal{P} \bar{\mathcal{Q}} \mathbf{u}_a^1 + (\mathbf{u}_a^2)^T diag(\psi) \mathbf{u}_a^2 + c^2(\bar{\mathcal{Q}}\mathbf{u}_a^1 - \mathcal{Q}\mathbf{u}_a^2), \quad (8.2)$$

where $\varphi := \begin{bmatrix} \varphi_1 \ldots \varphi_m \end{bmatrix}$ and $\psi := \begin{bmatrix} \psi_1 \ldots \psi_n \end{bmatrix}$ are row matrices with each element to be positive. $\mathcal{Q}$, which is the nonnegative matrix with each diagonal element greater than 1, models the vulnerability to attacks. $\bar{\mathcal{Q}}$ is a matrix with elements to be 1 or 0 and correlates IDS responses with attacks. $\bar{\mathcal{P}}$ represents how well the attacks are detected by the sensor network on average and $[\mathcal{P}]_{ij} = -[\bar{\mathcal{P}}]_{ij}$, $i = j$, and $[\mathcal{P}]_{ij} = [\bar{\mathcal{P}}]_{ij}$, $i \neq j$. c^1 and c^2 are both row vectors with each element to be positive, $r(\mathbf{u}_a^2)^T \mathcal{P} \bar{\mathcal{Q}} \mathbf{u}_a^1$, $r > 0$ represents the cost of false alarm for IDSs, $-r(\mathbf{u}_a^2)^T \mathcal{P} \bar{\mathcal{Q}} \mathbf{u}_a^1$ is the benefit of deception for the attacker, $(\mathbf{u}_a^1)^T diag(\varphi) \mathbf{u}_a^1$ and $(\mathbf{u}_a^2)^T diag(\psi) \mathbf{u}_a^2$ are the cost of detection for IDSs and cost of attacks for the attacker, respectively. $c^1(\mathcal{Q}\mathbf{u}_a^2 - \bar{\mathcal{Q}}\mathbf{u}_a^1)$ is the cost of the IDSs for successful attacks, and $c^2(\bar{\mathcal{Q}}\mathbf{u}_a^1 - \mathcal{Q}\mathbf{u}_a^2)$ is the cost of attacker for successful defense.

For the second stage game $\mathcal{G}_2$, $\mathbf{P_{b_1}}$ represents the **FI** H_∞ minimax controller that determines the control strategy, and $\mathbf{P_{b_2}}$ is the external disturbance. The system dynamics are

$$\begin{aligned} x_{k+1} &= A_k x_k + B_k u_k^a + D_k w_k, \\ u_k^a &= \alpha_k u_k. \end{aligned} \quad (8.3)$$

where A_k, B_k, and D_k, $k \in \mathbf{K} := \{1, 2, \ldots, N-1\}$ are time-varying matrices with appropriate matrices, x_k represents the state variable, u_k is the control command sent out by the controller, and u_k^a is the control input received by the controller after DoS attacks, w_k denotes the external disturbance. α_k is a stochastic variable describing the package drops at k. The control command is received by the actuator at k when $\alpha_k = 1$ and unreceived if $\alpha_k = 0$. α_k is an independently and identically distributed (i.i.d.) Bernoulli process with distribution as

$$\mathbb{P}\{\alpha_k = 1\} = \alpha, \qquad \mathbb{P}\{\alpha_k = 0\} = 1 - \alpha, \ \forall k \in \mathbf{K}. \quad (8.4)$$

Here transmission control protocol (TCP) is applied where each package is acknowledged, and the information set is $\mathcal{I}_0^1 = \{x_0\}$, $\mathcal{I}_k^1 = \{x_1, x_2, \ldots, x_k, \alpha_0, \alpha_1, \ldots, \alpha_{k-1}\}$ for the disturbance and $\mathcal{I}_k^2 = \{\mathcal{I}_k^1, w_k\}$ for the controller. We can see from the information sets that the **FI** H_∞ minimax control suggests that the controller has access to both the plant states and disturbances. $\mathcal{M}$ and $\mathcal{N}$ are introduced, which are the sets of the control and disturbance strategies, respectively. For admissible strategies μ_k and ν_k, we have $u_k = \mu_k(\mathcal{I}_k^2)$ and

$w_k = \nu_k(\mathcal{I}_k^1)$. Let us further define

$$\mu = \{\mu_0, \ldots, \mu_k, \ldots \mu_{N-1}\}, \quad \nu = \{\nu_0, \ldots, \nu_k, \ldots \nu_{N-1}\}. \tag{8.5}$$

According to [152], DoS attacks can degrade the channel quality which lowers package delivery rate (PDR) α. Let us define a function $\Lambda : \mathbb{R} \to [0, 1]$ mapping J_a^1 to PDR α. We assume here that Λ is a nonincreasing function of the cost of the defender (gains of the attacker) $J_a^1(\mathbf{u}_a^{1*}, \mathbf{u}_a^{2*})$, which suggests that the attacker's gain from attacking will increase if more control signals are intercepted.

Thus, PDR α is also parameterized by $\mathbf{u}_a^1$ and $\mathbf{u}_a^2$ and denoted as $\alpha(\mathbf{u}_a^1, \mathbf{u}_a^2)$. The associated performance index of $\mathcal{G}_2$ is hence determined by strategies of both $\mathbf{P}_{\mathbf{a}_i}$ and $\mathbf{P}_{\mathbf{b}_j}$, $i, j \in \{1, 2\}$

$$\begin{aligned} J_b(\mu, \nu, \mathbf{u}_a^1, \mathbf{u}_a^2) = \mathbb{E}_{\alpha_k}\{\|x_N\|_{Q_N}^2 \\ + \sum_{k=1}^{N-1} \{\|x_k\|_{Q_k}^2 + \alpha_k(\mathbf{u}_a^1, \mathbf{u}_a^2) \|u_k\|_{R_k}^2 - \gamma^2 \|w_k\|^2\}\}, \end{aligned} \tag{8.6}$$

where $\|x\|_S^2 = x^T S x$, and the scalar γ represents the disturbance attenuation level. We assume that $Q_N > 0$, $Q_k > 0$ and $R_k > 0$ for all $k \in \mathbf{K}$. $\mathbb{E}_{\alpha_k}\{\cdot\}$ is simplified as $\mathbb{E}\{\cdot\}$ henceforth.

8.2.2 Design Objective

Note that for $\mathcal{G}_1$ and $\mathcal{G}_2$, the information possessed by the players is asymmetric, since the intelligent attacker and controller are aware of their opponents' strategies and can adjust their own strategies accordingly. Thus, we introduce the concept of Stackelberg equilibrium solution here which is a natural formalism for describing systems with multi-levels in decision making. If the rational players act sequentially, they will reach Stackelberg equilibrium at last. In a Stackelberg game, we have the leader and the follower. The leader is defined as the players who hold powerful positions such that their strategies can be enforced to the follower. The roles of the players in this chapter are shown in Table 8.1.

Table 8.1 The roles of the players

Leader	Follower
$\mathbf{P}_{\mathbf{a}_1}$ (IDSs)	$\mathbf{P}_{\mathbf{a}_2}$ (Intelligent Attacker)
$\mathbf{P}_{\mathbf{b}_2}$ (External Disturbance)	$\mathbf{P}_{\mathbf{b}_1}$ (Controller)

Before presenting our design objective, the following definitions are provided.

Definition 8.1. (Stackelberg Configuration Strategy) The pair $(\mathbf{u}_a^{1*}, \mathbf{u}_a^{2*})$ is the Stackelberg configuration strategy for $\mathcal{G}_1$, if

$$J_a^{1*}(\mathbf{u}_a^{1*}, \mathbf{u}_a^{2*}) := J_a^1(\mathbf{u}_a^{1*}, R_2(\mathbf{u}_a^{1*})) = \min_{\mathbf{u}_a^1 \in U_a^1} J_a^1(\mathbf{u}_a^1, R_2(\mathbf{u}_a^1)), \tag{8.7}$$

where the singleton $R_2(\mathbf{u}_a^1) = \{\varpi \in U_a^2 : J_a^2(\mathbf{u}_a^1, \varpi) \leq J_a^2(\mathbf{u}_a^1, \mathbf{u}_a^2), \forall \mathbf{u}_a^2 \in U_a^2\}$ can be regarded as the rational reaction set of $\mathbf{P}_{\mathbf{a}_2}$ to the strategy $\mathbf{u}_a^1 \in U_a^1$ of $\mathbf{P}_{\mathbf{a}_1}$.

As mentioned before, games $\mathcal{G}_1$ and $\mathcal{G}_2$ are played sequentially and PDR $\alpha(\mathbf{u}_a^1, \mathbf{u}_a^2)$ will have a definite value once the inner game $\mathcal{G}_1$ is solved. Based on the rationality of the players, the estimated PDR will take the value $\alpha(\mathbf{u}_a^{1*}, \mathbf{u}_a^{2*})$. The Stackelberg strategy of $\mathcal{G}_2$ is defined as follows:

Definition 8.2. (Stackelberg Control Strategy) For cyber strategy $(\mathbf{u}_a^{1*}, \mathbf{u}_a^{2*})$, the pair of policies $(\mu^*, \nu^*) \in \mathcal{M} \times \mathcal{N}$ constitutes the Stackelberg control strategy for $\mathcal{G}_2$, if

$$\begin{aligned} J_b^*(\mu^*, \nu^*, \mathbf{u}_a^{1*}, \mathbf{u}_a^{2*}) &:= J_b(R(\nu^*), \nu^*, \mathbf{u}_a^{1*}, \mathbf{u}_a^{2*}) \\ &= \max_{\nu \in \mathcal{N}} J_b(R(\nu), \nu, \mathbf{u}_a^{1*}, \mathbf{u}_a^{2*}), \end{aligned} \tag{8.8}$$

where the singleton $R(\nu) = \{\varrho \in \mathcal{M} : J_b(\varrho, \nu, \mathbf{u}_a^{1*}, \mathbf{u}_a^{2*}) \leq J_b(\mu, \nu, \mathbf{u}_a^{1*}, \mathbf{u}_a^{2*}), \forall \mu \in \mathcal{M}\}$ can be regarded as the rational reaction set of $\mathbf{P}_{\mathbf{b}_1}$ to the strategy $\nu \in \mathcal{N}$ of $\mathbf{P}_{\mathbf{b}_2}$.

Since only μ^* and $\mathbf{u}_a^{1*}$ are implementable on the defense side, we define the pair $(\mu^*, \mathbf{u}_a^{1*})$ as the joint optimal defense strategy for RCS in the face of the intelligent DoS attacker. The main objective of this chapter is to develop a coupled design methodology such that the joint optimal defense strategy $(\mu^*, \mathbf{u}_a^{1*})$ can be found.

8.3 Main Contents

In this section, we find the analytical solutions for both the Stackelberg configuration strategy in the cyber layer and Stackelberg control strategy in the physical layer. The coupled design method is proposed to find the joint optimal defense strategy when the cost function of the attacker is unknown on the defense side.

8.3.1 Stackelberg Configuration Strategy for $\mathcal{G}_1$

In this subsection, we provide conditions such that the inner solution of the Stackelberg game $\mathcal{G}_1$ is unique, and present the analytical form of the solution.

Theorem 8.3. *There exists a unique Stackelberg equilibrium solution for $\mathcal{G}_1$. Furthermore, if we have*

$$r < \frac{\min\limits_i \bar{\mathcal{Q}}^T (c^1)^T}{\bar{\mathcal{Q}}^T \mathcal{P}^T \max\limits_i (diag(2\psi)^{-1} \mathcal{Q} (c^1)^T + (\theta^2)^T)}, \tag{8.9}$$

and $\mathcal{P}\bar{\mathcal{Q}} > 0$, the solutions will satisfy $\mathbf{u}_a^{1*} > 0$, $\mathbf{u}_a^{2*} > 0$ and are given by

$$\begin{aligned} \mathbf{u}_a^{1*} &= [diag(2\varphi) + r^2 \bar{\mathcal{Q}}^T \mathcal{P}^T [diag(\psi)]^{-1} \mathcal{P}\bar{\mathcal{Q}}]^{-1} [\bar{\mathcal{Q}}^T (c^1)^T \\ &\quad - r\bar{\mathcal{Q}}^T \mathcal{P}^T [diag(2\psi)]^{-1} \mathcal{Q}(c^1)^T - r\bar{\mathcal{Q}}^T \mathcal{P}^T \theta^2], \end{aligned} \tag{8.10}$$

$$\mathbf{u}_a^{2*} = \theta^2 + r[diag(2\psi)]^{-1} \mathcal{P}\bar{\mathcal{Q}} \mathbf{u}_a^{1*}. \tag{8.11}$$

where

$$\begin{aligned} \theta^1 &= \left[[c^1 \bar{\mathcal{Q}}]_{11}/(2\varphi_1), \ldots, [c^1 \bar{\mathcal{Q}}]_{1m}/(2\varphi_m) \right]^T, \\ \theta^2 &= \left[[c^2 \mathcal{Q}]_{11}/(2\psi_1), \ldots, [c^2 \mathcal{Q}]_{1n}/(2\psi_n) \right]^T. \end{aligned}$$

Proof. We first note that the cost function J_a^i is strictly convex in $\mathbf{u}_a^i$ for all $i \in \{1, 2\}$. From [14], we confine the search of an equilibrium solution to the class of pure strategies. To every announced strategy $\mathbf{u}_a^1$ of the defender, the attacker's unique response will be given by

$$\mathbf{u}_a^2 = \theta^2 + r[diag(2\psi)]^{-1} \mathcal{P}\bar{\mathcal{Q}} \mathbf{u}_a^1. \tag{8.12}$$

Towards this end, we are in a position to determine the Stackelberg strategy for the defender. Cost function $J_a^1(\mathbf{u}_a^1, \mathbf{u}_a^2)$ has to be minimized over U_a^1 subject to the constraint of the reaction of the attacker. Since the attacker's reaction is unique, we substitute (8.12) into $J_a^1(\mathbf{u}_a^1, \mathbf{u}_a^2)$ which yields

$$\begin{aligned} \tilde{J}_a^1 &= r(\theta^2 + r[diag(2\psi)]^{-1} \mathcal{P}\bar{\mathcal{Q}} \mathbf{u}_a^1)^T \mathcal{P}\bar{\mathcal{Q}} \mathbf{u}_a^1 + (\mathbf{u}_a^1)^T diag(\varphi) \mathbf{u}_a^1 \\ &\quad + c^1 (\mathcal{Q}[\theta^2 + r[diag(2\psi)]^{-1} \mathcal{P}\bar{\mathcal{Q}} \mathbf{u}_a^1] - \bar{\mathcal{Q}} \mathbf{u}_a^1). \end{aligned} \tag{8.13}$$

For the Stackelberg solution of the defender to be unique, we impose a strict convexity condition on $\tilde{J}_a^1$, which is

$$r^2 \bar{\mathcal{Q}}^T \mathcal{P}^T [diag(2\psi)]^{-1} \mathcal{P}\bar{\mathcal{Q}} + diag(\varphi) > 0. \tag{8.14}$$

This condition is easy to be verified since we have $\varphi, \psi > 0$. Then, the unique Stackelberg solution can be obtained by taking the first derivative of $\tilde{J}_a^1$ over $\mathbf{u}_a^1$

$$\begin{aligned} & r(\theta^2)^T \mathcal{P}\bar{\mathcal{Q}} + r^2 (\mathbf{u}_a^1)^T \bar{\mathcal{Q}}^T \mathcal{P}^T [diag(\psi)]^{-1} \mathcal{P}\bar{\mathcal{Q}} + (\mathbf{u}_a^1)^T diag(2\varphi) \\ & + rc^1 \mathcal{Q}[diag(2\psi)]^{-1} \mathcal{P}\bar{\mathcal{Q}} - c^1 \bar{\mathcal{Q}} = 0. \end{aligned} \tag{8.15}$$

By solving (8.15), we obtain $\mathbf{u}_a^{1*}$ as

$$\mathbf{u}_a^{1*} = [diag(2\varphi) + r^2\bar{\mathcal{Q}}^T\mathcal{P}^T[diag(\psi)]^{-1}\mathcal{P}\bar{\mathcal{Q}}]^{-1}[\bar{\mathcal{Q}}^T(c^1)^T \\ -r\bar{\mathcal{Q}}^T\mathcal{P}^T[diag(2\psi)]^{-1}\mathcal{Q}(c^1)^T - r\bar{\mathcal{Q}}^T\mathcal{P}^T\theta^2]. \tag{8.16}$$

From (8.12), the attacker's strategy yields

$$\mathbf{u}_a^{2*} = \theta^2 + r[diag(2\psi)]^{-1}\mathcal{P}\bar{\mathcal{Q}}\mathbf{u}_a^{1*}. \tag{8.17}$$

It is easy to see that if (8.9) holds, we have $\mathbf{u}_a^{1*} > 0$ and $\mathbf{u}_a^{2*} > 0$. Furthermore, there exist no boundary solutions since the Stackelberg equilibrium solution is unique. The proof is completed.

8.3.2 Stackelberg Control Strategy for $\mathcal{G}_2$

Once the upper layer security game $\mathcal{G}_1$ is solved, the PDR will take the estimated value $\alpha(\mathbf{u}_a^{1*}, \mathbf{u}_a^{2*})$. Towards this end, we will find the Stackelberg equilibrium solution for $\mathcal{G}_2$. In the sequel, we provide the conditions such that the cost function $J_b^*(\mu^*, \nu^*, \mathbf{u}_a^{1*}, \mathbf{u}_a^{2*})$ is upper bounded and the Stackelberg equilibrium solution can be found. Note that we use a shorthand notation of α^* in place of $\alpha(\mathbf{u}_a^{1*}, \mathbf{u}_a^{2*})$ henceforth.

Theorem 8.4. *For the linear-quadratic dynamic game (8.3), the estimated PDR α^* is given and $\boldsymbol{P}_{b_1}$'s control is allowed to depend on the current value of the action of $\boldsymbol{P}_{b_2}$. Then*

i *The dynamic game admits a unique Stackelberg equilibrium solution if*

$$M_k = \gamma^2 I - D_k^T(I - \alpha^* Z_{k+1}B_k(R_k \\ +B_k^T Z_{k+1}B_k)^{-1}B_k^T)Z_{k+1}D_k > 0, \quad k \in \mathbf{K}. \tag{8.18}$$

where the nonincreasing sequence $Z_k > 0$ is generated by Riccati recursion (8.19)

ii *The Riccati recursion is $Z_N = Q_N$ and*

$$Z_k = Q_k + P_{u_k}^T(\alpha^* R_k + \bar{\alpha}^* B_k^T Z_{k+1}B_k)P_{u_k} \\ -\gamma^2 P_{w_k}^T P_{w_k} + H_k^T Z_{k+1}H_k, \tag{8.19}$$

where

$$\begin{aligned}
\bar{\alpha}^* &= \alpha^*(1-\alpha^*), \\
H_k &= A_k - \alpha^* B_k P_{u_k} + D_k P_{w_k}, \\
P_{u_k} &= (R_k + B_k^T Z_{k+1}B_k)^{-1}B_k^T Z_{k+1}(D_k P_{w_k} + A_k), \\
P_{w_k} &= (\gamma^2 I - K_{w_k} Z_{k+1}D_k)^{-1}K_{w_k} Z_{k+1}A_k, \\
K_{w_k} &= D_k^T(I - \alpha^* Z_{k+1}B_k(R_k + B_k^T Z_{k+1}B_k)^{-1}B_k^T).
\end{aligned}$$

iii *Under condition* (**i**), *the Stackelberg solution of the leader is*

$$u_k^* = \mu_k^*(\mathcal{I}_k^2) = L_k w_k^* + F_k x_k, \tag{8.20}$$

where

$$L_k = -(R_k + B_k^T Z_{k+1} B_k)^{-1} B_k^T Z_{k+1} D_k,$$
$$F_k = -(R_k + B_k^T Z_{k+1} B_k)^{-1} B_k^T Z_{k+1} A_k,$$

whereas that of the follower is $w_k^* = \nu_k^*(\mathcal{I}_k^1) = P_{w_k} x_k$.

iv *Under condition* (**i**), *the value of the game is given by*

$$J_b^*(\mu^*, \nu^*, \boldsymbol{u}_a^{1*}, \boldsymbol{u}_a^{2*}) = x_1^T Z_1 x_1, \tag{8.21}$$

where x_1 *is the initial value of the state.*

Proof. From [14], the following is obtained by employing the dynamic programming

$$V_N(x_N) = x_N^T Q_N x_N, \tag{8.22}$$
$$V_{N-1}(x_{N-1}) = \max_{w_{N-1}} \min_{u_{N-1}} \mathbb{E}\{\|x_{N-1}\|^2_{Q_{N-1}} + \alpha_{N-1} \|u_{N-1}\|^2_{R_{N-1}} - \gamma^2 \|w_{N-1}\|^2 + x_N^T Q_N x_N\}. \tag{8.23}$$

The cost-to-go function V_k turns out to be quadratic which can be verified by induction method. Starting with $k = N$, the following can be written for $k = N - 1$.

$$\begin{aligned} V_{N-1}(x_{N-1}) = &\max_{w_{N-1}} \min_{u_{N-1}} \mathbb{E}\{\|x_{N-1}\|^2_{Q_{N-1}} + \alpha_{N-1} \|u_{N-1}\|^2_{R_{N-1}} \\ &-\gamma^2 \|w_{N-1}\|^2 + (A_{N-1}x_{N-1} + \alpha_{N-1}B_{N-1}u_{N-1} \\ &+D_{N-1}w_{N-1})^T Q_N (A_{N-1}x_{N-1} + \alpha_{N-1}B_{N-1}u_{N-1} \\ &+D_{N-1}w_{N-1})\} \qquad (8.24a) \\ = &\mathbb{E}\{\|x_{N-1}\|^2_{Q_{N-1}} + \alpha_{N-1} \|u_{N-1}^*\|^2_{R_{N-1}} - \gamma^2 \|w_{N-1}^*\|^2 \\ &+(A_{N-1}x_{N-1} + \alpha_{N-1}B_{N-1}u_{N-1}^* + D_{N-1}w_{N-1}^*)^T Q_N \\ &(A_{N-1}x_{N-1} + \alpha_{N-1}B_{N-1}u_{N-1}^* + D_{N-1}w_{N-1}^*)\}. \qquad (8.24b) \end{aligned}$$

Taking the first derivative of the right and side of (8.24a) with respect to u_{N-1}, we can obtain the follower's best response to every announced strategy w_{N-1} of the leader.

$$u_{N-1} = -(R + B^T Q_N B)^{-1} B^T Q_N (A x_{N-1} + D w_{N-1}), \tag{8.25}$$

which suggests that u_{N-1} is a linear combination of x_{N-1} and w_{N-1}. Thus, we denote $u_{N-1} = L_{N-1} w_{N-1} + F_{N-1} x_{N-1}$. Since $R_{N-1} + B_{N-1}^T Q_N B_{N-1} > 0$, the follower's response to every announced strategy of the leader is unique. By substituting u_{N-1} into (8.24a), we have

$$\begin{aligned}\tilde{V}_{N-1}(x_{N-1}) = &\alpha^*(L_{N-1}w_{N-1} + F_{N-1}x_{N-1})^T R_{N-1}(L_{N-1}w_{N-1}\\ &+F_{N-1}x_{N-1}) + \alpha^* x_{N-1}^T A_{N-1}^T Q_N B_{N-1}(L_{N-1}w_{N-1}\\ &+F_{N-1}x_{N-1}) + x_{N-1}^T A_{N-1}^T Q_N D_{N-1}w_{N-1}\\ &+\alpha^*(L_{N-1}w_{N-1} + F_{N-1}x_{N-1})^T B_{N-1}^T Q_N\\ &\times B_{N-1}(L_{N-1}w_{N-1} + F_{N-1}x_{N-1}) + \alpha^*(L_{N-1}w_{N-1}\\ &+F_{N-1}x_{N-1})^T B_{N-1}^T Q_N D_{N-1}w_{N-1}\\ &+w_{N-1}^T D_{N-1}^T Q_N A_{N-1}x_{N-1}\\ &+\alpha^* w_{N-1}^T D_{N-1}^T Q_N B_{N-1}(L_{N-1}w_{N-1}\\ &+F_{N-1}x_{N-1}) + x_{N-1}^T A_{N-1}^T Q_N A_{N-1}x_{N-1}\\ &+\alpha^*(L_{N-1}w_{N-1} + F_{N-1}x_{N-1})^T B_{N-1}Q_N\\ &\times A_{N-1}x_{N-1} + w_{N-1}^T D_{N-1}^T Q_N D_{N-1}w_{N-1}\\ &+x_{N-1}^T Q_{N-1}x_{N-1} - \gamma^2 w_{N-1}^T w_{N-1}. \end{aligned} \tag{8.26}$$

For the minimum of $\tilde{V}_{N-1}(x_{N-1})$ to be unique with respect to w_{N-1} , we have to pose a strictly concavity condition on $\tilde{V}_{N-1}(x_{N-1})$. which is actually $M_{N-1} > 0$. Under this condition, the unique minimizing solution can be obtained by setting the gradient of $\tilde{V}_{N-1}(x_{N-1})$ equal to zero, which yields w_{N-1}^*. By substituting u_{N-1}^* and w_{N-1}^* back into (8.24b), we get $V_{N-1} = x_{N-1}^T Z_{N-1}x_{N-1}$, which implies that the cost function has a linear quadratic structure with $k = N-1$. Since V_{N-1} is a linear quadratic function, we can proceed similarly and obtain the Stackelberg equilibrium solution (u_{N-2}^*, w_{N-2}^*) and the value function V_{N-2}. This procedure can also be carried out recursively for all $k \leq N-2$ and the value of the game yields $x_1^T Z_1 x_1$ in the end.

The positive definiteness of nonincreasing sequence Z_k, $k \in \mathbf{K}$ in (**i**) can be proved by the fact that, for $k = N-1$, we have

$$\begin{aligned}V_{N-1}(x_{N-1}) &= x_{N-1}^T Z_{N-1}x_{N-1}\\ &= \max_{w_{N-1}} \min_{u_{N-1}} \mathbb{E}\{\|x_{N-1}\|_{Q_{N-1}}^2 + \alpha_{N-1}\|u_{N-1}\|_{R_{N-1}}^2\\ &\quad -\gamma^2\|w_{N-1}\|^2 + x_N^T Z_N x_N\}\\ &= \max_{w_{N-1}} \mathbb{E}\{\|x_{N-1}\|_{Q_{N-1}}^2\\ &\quad +\alpha_{N-1}\|L_{N-1}w_{N-1} + F_{N-1}x_{N-1}\|_{R_{N-1}}^2\\ &\quad -\gamma^2\|w_{N-1}\|^2 + x_N^T Z_N x_N\}\\ &\geq \mathbb{E}\{\|x_{N-1}\|_{Q_{N-1}}^2 + \alpha_{N-1}\|F_{N-1}x_{N-1}\|_{R_{N-1}}^2\\ &\quad +x_N^T Z_N x_N \,| w_{N-1} = 0\} > 0. \end{aligned} \tag{8.27}$$

Since x_{N-1} is an arbitrary vector, we have $Z_{N-1} > 0$. Furthermore, it is easy to obtain $Z_{N-1} - Z_N \geq 0$ from (8.27). The same procedure can be repeated for all $k \leq N-2$.

Remark 8.5. If M_k is negative for some $\bar{k} \in \mathbf{K}$, the maximizer has the choice such that the upper value of the corresponding static game becomes unbounded [14]. Thus, (8.18) is also a necessary condition such that $\mathcal{G}_2$ is upper bounded.

The optimal disturbance attenuation level γ^* can be introduced as $\gamma^* := \inf\{\gamma : \gamma \in \Gamma\}$, where the set Γ is defined as $\Gamma := \{\gamma > 0 :$ condition (8.18)
holds$\}$.

8.3.3 Coupled Design of RCS

The process of the coupled design of RCS is summarized in this subsection. The inner game $\mathcal{G}_1$ is solved first, and then its solution α^* is propagated to $\mathcal{G}_2$ where the FI H_∞ minimax controller is found. The sequence of the solution of the game is consistent with the actual situation. In practice, the defender is usually unaware of the attacker's cost function J_a^2. Thus, we can not calculate $\mathbf{u}_a^{1*}$ and J_a^{1*} via (8.10) and (8.1) anymore. The process of finding the joint optimal strategy in the absence of the attacker's cost function is shown in Algorithm 9.

Remark 8.6. The detailed description of GA and the feasibility of using GA to solve the Stackelberg game can be found in [15].

8.4 Numerical Case

In this section, the proposed method is applied to the voltage regulation of UPS, which can provide high quality and reliable power for critical systems, such as life supporting system, emergency system and data storage system. Due to the importance of UPS, resilience becomes vital such that AC voltage can maintain at a desired setting regardless of the DoS attack and the load disturbance.

8.4.1 Dynamic Model

The dynamic model of the UPS is from [140], whose parameters are shown in Table 8.2.

Algorithm 9 Coupled design for RCS

Given: r, φ, $\mathcal{P}$, $\bar{\mathcal{Q}}$, $\mathcal{Q}$, c^1, γ, $\{Q_k\}_{k=1}^{N}$, $\{R_k\}_{k=1}^{N-1}$, function Λ and terminal period $\bar{T}$

Output: $(\mathbf{u}_a^{1*}, \{u_k^*\}_{k=1}^{N-1})$, J_a^{1*} and J_b^*

1. Choose a population of $\mathfrak{n}$ chromosomes.
2. Each chromosome represents an action: $(\mathbf{u}_{a1}^1, \mathbf{u}_{a2}^1, \ldots, \mathbf{u}_{a\mathfrak{n}}^1)$.
3. For $i = 1$ to $\mathfrak{n}$

   a $\mathbf{P}_{\mathbf{a}_1}$ takes the action $\mathbf{u}_{ai}^1$,
   b $\mathbf{P}_{\mathbf{a}_2}$ takes best response $\mathbf{u}_{ai}^2 = R_2(\mathbf{u}_{ai}^1)$
   c The fitness function is then given by $\frac{1}{J_a^1(\mathbf{u}_{ai}^1, \mathbf{u}_{ai}^2)}$

4. Generate the next generation of chromosomes by classical phases of GA: reproduction, crossover, and mutation.
5. Repeat (1)-(4) until the terminal period $\bar{T}$ is reached.
6. Obtain the estimated package drop $\alpha^* = \Lambda(J_a^{1*}(\mathbf{u}_a^{1*}, \mathbf{u}_a^{2*}))$.
7. Substitute α^* into (8.19) and calculate Z_k iteratively. u_k^* can be obtained correspondingly from (8.20) and the optimal control performance J_b^* is obtained from (8.21) in the end.

Table 8.2 Parameters of UPS

Symbol	Meaning
$L_f = 1mH$	Inductance
$R_{Lf} = 15m\Omega$	Resistance
$C_f = 300\mu F$	Capacity
$Y_o = 1S$	Admissible load admittance
$T_s = 0.0008s$	Sampling interval

The matrices and vectors are given by

$$A = \begin{bmatrix} 1 - \frac{T_s R_{Lf}}{L_f} & \frac{T_s}{L_f} & 0 \\ -\frac{T_s}{C_f} & 1 - \frac{T_s Y_o}{C_f} & 0 \\ 0 & T_s & 1 \end{bmatrix}, \quad B = \begin{bmatrix} \frac{T_s}{L_f} \\ 0 \\ 0 \end{bmatrix}, \quad D = \begin{bmatrix} 0 & \frac{T_s}{C_f} & 0 \\ -\frac{T_s}{L_f} & \frac{T_s Y_o}{C_f} & 0 \end{bmatrix}^T,$$

$$x_k = \begin{bmatrix} x_{1k} \; x_{2k} \; x_{3k} \end{bmatrix}^T, \quad w_k = \begin{bmatrix} i_k^o \; v_k^{rms} \end{bmatrix}^T, \tag{8.28}$$

where $\begin{bmatrix} x_{1k} \; x_{2k} \end{bmatrix}^T = \begin{bmatrix} i_k^{L_f} \; (v_k^{rms} - v_k^{orms}) \end{bmatrix}^T$, $x_{3k+1} = T_s x_{2k} + x_{3k}$, v_k^{orms} is the Root Mean Square (RMS) value of capacitor voltage, v_k^{rms} represents the RMS value of the sinusoidal reference, $i_k^{L_f}$ is the inductor current, i_k^o denotes the output disturbance. The associated performance index is

$$\mathbb{E}\{\|x_{201}\|_{20}^2 + \sum_{k=1}^{200} \{\|x_k\|_{20}^2 + \alpha_k \|u_k\|^2 - \gamma^2 \|w_k\|^2\}\}.$$

The parameters of the cost functions of the IDSs and the intelligent attacker are chosen to be $c^1 = 50$, $c^2 = 10$, $\mathcal{P} = 0.1$, $\varphi = 10$, $\psi = 10$, $r = 5$, $\mathcal{Q} = 1$ and $\bar{\mathcal{Q}} = 0.5$.

8.4.2 Simulation Results

First, let us check whether there exists a unique inner solution for Stackelberg game $\mathcal{G}_1$. Since $\mathcal{PQ} = 0.1 > 0$ and $r = 5 < 166$, the conditions in Theorem 8.3 are satisfied and we have $\mathbf{u}_a^{1*} = 1.2121$ and $\mathbf{u}_a^{2*} = 0.5152$ by 8.10 and 8.11. The corresponding cost functions $J_a^{1*} = 10.3030$ and $J_a^{2*} = 3.4068$ can be obtained by (8.1) and (8.2). we can also suppose that the attacker's cost function J_a^2 is unavailable to the defender and apply Algorithm 9. The mutation and crossover probability are chosen to be 0.25 and 0.75, respectively. Other parameters are chosen as $\mathfrak{n} = 80$ and $\bar{T} = 20$. Actions $\{\mathbf{u}_{ai}^1\}_{i=1}^{\mathfrak{n}}$ are chosen within the interval $[0, 10]$. The evolution process is shown in Figure 8.2, where we can see that the algorithm is convergent very fast. The strategies $\mathbf{u}_a^{1*} = 1.212$ and $J_a^{1*} = 10.300$ can be obtained by using GA, and the results are almost the same as the theoretical values, which demonstrates the effectiveness.

In the sequel, we choose the function $\Lambda(\mathbf{x}) = 0.8e^{-0.01\mathbf{x}}$, the initial value $x_1 = \begin{bmatrix} 1 \ 1 \ 0 \end{bmatrix}^T$, and $\gamma = 100$. The correlation between the IDS configuration strategy $\mathbf{u}_a^1$ and the underlying control performance J_b^* is shown in Figure 8.3. We also apply the control design method in [100] to 8.28. The comparison of γ^* between FI H_∞ minimax control and Moon et al. (2013) is shown in Figure 8.4.

8.4.3 Discussions

From Figure 8.3, we find that the proposed IDS configuration strategy $\mathbf{u}_a^{1*} = 1.21$ can minimize the optimal control performance J_b^* if the intelligent attacker is rational and takes the optimal response to the IDS's configuration strategy. The possible reason is that the optimal control performance index J_b^* is a nonincreasing function of α. This is quite intuitive that the optimal control performance J_b^* will not get better if more control commands are intercepted by the attacker, but this fact still needs to be proved. We leave it to further exploration of the coupled design of RCS.

From Figure 8.4, it is evident that the proposed method can get better optimal disturbance attenuation level γ^* than [100], since the FI H_∞ minimax controller can get direct access to the disturbance and is hence less conservative.

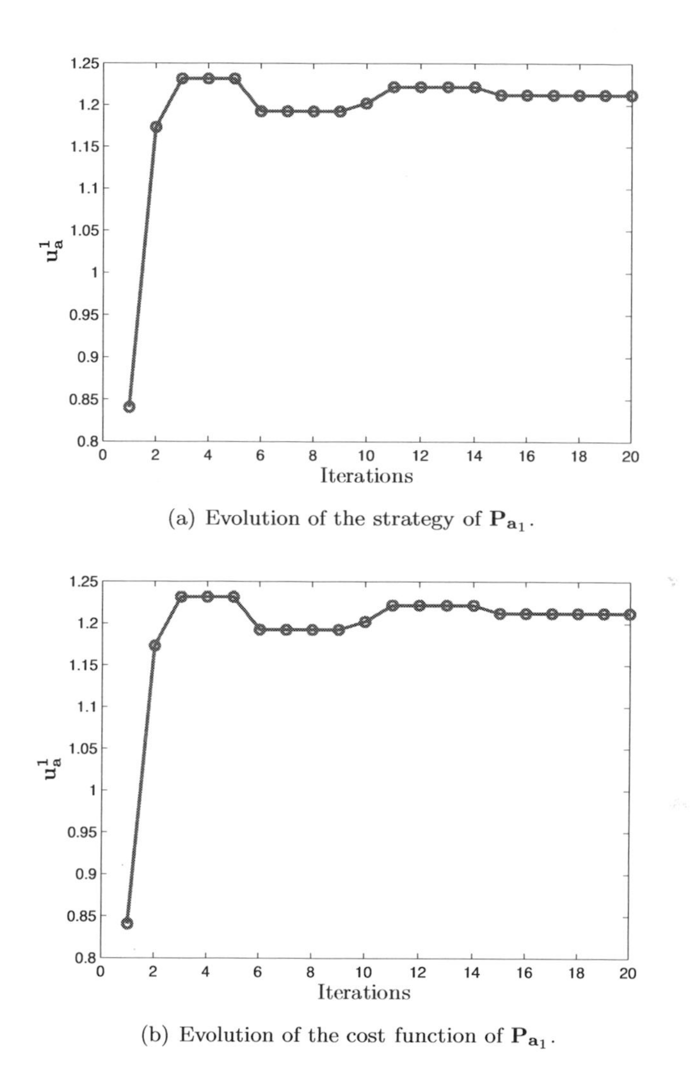

(a) Evolution of the strategy of $\mathbf{P}_{\mathbf{a}_1}$.

(b) Evolution of the cost function of $\mathbf{P}_{\mathbf{a}_1}$.

Figure 8.2 Evolution of the leader's strategy and cost in $\mathcal{G}_1$.

8.5 Conclusion

The exposure to public network has created new challenges for modern control system. Both the physical disturbance and cyber attacks have to be considered in the design of RCS. In this chapter, we have proposed a Stackelberg-in-Stackelberg theoretical structure to model the RCS that integrates cyber security into the control system. The Stackelberg configuration strategy for IDSs in face of the intelligent attacker has been provided in the cyber layer, and the Stackelberg control strategy for the FI H_∞ minimax control with

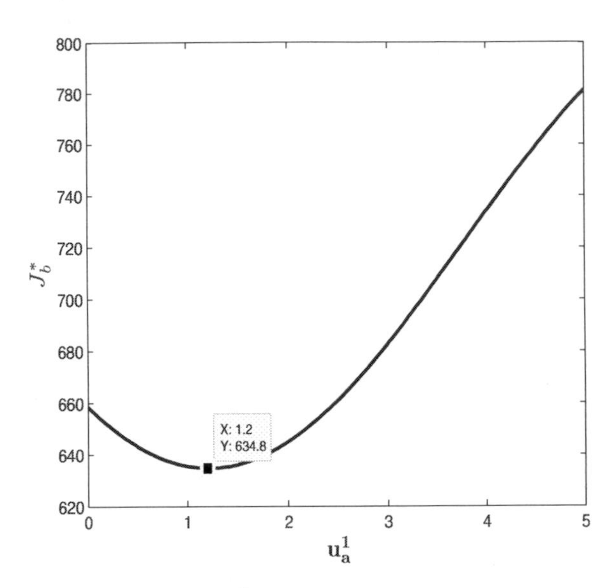

Figure 8.3 The relation curve of $\mathbf{u}_a^1$ and J_b^*.

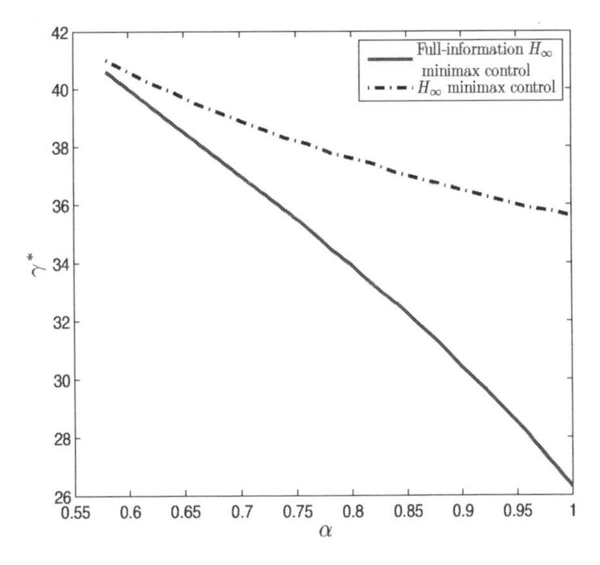

Figure 8.4 The comparison of γ^* between FI H_∞ minimax control and Moon et al. (2013).

package drops has been developed in the physical layer. Then, the coupled design method has been proposed. GA has been employed when the cost function of the intelligent attacker is unknown on the defense side. The advantage of the proposed method has been verified by a numerical example.

Chapter 9
Multitasking Optimal Control of NCSs

9.1 Introduction

NCSs have recently stirred much interest in the control community [144, 182, 166, 56].Much literature has exploited various aspects of NCSs, such as time delays [179], [110], packets dropout [75], packets disorder [88], quantization [153, 108], cyber attacks [171], and so on. Among these network-induced factors, two significant challenges are time delays and packets dropout, which lead to degradation of the control performance or even destabilize the whole controlled system [111]. On the other hand, it has been well known that disturbance exists in many control processes due to environment noise, measurement errors, and load variations [80, 25]. However, in reported results concerning the optimal control problem, the disturbances are usually assumed to be fully estimated or compensated, which is impractical in most cases. For systems with immeasurable and unpredictable disturbances, pure NE cannot be used as the final outcome of the game, since it will also be affected by such disturbances. Very recently, the so-called ϵ-NE has been proposed in [68] and [68], where the impact from the disturbances has been described by the ϵ-level.

In this chapter, we aim to investigate the multitasking optimal control problem of the NCS in the delta-domain. It is worth mentioning that the dynamics of the NCS is quite comprehensive to cover long time delays, packets dropout, and disturbances, thereby reflecting the reality closely. The problem to be dealt with represents the first of a few attempts to tackle the multitasking optimal control of NCS with disturbances. The challenges lie in how to define a comprehensive NCS model for the multitasking optimal control problem and how to quantify the influences of disturbance on NE. More specifically, the contributions of this chapter are mainly threefold. A novel delta-domain model is proposed to describe the NCS with random long time delays and packets dropout. Then, the multitasking optimal control problem is addressed for the proposed model in delta domain. Last, for the obtained control strategies, the ϵ-NE is proposed to describe the impacts from the disturbances, and the upper bound for the ϵ-level is provided explicitly.

The rest of the chapter is organized as follows. Section 10.2 is the problem formulation. Section 10.3 is the main results, where Nash strategies in delta-domain are provided and ϵ-level of ϵ-NE is provided explicitly. In Section 10.4, the numerical simulations are given. Conclusions are drawn in Section 10.5.

9.2 Problem Formulation

9.2.1 Delta-Domain Model of NCS

The block diagram of NCS in this chapter is shown in Figure 9.1, where we can see that time delays and packets dropout phenomenons are taken into account in the controller-to-actuator channel. The time delay is denoted by $\tau(t)$ and the packets dropout is described by $\alpha(t)$, which can also be regarded as an indicator of the packet received. Before proceeding, the following assumption

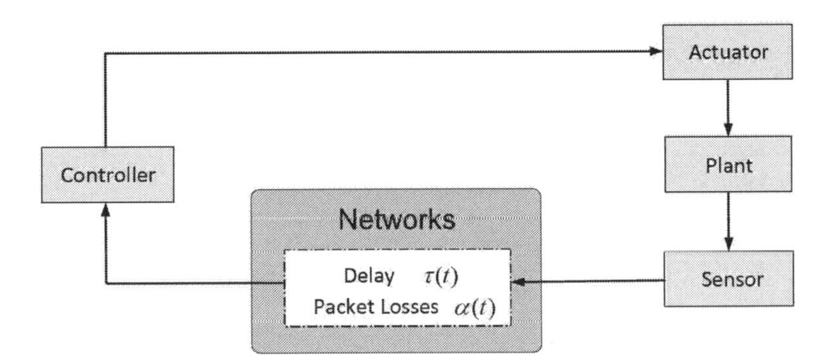

Figure 9.1 Structure of NCS with time delays and packets dropout.

is presented:

Assumption 9.1 *i* *Suppose that the sensor is time-driven, and the controller/actuator is event driven [75, 43, 161, 59].*

ii *Suppose that the time delay is upper bounded, i.e.,* $0 \leq \tau(t) < \sum_{l=1}^{h} T_{k-l}$, *where* h *is a positive scalar representing the maximum time delay steps [43, 59].*

Consider the continuous-domain system model of NCS

$$\dot{x}(t) = A_s x(t) + \sum_{i=1}^{S} \alpha^i(t) B_s^i u^i(t - \tau(t)) + D_s w(t), \tag{9.1}$$

where $x(\cdot) \in \mathbb{R}^n$ is the state vector, $u^i(\cdot) \in \mathbb{R}^m$ denotes the input of the ith player, and $w(\cdot) \in \mathbb{R}^m$ represents the disturbance term.

$A_s \in \mathbb{R}^{n \times n}$, $B_s^i \in \mathbb{R}^{n \times m}$, and $D_s \in \mathbb{R}^{n \times n}$ are continuous-domain matrices. The scalar $\alpha^i(t)$ is the indicator of packet received in the ith control channel and

$$\alpha^i(t) = \begin{cases} \mathbf{I}_{n \times n} \\ \mathbf{0}_{n \times n} \end{cases}. \tag{9.2}$$

Where $\mathbf{I}_{n \times n}$ means if the control input at the ith control channel is received at time; $\mathbf{0}_{n \times n}$ means if the control input at the ith control channel is lost at time.

Now, let us focus on a sampling interval $[t_k, t_{k+1}), \forall k \in \mathbf{K} := \{1, 2, \cdots, K\}$. The control input $u^i(\cdot)$ is piecewise constant to the plant, and hence there are at most h current and previous control input values that can be received at the actuator. If several control inputs are received at the same time, only the newest one acts on the controlled plant, and other previous inputs are dropped. According to Assumption 1, the controller is event-driven, and the plant will implement control input at the time instant $\sum_{l=0}^{k-1} \mathtt{T}_l + \Delta t_k^j$, $j \in \mathbf{H} := \{0, 1, 2, \ldots, h\}$, where $\Delta t_k^j = \tau_{k-j} - \sum_{l=1}^{j} \mathtt{T}_{k-l} \phi(j-1)$. Note that the superscript of Δt_k^j means that the sensor packet is sent at the time instant t_{k-j}, and the symbol τ_{k-j} is the time delays of the sensor packet sent at the time instant t_{k-j}. The timing diagram of signals on NCS can be seen in Figure 9.2.

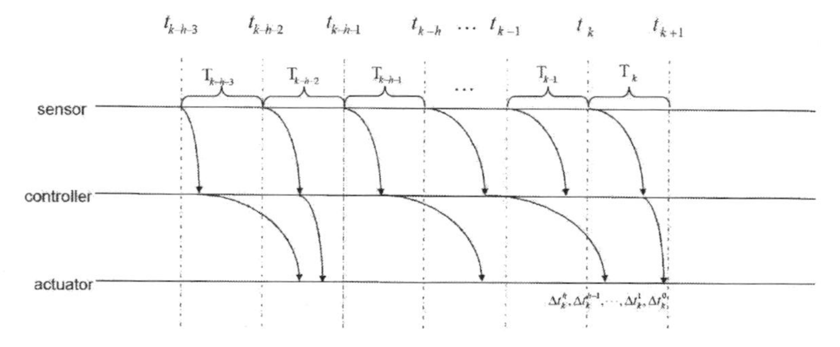

Figure 9.2 Timing diagram of signals transmitting in NCS.

From the above analysis and [43], system (9.1) can be discretized as the following system with nonuniform sampling period $\mathtt{T}_k$

$$x(t_{k+1}) = A_z(t_k)x(t_k) + \sum_{i=1}^{S}\left(\sum_{j=0}^{h} B_z^{ij}(t_k)u_\alpha^i(t_{k-j})\right) + D_z(t_k)w(t_k), \quad (9.3)$$

where

$$u_\alpha^i(t_{k-j}) = \alpha_{t_{k-j}}^i u^i(t_{k-j}),\ \forall i \in \mathbf{S} := \{1, 2, \cdots, S\},\ \forall j \in \mathbf{H},$$

$$\alpha_{t_{k-j}}^i = \begin{cases} \mathbf{I}_{n\times n} & \text{if the control input } i \text{ is received during } [t_k, t_{k+1}) \\ \mathbf{0}_{n\times n} & \text{if the control input } i \text{ is lost during } [t_k, t_{k+1}), \end{cases}$$

$$B_z^{ij}(t_k) = \int_{\tau_{k-j}-\sum_{l=1}^{j}\mathtt{T}_{k-l}}^{\tau_{k-(j-1)}-\sum_{l=1}^{j-1}\mathtt{T}_{k-l}} e^{A_s(\mathtt{T}_k - s)}B_s^i ds \cdot \phi(\mathtt{T}_{k-j} + \tau_{k-(j-1)} - \tau_{k-j}),$$
$$\cdot\, \phi(\tau_{k-j} - \sum\nolimits_{l=1}^{j} \mathtt{T}_{k-l}),\ \forall i \in \mathbf{S},\ \forall j \in \mathbf{H}\backslash\{0\},$$

$$B_z^{i0}(t_k) = \int_{\tau_k}^{\mathtt{T}_k} e^{A_s(\mathtt{T}_k - s)}B_s^i ds \cdot \phi(\mathtt{T}_k - \tau_k),\ \forall i \in \mathbf{S},$$

$$D_z(t_k) = \int_0^{\mathtt{T}_k} e^{A_s(\mathtt{T}_k - s)}D_s ds.$$

Note that the vector $u^i(\cdot)$ is the control input sent by controller i and $u_\alpha^i(\cdot)$ is the control input received at the corresponding actuator. Let $\alpha_{t_k}^i$ be the random variable and distributed according to the Bernoulli distribution. Suppose that $\alpha_{t_k}^i$ is i.i.d and, for $i \neq j,\ i, j \in \mathbf{S}$, $\alpha_{t_k}^i$ is independent of $\alpha_{t_k}^j$. Let us denote:

$$\mathbb{P}\{\alpha_{t_k}^i = 1\} = \alpha^i,\ \ \mathbb{P}\{\alpha_{t_k}^i = 0\} = 1 - \alpha^i,\ \forall i \in \mathbf{S},\ k \in \mathbf{K}.$$

Define the following vector:

$$\delta z(t_k) = \left[\delta x^T(t_k)\ \delta\mathbf{u}^T(t_{k-1})\ \delta\mathbf{u}^T(t_{k-2}) \cdots\ \delta\mathbf{u}^T(t_{k-h+1})\ \delta\mathbf{u}^T(t_{k-h})\right]^T (9.4)$$

where $z(t_k) \in \mathbb{R}^{n+mSh}$, and $\mathbf{u}(\cdot) := \left[u^{1^T}(\cdot)\ \ u^{2^T}(\cdot)\ \ \ldots\ \ u^{S^T}(\cdot)\right]^T$. With the augmented state vector $z(t_k)$ and definition of delta operator, system (9.3) can be rewritten as

$$\Sigma_e:\ \ \delta z(t_k) = \mathcal{A}_\delta(t_k)z(t_k) + \sum_{i=1}^{S} \mathcal{B}_\delta^i(t_k)u^i(t_k) + \mathcal{D}_\delta(t_k)w(t_k), \quad (9.5)$$

where

$$\mathcal{A}_\delta(t_k) = \begin{bmatrix} A_\delta(t_k) & \mathbf{B}_\delta^1(t_k) & \mathbf{B}_\delta^2(t_k) & \cdots & \mathbf{B}_\delta^{h-1}(t_k) & \mathbf{B}_\delta^h(t_k) \\ \mathbf{0}_{Sm\times n} & -\frac{1}{\mathrm{T}_k}\mathbf{I}_{Sm\times Sm} & \mathbf{0}_{Sm\times Sm} & \cdots & \mathbf{0}_{Sm\times Sm} & \mathbf{0}_{Sm\times Sm} \\ \mathbf{0}_{Sm\times n} & \frac{1}{\mathrm{T}_k}\mathbf{I}_{Sm\times Sm} & -\frac{1}{\mathrm{T}_k}\mathbf{I}_{Sm\times Sm} & \cdots & \mathbf{0}_{Sm\times Sm} & \mathbf{0}_{Sm\times Sm} \\ \vdots & \vdots & \vdots & \ddots & \vdots & \vdots \\ \mathbf{0}_{Sm\times n} & \mathbf{0}_{Sm\times Sm} & \mathbf{0}_{Sm\times Sm} & \cdots & -\frac{1}{\mathrm{T}_k}\mathbf{I}_{Sm\times Sm} & \mathbf{0}_{Sm\times Sm} \\ \mathbf{0}_{Sm\times n} & \mathbf{0}_{Sm\times Sm} & \mathbf{0}_{Sm\times Sm} & \cdots & \frac{1}{\mathrm{T}_k}\mathbf{I}_{Sm\times Sm} & -\frac{1}{\mathrm{T}_k}\mathbf{I}_{Sm\times Sm} \end{bmatrix},$$

$$\mathcal{B}_\delta^i(t_k) = \begin{bmatrix} \alpha_{t_k}^i B_\delta^{i0}(t_k) \\ \frac{1}{\mathrm{T}_k}\tilde{\mathbf{I}}_{Sm\times m}^i \\ \mathbf{0}_{mS(h-1)\times m} \end{bmatrix}, \quad \mathcal{D}_\delta(t_k) = \begin{bmatrix} D_\delta(t_k) \\ \mathbf{0}_{mSh\times n} \end{bmatrix},$$

$$\mathbf{B}_\delta^j(t_k) = \begin{bmatrix} \alpha_{t_{k-j}}^1 B_\delta^{1j}(t_k) & \alpha_{t_{k-j}}^2 B_\delta^{2j}(t_k) & \cdots & \alpha_{t_{k-j}}^S B_\delta^{Sj}(t_k) \end{bmatrix}, \ \forall j \in \mathbf{H}/0,$$

$$\tilde{\mathbf{I}}_{Sm\times m}^i = \begin{bmatrix} \mathbf{0}_{(i-1)m\times m} \\ \mathbf{I}_{m\times m} \\ \mathbf{0}_{(S-i)m\times m} \end{bmatrix}, \ \forall i \in \mathbf{S}.$$

with $A_\delta(t_k) = (e^{A_s\mathrm{T}_k} - I)/\mathrm{T}_k$, $B_\delta^{ij}(t_k) = \frac{1}{\mathrm{T}_k}B_z^{ij}(t_k)$, $B_\delta^{i0}(t_k) = \frac{1}{\mathrm{T}_k}B_z^{i0}(t_k)$, $D_\delta(t_k) = \frac{1}{\mathrm{T}_k}D_z(t_k)$.

9.2.2 Design Objective for Multitasking NCS

System (Σ_e) is said to be the nominal system (Σ_n) if $w(t_k) \equiv 0$. In the situation where the central coordination is infeasible [91], it is more practical that each controller only optimized its individual objective function. The individual cost function of system (Σ_n) yields

$$J^i = \mathbb{E}\Big\{ z^T(t_K)\mathcal{Q}_\delta^{iK} z(t_K) + \sum_{k=0}^{K-1} \mathrm{T}_k\big(z^T(t_k)\mathcal{Q}_\delta^i z(t_k) + \Phi(t_k)\big)\Big\}, \ i \in \mathbf{S}, \quad (9.6)$$

where $\Phi(t_k) = \alpha_{t_k}^i {u^i}^T(t_k)\mathcal{R}_\delta^{ii} u^i(t_k) + \sum_{j=1,j\neq i}^S \varepsilon_j {u^j}^T(t_k)\mathcal{R}_\delta^{ij} u^j(t_k)$. We assume that $\mathcal{Q}_\delta^{iK} \geq 0$, $\mathcal{Q}_\delta^i \geq 0$, $\mathcal{R}_\delta^{ii} > 0$ and $\mathcal{R}_\delta^{ij} > 0$ for $\forall i \in \mathbf{S}$, $\forall j \in \mathbf{S}$. Parameter ε_j is used to characterize the 'fairness' of the game, which is a very small positive scalar implying that the ith player needn't be punished too much by others' faults. Under the network environment, we adopt transmission control protocol [119] where each package is acknowledged, and the information set is $\mathcal{I}_{t_0} = \{z(t_0)\}$, $\mathcal{I}_{t_k} = \{z(t_1), z(t_2), \ldots, z(t_k), \alpha_{t_{k-1}}^1, \ldots, \alpha_{t_{k-1}}^S, \ldots, \alpha_{t_0}^1, \ldots, \alpha_{t_0}^S\}$. Thus, the control input can be expressed as $u^i(t_k) = \mu_{t_k}(\mathcal{I}_{t_k})$, where μ_{t_k} is the admissible strategy. Based on the rationality of all players, for system (Σ_n), pure NE can be employed as the final outcome of the game, which is $J^i(u^{i*}, u^{-i*}) \leq J^i(u^i, u^{-i*})$ [16]. Nevertheless, it is worth mentioning that there exists disturbance $w(t_k)$ in system (Σ_e), which can perturb the pure NE. To tackle with system (Σ_e), we introduce the concept of ϵ-NE as follows:

$$\hat{J}^i(u^{i*}, u^{-i*}) \leq \inf_{u^i \in U^i_{admis}} \hat{J}^i(u^i, u^{-i*}) + \epsilon^i, \ i \in \mathbf{S}, \tag{9.7}$$

where $\hat{J}^i(u^{i*}, u^{-i*})$ has same structure as $J^i(u^{i*}, u^{-i*})$, but includes the impacts from the disturbance. The scalar $\epsilon^i \geq 0$ is employed to describe how far the ϵ-NE deviates from the pure NE. Furthermore, the following assumption is presented.

Assumption 9.2 *Suppose that the disturbance $\omega(t_k)$ and control input $u^i(t_k)$ are both quadratically bounded, i.e., $\|w(t_k)\|^2 \leq \gamma$ and $\|u^i(t_k)\| \leq \theta^i$.*

In this chapter, we aim to design feedback Nash strategies for system Σ_n such that the cost function (9.6) can be minimized individually. In addition, for the obtained Nash strategies u^{i*}, the ϵ-level of the ϵ-NE is provided explicitly. Before ending this section, the following two lemmas are provided which will be used subsequently.

Lemma 9.1. *[162] The property of delta operator, for any time function $x(t_k)$ and $y(t_k)$, there exists*

$$\delta(x(t_k)y(t_k)) = y(t_k)\delta x(t_k) + x(t_k)\delta y(t_k) + T_k \delta x(t_k)\delta y(t_k),$$

where T_k is the sampling period.

Lemma 9.2. *[68] Let x, y be any n_1-dimensional real vectors, and Π be an $n_1 \times n_1$ symmetric positive semi-definite matrix. Then, we have $x^T y + y^T x \leq x^T \Pi x + y^T \Pi^{-1} y$.*

Remark 9.3. i The so-called multitasking control can describe the case where each controller pursues its own target without cooperation. It should be noticed that all the system states should be available to the controller, which can be seen from the information set $\mathcal{I}_{t_k}$. On the other hand, the graphic difference/differential game can describe the case where the controller only needs to get access to part of the system states [1]. Thus, the graphic difference/differential game is capable of describing the multi-agent systems with certain topologies that is not within the scope of this chapter.

ii It can be observed from the information set $\mathcal{I}_{t_k}$ that the proposed control strategy requires the current system state as well as past control input sequence $\mathbf{u}(t_{k-1}), ..., \mathbf{u}(t_{k-h})$. In practice, the buffer located at the side of the controller has a limited capacity for storage and hence the proposed control scheme cannot tolerant overlong time delays. In other words, even if the h in Assumption 9.1 can be a arbitrarily large scaler in theory, the proposed control scheme has limitation in practice.

iii In this chapter, the NE has been employed as a performance index to describe the situation where the central coordination is difficult for all the controllers. It should be pointed out that the performance obtained by NE solution is worse than or equal to that obtained by socially optimal solution, and such performance degradation can be quantified by the concept of 'price-of-anarchy' [17]. However, for the complex and large scaled systems, it makes more sense to use NE to describe the practical situation. For example, in classical power grids, the overall system can be optimized by solving a system-level optimization problem by using a centralized objective function. However, as the scale of the power grids grows, a networked group of distributed energy sources can locate at the distribution networked of geographical areas. Therefore, the distributed operation of the smart grid nodes is required and it makes practical sense to define a specific objective function to meet consumers' individual satisfaction in each control area. Furthermore, for the micro-grid, which can operate autonomously in the isolated mode having no connection to the coordination centre, it's more natural to adopt distributed analytical techniques such as game theoretic tool [116]. Due to the above discussions, noncooperative dynamic games have been extensively exploited in the coordinated control problem of multi-agent systems [1], and control problem of power systems, such as the heating, ventilation, and air conditioning system [91], and load frequency control system [138, 137].

9.3 Main Results

In this section, the Nash strategies and Riccati-like recursions are provided for the multitasking optimal control problem of NCS in delta domain. Furthermore, the explicit expression of the ϵ-NE is provided.

9.3.1 Design of the Control Strategy

In the following, the sufficient conditions are provided such that, for system Σ_n, the individual cost function (9.6) can be minimized in the delta-domain. Also, the explicit expression for the feedback Nash strategy is given.

Theorem 9.4. *Consider the system Σ_n with the cost function* (9.6). *The following conclusions are presented.*

- *There exists the unique NE if*

$$\alpha^i \mathcal{R}_\delta^{ii} + T_k \mathbb{E}\{\mathcal{B}_\delta^{iT}(t_k) P^i(t_{k+1}) \mathcal{B}_\delta^i(t_k)\} > 0, \tag{9.8}$$

and the matrix $\Theta(t_k)$ is invertible with

$$\Theta^{ii}(t_k) = \alpha^i \mathcal{R}_\delta^{ii} + T_k \mathbb{E}\{\mathcal{B}_\delta^{iT}(t_k) P^i(t_{k+1}) \mathcal{B}_\delta^i(t_k)\},\ i \in \boldsymbol{S},$$
$$\Theta^{ij}(t_k) = T_k \mathbb{E}\{\mathcal{B}_\delta^{iT}(t_k) P^i(t_{k+1}) \mathcal{B}_\delta^j(t_k)\},\ i, j \in \boldsymbol{S},\ i \neq j,$$

- *Under condition 1, the NE strategies are given as*

$$u^{i*}(t_k) = -L^i(t_k) z(t_k),\ \forall i \in \boldsymbol{S}, \tag{9.9}$$

where

$$\begin{aligned} L^i(t_k) = &\left\{\alpha^i \mathcal{R}_\delta^{ii} + T_k \mathbb{E}\left\{\mathcal{B}_\delta^{iT}(t_k) P^i(t_{k+1}) \mathcal{B}_\delta^i(t_k)\right\}\right\}^{-1} \\ &\Bigg\{\mathbb{E}\left\{\mathcal{B}_\delta^{iT}(t_k) P^i(t_{k+1})(T_k \mathcal{A}_\delta(t_k) + I)\right\} \\ &- T_k \sum_{j=1, j\neq i}^{S} \mathbb{E}\left\{\mathcal{B}_\delta^{iT}(t_k) P^i(t_{k+1}) \mathcal{B}_\delta^j(t_k)\right\} L^j(t_k)\Bigg\}. \end{aligned} \tag{9.10}$$

- *The backward iterations are carried out with $P^i(t_K) = \mathcal{Q}_\delta^{iK}$, and*

$$\begin{aligned} -\delta P^i(t_k) = &\mathcal{Q}_\delta^i + \alpha^i L^{iT}(t_k) \mathcal{R}_\delta^{ii} L^i(t_k) + \sum_{j=1, j\neq i}^{S} \varepsilon_j L^{jT}(t_k) \mathcal{R}_\delta^{ij} L^j(t_k) \\ &+ T_k \mathbb{E}\{F^T(t_k) P^i(t_{k+1}) F(t_k)\} + \mathbb{E}\{F(t_k)\}^T P^i(t_{k+1}) \\ &+ P^i(t_{k+1}) \mathbb{E}\{F(t_k)\}, \end{aligned} \tag{9.11}$$
$$P^i(t_k) = P^i(t_{k+1}) - T_k \delta P^i(t_k), \tag{9.12}$$

where $F(t_k) = \mathcal{A}_\delta(t_k) - \sum_{i=1}^{S} \mathcal{B}_\delta^i(t_k) L^i(t_k)$.

- *Under condition 1, the NE value for player i is $J^{i*} = z^T(t_0) P^i(t_0) z(t_0)$, $i \in \boldsymbol{S}$, where $z(t_0)$ is the initial value.*

Proof: Let us choose the delta-domain cost function as $V^i(z(t_{k+1})) = z^T(t_{k+1}) P^i(t_{k+1})$ with $P^i(t_{k+1}) > 0$. It follows from Lemma 9.1 that

$$\begin{aligned} V^i(z(t_{k+1})) = &\mathrm{T}_k \delta(z^T(t_k) P^i(t_k) z(t_k)) + z^T(t_k) P^i(t_k) z(t_k) \\ = &\mathrm{T}_k \delta z^T(t_k) P^i(t_{k+1}) z(t_k) + \mathrm{T}_k z^T(t_k) P^i(t_{k+1}) \delta z(t_k) \\ &+ \mathrm{T}_k^2 \delta z^T(t_k) P^i(t_{k+1}) \delta z(t_k) + z^T(t_k) P^i(t_{k+1}) z(t_k). \end{aligned}$$

In virtue of the dynamic programming method [16], we have

$$
\begin{aligned}
V^i(z(t_k)) &= \min_{u^i(t_k)} \mathbb{E}\Big\{\mathrm{T}_k z^T(t_k)\mathcal{Q}_\delta^i z(t_k) + \alpha_{t_k}^i \mathrm{T}_k u^{iT}(t_k)\mathcal{R}_\delta^{ii} u^i(t_k) \\
&\quad + \mathrm{T}_k \sum_{j=1, j\neq i}^{S} \varepsilon_j u^{jT}(t_k)\mathcal{R}_\delta^{ij} u^j(t_k) + V^i(z(t_{k+1}))\Big\} \\
&= \min_{u^i(t_k)} \mathbb{E}\Big\{\mathrm{T}_k z^T(t_k)\mathcal{Q}_\delta^i z(t_k) + \alpha_{t_k}^i \mathrm{T}_k u^{iT}(t_k)\mathcal{R}_\delta^{ii} u^i(t_k) \\
&\quad + \mathrm{T}_k \sum_{j=1, j\neq i}^{S} \varepsilon_j u^{jT}(t_k)\mathcal{R}_\delta^{ij} u^j(t_k) \\
&\quad + \mathrm{T}_k \Big(\mathcal{A}_\delta(t_k)z(t_k) + \sum_{i=1}^{S} \mathcal{B}_\delta^i(t_k)u^i(t_k)\Big)^T P^i(t_{k+1})z(t_k) \\
&\quad + \mathrm{T}_k z^T(t_k)P^i(t_{k+1})\Big(\mathcal{A}_\delta(t_k)z(t_k) + \sum_{i=1}^{S} \mathcal{B}_\delta^i(t_k)u^i(t_k)\Big) \\
&\quad + z^T(t_k)P^i(t_{k+1})z(t_k) \\
&\quad + \mathrm{T}_k^2 \Big(\mathcal{A}_\delta(t_k)z(t_k) + \sum_{i=1}^{S} \mathcal{B}_\delta^i(t_k)u^i(t_k)\Big)^T P^i(t_{k+1}) \\
&\quad \times \Big(\mathcal{A}_\delta(t_k)z(t_k) + \sum_{i=1}^{S} \mathcal{B}_\delta^i(t_k)u^i(t_k)\Big)\Big\}. \qquad (9.13)
\end{aligned}
$$

Since the second derivative of equation (9.13) yields

$$
\alpha^i \mathcal{R}_\delta^{ii} + \mathrm{T}_k \mathbb{E}\{\mathcal{B}_\delta^{iT}(t_k)P^i(t_{k+1})\mathcal{B}_\delta^i(t_k)\} > 0
$$

it is concluded that the cost function $V^i(z(t_k))$ is strictly convex of $u^i(t_k)$. Therefore, the minimizer is obtained by solving

$$
\begin{aligned}
&\partial V^i(z(t_k))/\partial u^i(t_k) \\
&= \alpha^i \mathcal{R}_\delta^{ii} u^i(t_k) + \mathbb{E}\{\mathcal{B}_\delta^{iT}(t_k)P^i(t_{k+1})(\mathrm{T}_k \mathcal{A}_\delta(t_k) + I)z(t_k)\} \\
&\quad + \mathrm{T}_k \mathbb{E}\{\mathcal{B}_\delta^{iT}(t_k)P^i(t_{k+1})\mathcal{B}_\delta^i(t_k)u^i(t_k)\} \\
&\quad + \mathrm{T}_k \mathbb{E}\Big\{\mathcal{B}_\delta^{iT}(t_k)P^i(t_{k+1}) \sum_{j=1, j\neq i}^{S} \mathcal{B}_\delta^j(t_k)u^j(t_k)\Big\} = 0. \qquad (9.14)
\end{aligned}
$$

The feedback Nash strategies are then obtained as $u^{i*}(t_k) = -L^i(t_k)z(t_k)$ with

$$L^i(t_k) = \left\{\alpha^i \mathcal{R}_\delta^{ii} + \mathtt{T}_k \mathbb{E}\left\{\mathcal{B}_\delta^{iT}(t_k) P^i(t_{k+1}) \mathcal{B}_\delta^i(t_k)\right\}\right\}^{-1} \left\{\mathbb{E}\{\mathcal{B}_\delta^{iT}(t_k) P^i(t_{k+1})(\mathtt{T}_k \mathcal{A}_\delta(t_k) + I)\} - \mathtt{T}_k \sum_{j=1, j\neq i}^{S} \mathbb{E}\{\mathcal{B}_\delta^{iT}(t_k) P^i(t_{k+1}) \mathcal{B}_\delta^j(t_k)\} L^j(t_k)\right\}. \tag{9.15}$$

Notice that (9.15) should set up for all players. Hence, it is required that $\Theta(t_k)\mathbf{L}(t_k) = \Pi(t_k)$, where $\Theta(t_k)$ is the invertible matrix, and

$$\begin{aligned} \Pi^{ii}(t_k) &= \mathbb{E}\{\mathcal{B}_\delta^{iT}(t_k) P^i(t_{k+1})(\mathtt{T}_k \mathcal{A}_\delta(t_k) + I)\}, \ i \in \mathbf{S}, \\ \Pi^{ij}(t_k) &= 0, \ i, j \in \mathbf{S}, \ i \neq j, \\ \mathbf{L}(t_k) &= \left[L^{1T}, L^{2T}, \cdots, L^{ST}\right]^T. \end{aligned}$$

Substituting $u^{i*}(t_k) = -L^i(t_k) z(t_k)$ into (9.13), one has that

$$\begin{aligned} 0 = {} & \mathcal{Q}_\delta^i + \alpha^i L^{iT}(t_k) \mathcal{R}_\delta^{ii} L^i(t_k) \\ & + \sum_{j=1, j\neq i}^{S} \varepsilon_j L^{jT}(t_k) \mathcal{R}_\delta^{ij} L^j(t_k) + \delta P^i(t_k) \\ & + \mathtt{T}_k \mathbb{E}\{F^T(t_k) P^i(t_{k+1}) F(t_k)\} \\ & + \mathbb{E}\{F(t_k)\}^T P^i(t_{k+1}) + P^i(t_{k+1}) \mathbb{E}\{F(t_k)\}, \end{aligned} \tag{9.16}$$

$$P^i(t_k) = P^i(t_{k+1}) - \mathtt{T}_k \delta P^i(t_k), \tag{9.17}$$

where $F(t_k) = \mathcal{A}_\delta(t_k) - \sum_{i=1}^{S} \mathcal{B}_\delta^i(t_k) L^i(t_k)$. The backward induction method can be carried out continuously and the NE value for player i is obtained as $J^{i*} = z^T(t_0) P^i(t_0) z(t_0)$. The proof is completed. □

Remark 9.5. The advantages of the delta-domain results in Theorem 9.4 can be summarized as follows:

1. The sampling interval in Theorem 9.4 becomes an explicit parameter and can be tuned according to the network environment [161];
2. The results in Theorem 9.4 can be seen as a unified form of the results for both discrete- and continuous-time systems, and hence the possible numerical stiffness at a high sampling rate can be circumvented [162]. To be specific, the delta-domain results in Theorem 9.4 can be easily converted to its discrete-domain analogue by using $\mathcal{A}_z(k) = \mathtt{T}_k \mathcal{A}_\delta(t_k) + I$ and $\mathcal{B}_z^i(k) = \mathtt{T}_k \mathcal{B}_\delta^i(t_k)$. On the other hand, the equivalence of the results in Theorem 9.4 to those of the continuous-time system can be verified as follows: It follows from $\mathtt{T}_k \to 0$ that $\mathcal{A}_\delta(t_k) \to \mathcal{A}_s(t)$ and $\mathcal{B}_\delta^i(t_k) \to \mathcal{B}_s^i(t)$. Furthermore, the feedback gain (9.10) becomes

$$L^i(t) = (\alpha^i \mathcal{R}_\delta^{ii})^{-1} \mathbb{E}\{\mathcal{B}_s^i(t)^T P^i(t)\}. \tag{9.18}$$

By using (9.18) and $\lim_{\mathtt{T}_k \to 0} \delta P^i(k) = \dot{P}^i(t)$, when $\mathtt{T}_k \to 0$, the Riccati backward iteration (9.11) becomes

$$
\begin{aligned}
-\dot{P}^i(t) = & \mathcal{Q}_\delta^i + \mathbb{E}\left\{P^i(t)\mathcal{B}_s^i(t)(\alpha^i(t)\mathcal{R}_\delta^{ii})^{-1}\mathcal{B}_s^{iT}(t)P^i(t)\right\} \\
& + \sum_{j=1, j\neq i}^{S} \epsilon_j \mathbb{E}\Big\{P^j(t)\mathcal{B}_s^j(t)(\alpha^j(t)\mathcal{R}_\delta^{jj})^{-1} \\
& \times \mathcal{R}_\delta^{ij}(\alpha^j(t)\mathcal{R}_\delta^{jj})^{-1}\mathcal{B}_s^{jT}(t)P^j(t)\Big\} \\
& + \mathbb{E}\{\mathcal{A}_s(t)^T P^i(t)\} + \mathbb{E}\{P^i(t)\mathcal{A}_s(t)\} \\
& - \mathbb{E}\Big\{\Big(\sum_{i=1}^{S} \mathcal{B}_s^i(t)(\alpha^i(t)\mathcal{R}_\delta^{ii})^{-1}\mathcal{B}_s^i(t)^T P^i(t)\Big)^T P^i(t)\Big\} \\
& - \mathbb{E}\Big\{P^i(t)\Big(\sum_{i=1}^{S} \mathcal{B}_s^i(t)(\alpha^i(t)\mathcal{R}_\delta^{ii})^{-1}\mathcal{B}_s^i(t)^T P^i(t)\Big)\Big\}. \quad (9.19)
\end{aligned}
$$

Note that (9.18) and (9.19) are consistent with the feedback Nash strategies and Riccati recursions in the continuous-time systems [16]. Therefore, it is concluded that Theorem 9.4 unifies the results in continuous- and discrete-time systems.

9.3.2 Robustness of ϵ-NE

So far, the feedback Nash gain $L^i(t_k), i \in \mathbf{S}$ has been obtained for system Σ_n. It should be pointed out that the disturbance $w(t_k)$ can lead to a certain deviation from the pure NE. In this subsection, system Σ_e will be considered and the estimation for the ϵ-level of ϵ-NE will be provided. We first present the definitions of the following dynamic process:

1. $\bar{z}(t_k)$ is the state vector for the system (Σ_n) if each player uses NE strategies.
2. $z(t_k)$ is the state vector for the system (Σ_e) if each player uses NE strategies.
3. $\tilde{z}(t_k)$ is the state vector for the system (Σ_n) if each player uses NE strategies except player i. Note that we still have $u^i(\tilde{x}(k)) \in U^i_{admis}$.
4. $\check{z}(t_k)$ is the state vector for the system (Σ_e) if each player uses NE strategies except player i.

It follows from the definition of U^i_{admis} that $\sum_{k=0}^{K} \|\bar{z}(t_k)\|^2 \leq \beta_3$ and $\sum_{k=0}^{K} \|\check{z}(t_k)\|^2 \leq \beta_2$. Before presenting the results, the following notations are given:

$$
\begin{aligned}
&\bar{\mathcal{Q}}_\delta^i(t_K) = \mathcal{Q}_\delta^{iK},\ \tilde{\mathcal{Q}}_\delta^i(t_K) = \mathcal{Q}_\delta^{iK},\\
&\bar{\mathcal{Q}}_\delta^i(t_k) = \mathrm{T}_k\mathcal{Q}_\delta^i + \mathrm{T}_k\sum_{j\neq i}^{S}\epsilon_j L^{jT}(t_k)\mathcal{R}_\delta^{ij}L^j(t_k) + \mathrm{T}_k\alpha_{t_k}^i L^{iT}(t_k)\mathcal{R}_\delta^{ii}L^i(t_k),\\
&\tilde{\mathcal{Q}}_\delta^i(t_k) = \mathrm{T}_k\mathcal{Q}_\delta^i + \mathrm{T}_k\sum_{j\neq i}^{S}\epsilon_j L^{jT}(t_k)\mathcal{R}_\delta^{ij}L^j(t_k),\\
&V_1(t_k) = \bar{e}^T(t_k)S_1(t_k)\bar{e}(t_k),\ V_2^i(t_k) = \tilde{e}^T(t_k)S_2^i(t_k)\tilde{e}(t_k),\\
&\bar{e}(t_k) = z(t_k) - \bar{z}(t_k),\ \tilde{e}(t_k) = \tilde{z}(t_k) - \check{z}(t_k),\\
&\tilde{S}_1 := \{\tilde{S}_1 > S_1(t_k), \forall k\in\mathbf{K}\},\ \tilde{S}_2^i := \{\tilde{S}_2^i > S_2^i(t_k), \forall k\in\mathbf{K}\},\\
&\Gamma_1 = (\mathcal{D}_\delta^{\mathrm{T}}\tilde{S}_1\Lambda_1^{-1}\tilde{S}_1^T\mathcal{D}_\delta + \mathcal{D}_\delta^{\mathrm{T}}\tilde{S}_1\mathcal{D}_\delta),\\
&\Gamma_2 = \mathcal{D}_\delta^{\mathrm{T}}\tilde{S}_2^i\Lambda_3^{-1}\tilde{S}_2^{iT}\mathcal{D}_\delta + \mathcal{D}_\delta^{\mathrm{T}}\tilde{S}_2^i\Lambda_4^{-1}\tilde{S}_2^{iT}\mathcal{D}_\delta + \mathcal{D}_\delta^{\mathrm{T}}\tilde{S}_2^i\mathcal{D}_\delta,\\
&\kappa_1 = 4\max_{\forall k\in\mathbf{K}}\Big(\mathrm{T}_k^2\Big\{\mathbb{E}\{\mathcal{B}_\delta^{iT}(t_k)\Lambda_2^{-1}\mathcal{B}_\delta^i(t_k)\} + \mathbb{E}\{\mathcal{B}_\delta^{iT}(t_k)\tilde{S}_2^i\mathcal{B}_\delta^i(t_k)\}\\
&\qquad +\mathbb{E}\{\mathcal{B}_\delta^{iT}(t_k)\tilde{S}_2^i\Lambda_4\tilde{S}_2^{iT}\mathcal{B}_\delta^i(t_k)\}\Big\}\Big).
\end{aligned}
$$

In the following theorem, the explicit expression for the upper bound of ϵ-NE is provided.

Theorem 9.6. *Consider the system Σ_e with the cost function* (9.6). *For given NE strategies u^{i*} and positive definite matrices $\{\Lambda_j\}_{j=1}^4$, $\mathcal{L}_1$, $\mathcal{L}_2^i$, $i\in\boldsymbol{S}$, if there exist positive definite matrices $S_1(t_k)$, $S_2^i(t_k)$, $\forall i\in\boldsymbol{S}$, $\forall k\in\boldsymbol{K}$ such that the following delta-domain Riccati-like recursions hold*

$$
\begin{aligned}
-\delta S_1(t_k) = {}& T_k\mathbb{E}\{\bar{\mathcal{A}}_\delta^T(t_k)S_1(t_{k+1})\bar{\mathcal{A}}_\delta(t_k)\}\\
&+\mathbb{E}\{\bar{\mathcal{A}}_\delta(t_k)\}^T S_1(t_{k+1}) + S_1(t_{k+1})\mathbb{E}\{\bar{\mathcal{A}}_\delta(t_k)\} \qquad (9.20)\\
&+\frac{1}{T_k}\left\{\mathbb{E}\{(T_k\bar{\mathcal{A}}_\delta(t_k)+I)^T\Lambda_1(T_k\bar{\mathcal{A}}_\delta(t_k)+I)\} + \mathcal{L}_1\right\},\\
S_1(t_k) = {}& S_1(t_{k+1}) - T_k\delta S_1(t_k),\ S_1(t_K) = 0, \qquad (9.21)
\end{aligned}
$$

$$
\begin{aligned}
-\delta S_2^i(t_k) = {}& T_k\mathbb{E}\{\tilde{\mathcal{A}}_\delta^{iT}(t_k)S_2^i(t_{k+1})\tilde{\mathcal{A}}_\delta^i(t_k)\} \qquad (9.22)\\
&+\mathbb{E}\{\tilde{\mathcal{A}}_\delta^i(t_k)\}^T S_2^i(t_{k+1}) + S_2^i(t_{k+1}) + \mathbb{E}\{\tilde{\mathcal{A}}_\delta^i(t_k)\}\\
&+\frac{1}{T_k}\Big\{\mathbb{E}\{(T_k\tilde{\mathcal{A}}_\delta^i(t_k)+I)^T\Lambda_2(T_k\tilde{\mathcal{A}}_\delta^i(t_k)+I)\}\\
&+\mathbb{E}\{(T_k\tilde{\mathcal{A}}_\delta^i(t_k)+I)^T\Lambda_3(T_k\tilde{\mathcal{A}}_\delta^i(t_k)+I)\} + \mathcal{L}_2^i\Big\},\\
S_2^i(t_k) = {}& S_2^i(t_{k+1}) - T_k\delta S_2^i(t_k),\ S_2^i(t_K) = 0,\ \forall i\in\boldsymbol{S}, \qquad (9.23)
\end{aligned}
$$

where

$$\bar{\mathcal{A}}_\delta(t_k) = \mathcal{A}_\delta(t_k) - \sum_{i=1}^{S} \mathcal{B}_\delta^i(t_k)L^i(t_k)$$

$$\tilde{\mathcal{A}}_\delta^i(t_k) = \mathcal{A}_\delta(t_k) - \sum_{j\neq i}^{S} \mathcal{B}_\delta^j(t_k)L^j(t_k)$$

then the Nash strategies (9.9) provides the ϵ-level of NE, i.e.,

$$\hat{J}^i(u^{i*}, u^{-i*}) \leq \hat{J}^i(u^i, u^{-i*}) + \epsilon^i, \tag{9.24}$$

where

$$\begin{aligned}\epsilon^i = &\max_{\forall k\in \boldsymbol{K}} \lambda_{\max}\left\{\mathbb{E}\{\bar{\mathcal{Q}}_\delta^i(t_k)\}\right\}\left(\Theta_1 + 2\sqrt{\beta_1(\Theta_1)}\right)\\ &+ \max_{\forall k\in \boldsymbol{K}} \lambda_{\max}\left\{\mathbb{E}\{\tilde{\mathcal{Q}}_\delta^i(t_k)\}\right\}\left(\Theta_2 + \Xi + 2\sqrt{\beta_2(\Theta_2+\Xi)}\right)\\ &+4\lambda_{\max}\{\mathcal{R}_\delta^{ii}\}K\theta^{i2}.\end{aligned} \tag{9.25}$$

with

$$\begin{aligned}\Theta_1 &= \max_{\forall k\in \boldsymbol{K}} T_k^2 K\lambda_{\min}^{-1}\{\mathcal{L}_1\}\lambda_{\max}\{\Gamma_1\}\gamma + \lambda_{\min}^{-1}\{\mathcal{L}_1\}V_1(t_0)\\ \Theta_2 &= \max_{\forall k\in \boldsymbol{K}} T_k^2 K\lambda_{\min}^{-1}\{\mathcal{L}_2^i\}\lambda_{\max}\{\Gamma_2\}\gamma\\ \Xi_2 &= \max_{\forall k\in \boldsymbol{K}} T_k K\lambda_{\min}^{-1}\{\mathcal{L}_2^i\}\theta^{i2}\kappa_1 + \lambda_{\min}^{-1}\{\mathcal{L}_2^i\}V_2^i(t_0)\end{aligned}$$

Proof: According to the definition of NE, the following inequality sets up

$$J^i(u^{i*}, u^{-i*}) \leq J^i(u^i, u^{-i*}). \tag{9.26}$$

Adding and subtracting $\hat{J}(u^{i*}, u^{-i*})$ and $\hat{J}(u^i, u^{-i*})$ to both sides of inequality (9.26), we have

$$\hat{J}^i(u^{i*}, u^{-i*}) \leq \hat{J}^i(u^i, u^{-i*}) + \epsilon^i, \tag{9.27}$$

where $\epsilon^i = \Delta J_1^i + \Delta J_2^i$, with $\Delta J_1^i = \hat{J}^i(u^{i*}, u^{-i*}) - J^i(u^{i*}, u^{-i*})$ and $\Delta J_2^i = J^i(u^i, u^{-i*}) - \hat{J}^i(u^i, u^{-i*})$. The estimation of ΔJ_1^i and ΔJ_2^i is provided, respectively.

(1) The estimation of ΔJ_1^i

The corresponding cost function using feedback Nash strategies without external disturbance is

$$J^i(u^{i*}, u^{-i*}) = \sum_{k=0}^{K} \bar{z}^T(t_k)\bar{\mathcal{Q}}_\delta^i(t_k)\bar{z}(t_k).$$

Considering the disturbance, one has

$$\hat{J}^i(u^{i*}, u^{-i*}) = \sum_{k=0}^{K} z^T(t_k)\bar{\mathcal{Q}}_\delta^i(t_k)z(t_k).$$

Then, ΔJ_1^i is calculated as

$$\begin{aligned}
&\Delta J_1^i = \hat{J}^i(u^{i*}, u^{-i*}) - J^i(u^{i*}, u^{-i*}) \\
&= \sum_{k=0}^{K} \{z^T(t_k)\mathbb{E}\{\bar{\mathcal{Q}}_\delta^i(t_k)\}z(t_k) - \bar{z}^T(t_k)\mathbb{E}\{\bar{\mathcal{Q}}_\delta^i(t_k)\}\bar{z}(t_k)\} \\
&\leq \max_{\forall k \in \mathbf{K}} \lambda_{\max}\left\{\mathbb{E}\{\bar{\mathcal{Q}}_\delta^i(t_k)\}\right\} \\
&\times \left(\sum_{k=0}^{K} \|\bar{e}(t_k)\|^2 + 2\sqrt{\sum_{k=0}^{K} \|\bar{e}(t_k)\|^2}\sqrt{\sum_{k=0}^{K} \|\bar{z}(t_k)\|^2}\right).
\end{aligned} \tag{9.28}$$

Considering the definition of $\bar{e}(t_k)$, we have

$$\delta\bar{e}(t_k) = \bar{\mathcal{A}}_\delta(t_k)\bar{e}(k) + \mathcal{D}_\delta w(t_k). \tag{9.29}$$

Denoting the energetic function $V_1(t_k)$ as $V_1(t_k) = \bar{e}^T(t_k)S_1(t_k)\bar{e}(t_k)$, one has that

$$\begin{aligned}
\delta V_1(t_k) &= \frac{1}{\mathrm{T}_k}\left(\mathbb{E}\left\{\bar{e}^T(t_{k+1})S_1(t_{k+1})\bar{e}(t_{k+1})\right\} - \bar{e}^T(t_k)S_1(t_k)\bar{e}(t_k)\right) \\
&= \frac{1}{\mathrm{T}_k}\left(\mathbb{E}\left\{((\mathrm{T}_k\bar{\mathcal{A}}_\delta(t_k) + I)\bar{e}(t_k) + \mathrm{T}_k\mathcal{D}_\delta w(t_k))^T S_1(t_{k+1})\right.\right. \\
&\quad \left.\left.((\mathrm{T}_k\bar{\mathcal{A}}_\delta(t_k) + I)\bar{e}(t_k) + \mathrm{T}_k\mathcal{D}_\delta w(t_k))\right\} - \bar{e}^T(t_k)S_1(t_k)\bar{e}(t_k)\right) \\
&\leq \frac{1}{\mathrm{T}_k}\bar{e}^T(t_k)\left(\mathbb{E}\{(\mathrm{T}_k\bar{\mathcal{A}}_\delta(t_k) + I)^T S_1(t_{k+1})(\mathrm{T}_k\bar{\mathcal{A}}_\delta(t_k) + I)\}\right. \\
&\quad \left.+\mathbb{E}\{(\mathrm{T}_k\bar{\mathcal{A}}_\delta(t_k) + I)^T \Lambda_1(\mathrm{T}_k\bar{\mathcal{A}}_\delta(t_k) + I)\} + \mathcal{L}_1 - S_1(t_k)\right)\bar{e}(t_k) \\
&\quad -\frac{1}{\mathrm{T}_k}\bar{e}^T(t_k)\mathcal{L}_1\bar{e}(t_k) + \mathrm{T}_k w^T(t_k)\left(\mathcal{D}_\delta^{\mathrm{T}} S_1(t_{k+1})\Lambda_1^{-1}S_1^{\mathrm{T}}(t_{k+1})\mathcal{D}_\delta\right. \\
&\quad \left.+\mathcal{D}_\delta^{\mathrm{T}} S_1(t_{k+1})\mathcal{D}_\delta\right) w(t_k),
\end{aligned} \tag{9.30}$$

where $\mathcal{L}_1$ is a positive define matrix. Let us select $S_1(t_k)$ such that the Riccati recursion (9.20) holds, and the following inequality is further obtained.

$$\delta V_1(t_k) \leq -\frac{1}{\mathrm{T}_k}\bar{e}^T(t_k)\mathcal{L}_1\bar{e}(t_k) + \mathrm{T}_k w^T(t_k)\Gamma_1 w(t_k). \tag{9.31}$$

Thereby, inequality (9.31) is rewritten as

$$\bar{e}^T(t_k)\mathcal{L}_1\bar{e}(t_k) \leq \mathrm{T}_k^2 w^T(t_k)\Gamma_1 w(t_k) - \mathrm{T}_k\delta V_1(t_k).$$

One has that

$$\lambda_{\min}\{\mathcal{L}_1\}\sum_{k=0}^{K}\|\bar{e}(t_k)\|^2 \le \max_{\forall k\in\mathbf{K}} \mathrm{T}_k^2\lambda_{\max}\{\Gamma_1\}\sum_{k=0}^{K}\|w(t_k)\|^2 \\ -(V_1(t_K)-V_1(t_0)),$$

which leads to

$$\sum_{k=0}^{K}\|\bar{e}(t_k)\|^2 \le \max_{\forall k\in\mathbf{K}} \mathrm{T}_k^2\lambda_{\min}^{-1}\{\mathcal{L}_1\}\lambda_{\max}\{\Gamma_1\}\sum_{k=0}^{K}\|w(t_k)\|^2 \\ +\lambda_{\min}^{-1}\{\mathcal{L}_1\}V_1(t_0).$$

With the constraint of bounded external disturbance, the following inequality sets up

$$\sum_{k=0}^{K}\|\bar{e}(t_k)\|^2 \le \max_{\forall k\in\mathbf{K}} \mathrm{T}_k^2 K\lambda_{\min}^{-1}\{\mathcal{L}_1\}\lambda_{\max}\{\Gamma_1\}\gamma+\lambda_{\min}^{-1}\{\mathcal{L}_1\}V_1(t_0). \quad (9.32)$$

Substituting (9.32) into (9.28), we have

$$\Delta J_1^i \le \max_{\forall k\in\mathbf{K}} \lambda_{\max}\left\{\mathbb{E}\{\bar{\mathcal{Q}}_\delta^i(t_k)\}\right\}\left(\Theta_1+2\sqrt{\beta_1(\Theta_1)}\right). \quad (9.33)$$

where

$$\Theta_1 = \max_{\forall k\in\mathbf{K}} \mathrm{T}_k^2 K\lambda_{\min}^{-1}\{\mathcal{L}_1\}\lambda_{\max}\{\Gamma_1\}\gamma+\lambda_{\min}^{-1}\{\mathcal{L}_1\}V_1(t_0)$$

(2) The estimation of ΔJ_2^i

The cost functions with and without the disturbances are shown as follows:

$$J^i(u^i,u^{-i*}) = \sum_{k=0}^{K}\mathbb{E}\{\tilde{z}^T(t_k)\tilde{\mathcal{Q}}_\delta^i(t_k)\tilde{z}(t_k)+\alpha_{t_k}^i u^{iT}(\tilde{z}(t_k))\mathcal{R}_\delta^{ii}u^i(\tilde{z}(t_k))\}.$$

$$\hat{J}^i(u^i,u^{-i*}) = \sum_{k=0}^{K}\mathbb{E}\{\check{z}^T(t_k)\tilde{\mathcal{Q}}_\delta^i(t_k)\check{z}(t_k)+\alpha_{t_k}^i u^{iT}(\check{z}(t_k))\mathcal{R}_\delta^{ii}u^i(\check{z}(t_k))\}.$$

Thus, ΔJ_2^i is calculated as

$$\begin{aligned}\Delta J_2^i &= J^i(u^i, u^{-i*}) - \hat{J}^i(u^i, u^{-i*}) \\ &= \sum_{k=0}^{K} \mathbb{E}\left\{\tilde{z}^T(t_k)\tilde{\mathcal{Q}}_\delta^i(t_k)\tilde{z}(t_k) + \alpha_{t_k}^i u^{iT}(\tilde{z}(t_k))\mathcal{R}_\delta^{ii}u^i(\tilde{z}(t_k))\right\} \\ &\quad - \sum_{k=0}^{K} \mathbb{E}\left\{\check{z}^T(t_k)\tilde{\mathcal{Q}}_\delta^i(t_k)\check{z}(t_k) + \alpha_{t_k}^i u^{iT}(\check{z}(t_k))\mathcal{R}_\delta^{ii}u^i(\check{z}(t_k))\right\} \\ &\le \max_{\forall k \in \mathbf{K}} \lambda_{\max}\left\{\mathbb{E}\{\tilde{\mathcal{Q}}_\delta^i(t_k)\}\right\}\left(\sum_{k=0}^{K}\|\tilde{e}(t_k)\|^2\right. \\ &\quad \left.+2\sqrt{\sum_{k=0}^{K}\|\tilde{e}(t_k)\|^2}\sqrt{\sum_{k=0}^{K}\|\check{z}(t_k)\|^2}\right) + 4\lambda_{\max}\{\mathcal{R}_\delta^{ii}\}K\theta^{i2}. \end{aligned} \tag{9.34}$$

The error equation of $\tilde{e}(t_k)$ in delta-domain is

$$\delta\tilde{e}(t_k) = \tilde{\mathcal{A}}_\delta^i(t_k)\tilde{e}(t_k) + \mathcal{B}_\delta^i(t_k)u^i(\tilde{z}(t_k)) - \mathcal{B}_\delta^i(t_k)u^i(\check{z}(t_k)) - \mathcal{D}_\delta w(t_k). \tag{9.35}$$

Denoting $V_2^i(t_k) = \tilde{e}^T(t_k)S_2^i(t_k)\tilde{e}(t_k)$, we have the positive matrix $S_2^i(t_k)$ selected to satisfy the Riccati equation (9.22). Let us induce $\tilde{S}_2^i$, which satisfies $\tilde{S}_2^i > S_2^i(t_k)$, $\forall k \in \mathbf{K}$, and the following inequality is obtained according to the Clarkson's inequality.

$$\delta V_2^i(t_k) \le -\frac{1}{\mathtt{T}_k}\tilde{e}^T(t_k)\mathcal{L}_2^i\tilde{e}(t_k) + \theta^{i2}\kappa_1 + \mathtt{T}_k w^T(t_k)\Gamma_2 w(t_k). \tag{9.36}$$

Furthermore, inequality (9.36) can be rewritten as

$$\begin{aligned}\lambda_{\min}\{\mathcal{L}_2^i\}\sum_{k=0}^{K}\|\tilde{e}(t_k)\|^2 &\le \max_{\forall k\in\mathbf{K}} \mathtt{T}_k^2\lambda_{\max}\{\Gamma_2\}\sum_{k=0}^{K}\|w(t_k)\|^2 \\ &\quad + \max_{\forall k\in\mathbf{K}} \mathtt{T}_k K\theta^{i2}\kappa_1 - (V_2^i(t_K) - V_2^i(t_0)),\end{aligned}$$

which further implies that

$$\begin{aligned}\sum_{k=0}^{K}\|\tilde{e}(t_k)\|^2 &\le \max_{\forall k\in\mathbf{K}} \mathtt{T}_k^2\lambda_{\min}^{-1}\{\mathcal{L}_2^i\}\lambda_{\max}\{\Gamma_2\}\sum_{k=0}^{K}\|w(t_k)\|^2 \\ &\quad + \max_{\forall k\in\mathbf{K}} \mathtt{T}_k K\lambda_{\min}^{-1}\{\mathcal{L}_2^i\}\theta^{i2}\kappa_1 + \lambda_{\min}^{-1}\{\mathcal{L}_2^i\}V_2^i(t_0).\end{aligned}$$

With the constraint of the bounded external disturbance, the following inequality sets up

$$\sum_{k=0}^{K} \|\tilde{e}(t_k)\|^2 \leq \max_{\forall k \in \mathbf{K}} \mathtt{T}_k^2 K \lambda_{\min}^{-1}\{\mathcal{L}_2^i\}\lambda_{\max}\{\Gamma_2\}\gamma$$
$$+ \max_{\forall k \in \mathbf{K}} \mathtt{T}_k K \lambda_{\min}^{-1}\{\mathcal{L}_2^i\}\theta^{i2}\kappa_1 + \lambda_{\min}^{-1}\{\mathcal{L}_2^i\}V_2^i(t_0). \tag{9.37}$$

Substituting (9.37) into (9.34), we have

$$\Delta J_2^i = \max_{\forall k \in \mathbf{K}} \lambda_{\max}\left\{\mathbb{E}\{\tilde{\mathcal{Q}}_\delta^i(t_k)\}\right\}\left(\Theta_2 + \Xi_2 + 2\sqrt{\beta_2(\Theta_2 + \Xi_2)}\right)$$
$$+4\lambda_{\max}\{\mathcal{R}_\delta^{ii}\}K\theta^{i2} \tag{9.38}$$

where

$$\Theta_2 = \max_{\forall k \in \mathbf{K}} \mathtt{T}_k^2 K \lambda_{\min}^{-1}\{\mathcal{L}_2^i\}\lambda_{\max}\{\Gamma_2\}\gamma$$
$$\Xi_2 = \max_{\forall k \in \mathbf{K}} \mathtt{T}_k K \lambda_{\min}^{-1}\{\mathcal{L}_2^i\}\theta^{i2}\kappa_1 + \lambda_{\min}^{-1}\{\mathcal{L}_2^i\}V_2^i(t_0)$$

Finally, the estimation of the ϵ-level of $\epsilon-$NE is obtained by $\epsilon^i = \Delta J_1^i + \Delta J_2^i$. The proof is completed. □

Remark 9.7. The Riccati-like recursions (9.20) and (9.22) in the delta-domain are actually a unified form of results in discrete- and continuous-time system. For instance, by using $\bar{\mathcal{A}}_z(k) = \mathtt{T}_k\bar{\mathcal{A}}_\delta(t_k) + I$, the delta-domain result (9.20) can be converted to its discrete-domain analogue as follows:

$$S_1(k) = \mathbb{E}\{\bar{\mathcal{A}}_z^T(k)S_1(k+1)\bar{\mathcal{A}}_z(k)\} + \mathbb{E}\{\bar{\mathcal{A}}_z^T(k)\Lambda_1\bar{\mathcal{A}}_z(k)\} + \mathcal{L}_1.$$

On the other hand, if we choose $\Lambda_1 = \mathtt{T}_k\bar{\Lambda}_1$ and $\mathcal{L}_1 = \mathtt{T}_k\bar{\mathcal{L}}_1$, the following continuous-domain results can be obtained as $\mathtt{T}_k \to 0$

$$\dot{S}_1(t) + \mathbb{E}\{\bar{\mathcal{A}}_s(t)\}^T S_1(t) + S_1(t)\mathbb{E}\{\bar{\mathcal{A}}_s(t)\} + \bar{\Lambda}_1 + \bar{\mathcal{L}}_1 = 0.$$

Furthermore, if we assign $w(t_k) = 0$ and $\bar{e}(t_0) = \tilde{e}(t_0) = 0$, the pure Nash strategies will be obtained with $\epsilon^i = 0$.

Up to now, an upper bound for the ϵ-NE has been provided explicitly. In the following, the upper bound for system (Σ_e) with an average cost is presented.

Corollary 9.8. *Consider the system (Σ_e) with the average cost defined by $\hat{J}_{av}^i(u^i, u^{-i}) = \frac{1}{K}\hat{J}^i(u^i, u^{-i})$. If the Nash strategies u^{i*} for any K yields an ϵ-NE of the delta-domain LQ game, we have*

$$\hat{J}_{av}^i(u^{i*}, u^{-i*}) \leq \hat{J}_{av}^i(u^i, u^{-i*}) + \epsilon_{av}^i, \tag{9.39}$$

where

$$\epsilon_{av}^{i} = \max_{\forall k \in \boldsymbol{K}} \lambda_{\max} \mathbb{E}\left\{\bar{\mathcal{Q}}_{\delta}^{i}(t_k)\right\} \max_{\forall k \in \boldsymbol{K}} T_k^2 \lambda_{\min}^{-1}\{\mathcal{L}_1\} \lambda_{\max}\{\Gamma_1\}\gamma$$
$$+ \max_{\forall k \in \boldsymbol{K}} \lambda_{\max} \mathbb{E}\left\{\tilde{\mathcal{Q}}_{\delta}^{i}(t_k)\right\} \Bigg(\max_{\forall k \in \boldsymbol{K}} T_k^2 \lambda_{\min}^{-1}\{\mathcal{L}_2^i\} \lambda_{\max}\{\Gamma_2\}\gamma$$
$$+ \max_{\forall k \in \boldsymbol{K}} T_k \lambda_{\min}^{-1}\{\mathcal{L}_2^i\} \theta^{i2} \kappa_1 \Bigg)$$
$$+4\lambda_{\max}\{\mathcal{R}_{\delta}^{ii}\} \bar{\delta}^{i2} + O\left(\frac{1}{\sqrt{K}}\right).$$

Proof: The results can be directly obtained from Theorem 9.2 and hence the proof process is omitted. □

Remark 9.9. From Theorem 9.6, it is evident that the estimation of ϵ-level is not dependent on the initial value $z(t_0)$ when $K \to \infty$. This is because costs incurred in early stages have little influence on the total cost. When $K \to \infty$, average cost of every stage in the early stages is reduced to zero for any fixed K_k, i.e.,

$$\lim_{K \to \infty} \frac{1}{\sum_{l=0}^{K} \mathrm{T}_l} \sum_{k=0}^{K_k - 1} \mathrm{T}_k \{ z^T(t_k) \mathcal{Q}_{\delta}^{i} z(t_k) + \alpha_{t_k}^{i} u^{iT} \mathcal{R}_{\delta}^{ii} u^{i}(t_k)$$
$$+ \sum_{j=1, j \neq i}^{S} \epsilon_j u^{jT}(t_k) \mathcal{R}_{\delta}^{ij} u^{j}(t_k) \} = 0.$$

9.4 Numerical Simulation

In this section, the proposed methodology is applied to a two-area Load Frequency Control (LFC) problem [96]. The two-area interconnected power system is presented as in Figure 9.3, and the two-area power system model yields

$$\dot{x}(t) = Ax(t) + Bu(t) + F\Delta P_d(t),$$

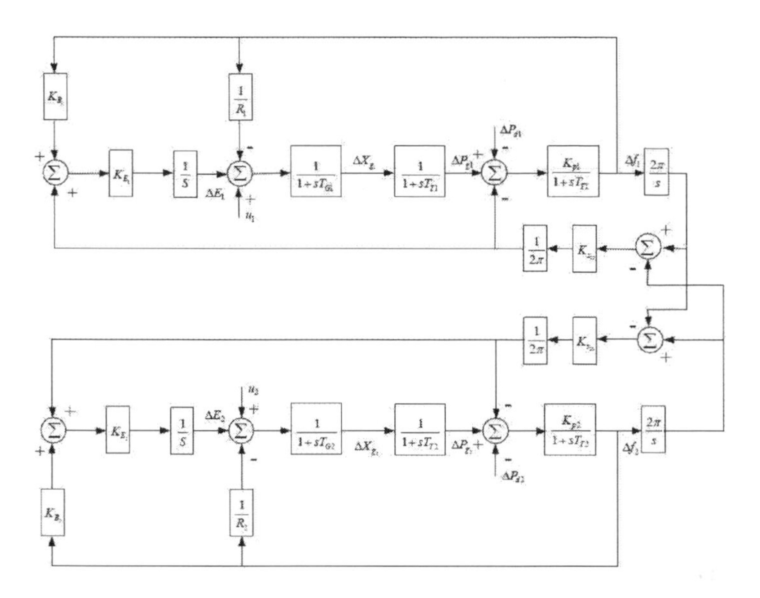

Figure 9.3 Structure of a two-area LFC system.

where

$$x(t) = \left[x^{1T}(t) \; x^{2T}(t) \right]^T, \; u(t) = \left[u^{1T}(t) \; u^{2T}(t) \right]^T,$$
$$\Delta P_d(t) = \left[\Delta P_{d_1}^T(t) \; \Delta P_{d_2}^T(t) \right]^T,$$
$$A = \begin{bmatrix} A^{11} & A^{12} \\ A^{21} & A^{22} \end{bmatrix}, \; B = \mathrm{diag}\left[B^1 \; B^2 \right], \; F = \mathrm{diag}\left[F^1 \; F^2 \right],$$
$$A^{ii} = \begin{bmatrix} -\frac{1}{\mathrm{T}_{p_i}} & \frac{K_{p_i}}{\mathrm{T}_{p_i}} & 0 & 0 & -\frac{K_{p_i}}{2\pi \mathrm{T}_{p_i}} \sum_{j \in \mathbf{S}, j \neq i} K_{s_{ij}} \\ 0 & -\frac{1}{\mathrm{T}_{\mathrm{T}_i}} & \frac{1}{\mathrm{T}_{\mathrm{T}_i}} & 0 & 0 \\ -\frac{1}{R_i \mathrm{T}_{G_i}} & 0 & -\frac{1}{\mathrm{T}_{G_i}} & \frac{1}{\mathrm{T}_{G_i}} & 0 \\ K_{E_i} K_{B_i} & 0 & 0 & 0 & \frac{K_{E_i}}{2\pi} \sum_{j \in \mathbf{S}, j \neq i} K_{s_{ij}} \\ 2\pi & 0 & 0 & 0 & 0 \end{bmatrix},$$
$$B^i = \left[0 \; 0 \; \frac{1}{\mathrm{T}_{G_i}} \; 0 \; 0 \right]^T, \; F^i = \left[\frac{K_{p_i}}{\mathrm{T}_{p_i}} \; 0 \; 0 \; 0 \; 0 \right]^T,$$
$$A^{ij} = \begin{bmatrix} 0\,0\,0\,0 & -\frac{K_{p_i}}{2\pi \mathrm{T}_{p_i}} K_{s_{ij}} \\ 0\,0\,0\,0 & 0 \\ 0\,0\,0\,0 & 0 \\ 0\,0\,0\,0 & \frac{K_{E_i}}{2\pi} K_{s_{ij}} \\ 0\,0\,0\,0 & 0 \end{bmatrix}, \; x^i(t) = \begin{bmatrix} \Delta f_i(t) \\ \Delta P_{g_i}(t) \\ \Delta X_{g_i}(t) \\ \Delta E_i(t) \\ \Delta \delta_i(t) \end{bmatrix}, \; i, j \in \{1, 2\}.$$

Variables $\Delta f_i(t)$, $\Delta P_{g_i}(t)$, $\Delta X_{g_i}(t)$, $\Delta E_i(t)$, and $\Delta \delta_i(t)$ are the changes of frequency, power output, governor valve position, integral control, and rotor angle deviation, respectively. $\Delta P_{d_i}(t)$ is the vector of load disturbance. Parameters $\mathtt{T}_{p_i}$, $\mathtt{T}_{\mathrm{T}_i}$, and $\mathtt{T}_{G_i}$ are time constants of power system, turbine, and governor, respectively. Constants K_{p_i}, K_{E_i}, K_{B_i} are power system gain, integral control gain, and frequency bias factor, and $K_{s_{ij}}$ is the interconnection gain between area i and j $(i \neq j)$. Parameter R_i is the speed regulation coefficient. Some basic parameters of the system are shown in Table 9.1 [96]. Considering the time delays and packets dropout, and assigning random sam-

Table 9.1 Parameters of Two-Area LFC

Parameters	$\mathtt{T}_{p_i}$	K_{p_i}	$\mathtt{T}_{\mathrm{T}_i}$	$\mathtt{T}_{G_i}$	R_i	K_{E_i}	K_{B_i}	$K_{s_{ij}}$
Area 1	20	120	0.3	0.08	2.4	10	0.41	0.55
Area 2	25	112.5	0.33	0.072	2.7	9	0.37	0.65

pling period $\mathtt{T}_k = 0.05s$, we have the following delta operator system model for LFC

$$\delta x(t_k) = A_\delta(t_k)x(t_k) + \sum_{i=1}^{2}\sum_{j=1}^{h}\alpha^i_{t_{k-j}}B^{ij}_\delta(t_k)u^i(t_{k-j}) + F_\delta \Delta P_d(t_k). \quad (9.40)$$

Suppose that the upper bound of time delays is $h = 2$, the packets dropout rate is 0.3, i.e., $\alpha^i = 0.7$, and the sampling interval is $\mathtt{T}_s = 0.05s$. The random time delays τ_k obeys uniform distribution in the interval $[0.04s,\ 0.1s]$. With these parameters, the augmented system model can be established in the delta-domain according to (9.5). The weighting matrices $\mathcal{Q}^i_\delta, i \in \mathbf{S}$ and $\mathcal{Q}^K_\delta$ are taken as identity matrices. We assign that $\mathcal{R}^{ij}_\delta = I$ for all $i \in \mathbf{S},\ j \in \mathbf{S}$, the scalar $c_j = 0.01$ for all $j \in \mathbf{S}$, and $K = 300$. The initial state is selected as $x(t_0) = [0.5\ 0\ 0\ 1\ 1\ 0.5\ 0\ 0\ 1\ 1]^T$. The resultant curves of state using multitasking control are shown in Figure 9.4(a) and Figure 9.4(b). The dash lines and solid lines represent the control resultant curves with and without external disturbance, respectively. It is evident that the resultant curves with disturbance perform worse, which can be described by the ϵ-level of the ϵ-NE.

Next, let us provide an upper bound for ϵ-NE. Parameters are chosen as $\Lambda_j = \mathbf{I}_{14\times14}$, for all $j \in \{1,2,3,4\}$, and $\mathcal{L}_1 = \mathcal{L}^1_2 = \mathcal{L}^2_2 = \mathbf{I}_{14\times14}$. The upper bounds in (9.25) are $\beta_1 = 5.17$, $\beta_2 = 22.5$, $\theta^i = 0.5$, and $\gamma = 0.01$, and the other related values in (9.25) are calculated as $\lambda_{\max}\{\mathbb{E}\{\bar{\mathcal{Q}}^1_\delta(t_k)\}\} = 2.5434$, $\lambda_{\max}\{\mathbb{E}\{\bar{\mathcal{Q}}^2_\delta(t_k)\}\} = 2.3573$, $\lambda_{\max}\{\mathbb{E}\{\tilde{\mathcal{Q}}^1_\delta(t_k)\}\} = 2.0044$, $\lambda_{\max}\{\mathbb{E}\{\tilde{\mathcal{Q}}^2_\delta(t_k)\}\} = 2.0066$. Thus, we have $\epsilon^1 = 155.1087$ and $\epsilon^2 = 131.4880$. The cost function values are $\hat{J}^1 = 1.4085 * 10^3$ and $\hat{J}^2 = 1.4021 * 10^3$. Therefore, the upper limit for a possible deviation of the ϵ-NE are 11.01% and 9.38% of the cost function's values, respectively.

In order to illustrate the factors impacting the ϵ-level, the following experimental results are presented. The relationship of ϵ^i with K and γ is shown in Figure 9.5, where we can see that the upper bound of ϵ^i becomes larger with

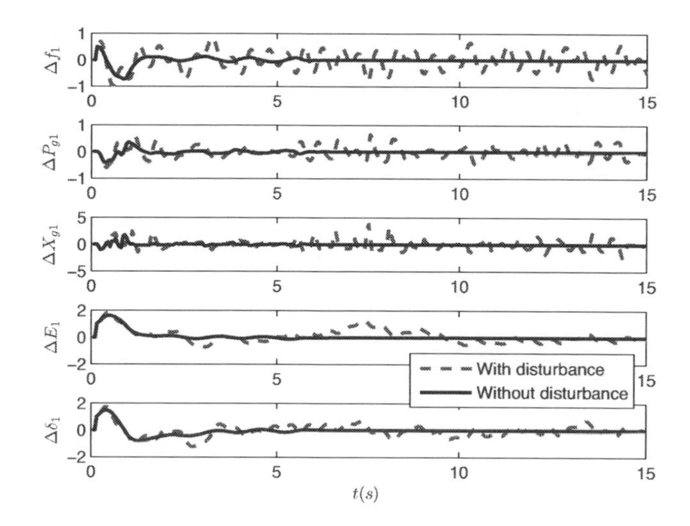

(a) Comparative results in Area 1.

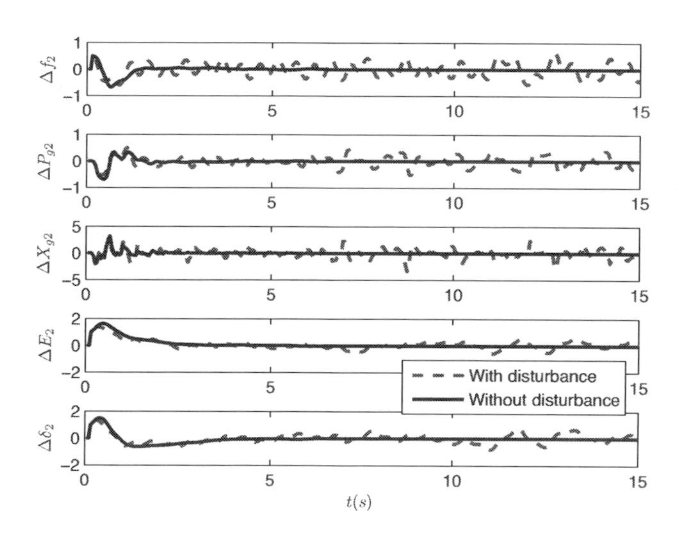

(b) Comparative results in Area 2.

Figure 9.4 The comparison of resultant curves with and without the disturbance: The symbols $\Delta f_i(t)$, $\Delta P_{g_i}(t)$, $\Delta X_{g_i}(t)$, $\Delta E_i(t)$ and $\Delta \delta_i(t)$, $\quad i = 1, 2$ denote the changes of frequency, power output, governor valve position, integral control, and rotor angle deviation, respectively.

the growth of the finite-horizon K and the upper bound of disturbance γ. The relationship of α^i and upper bound of ϵ^i is depicted in Figure 9.6, which shows that larger packets dropout rate leads to larger ϵ^i.

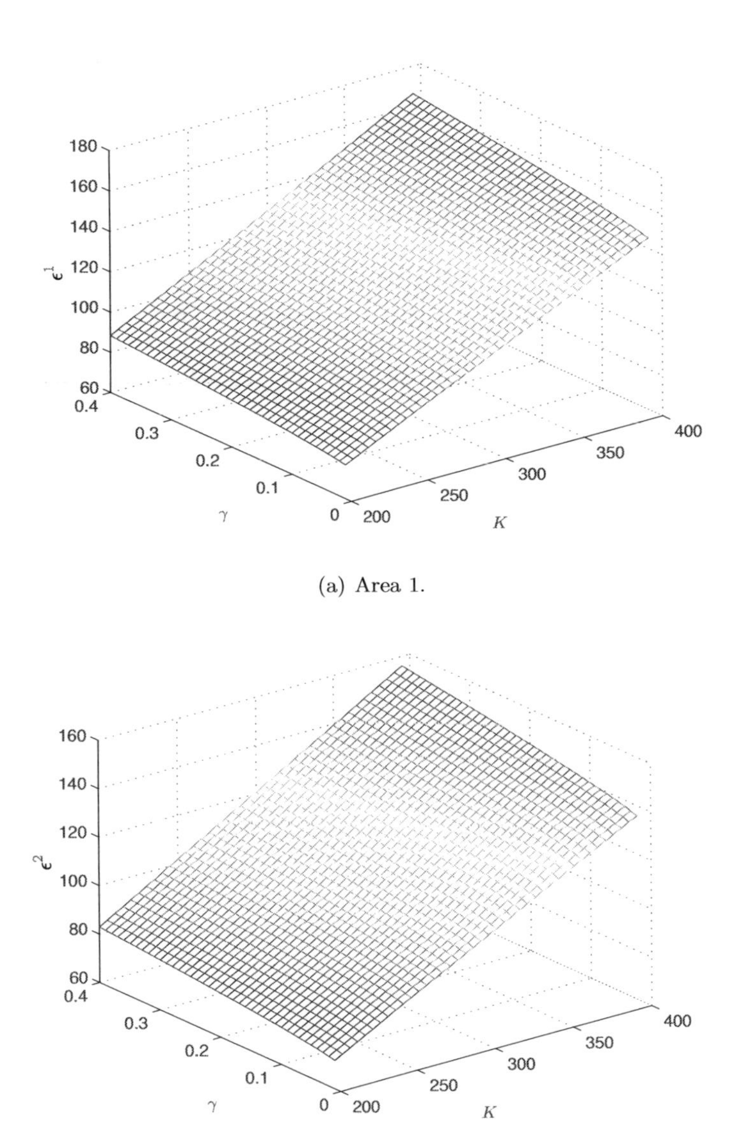

(a) Area 1.

(b) Area 2.

Figure 9.5 The relationship of K, γ, and upper bound of ϵ^i.

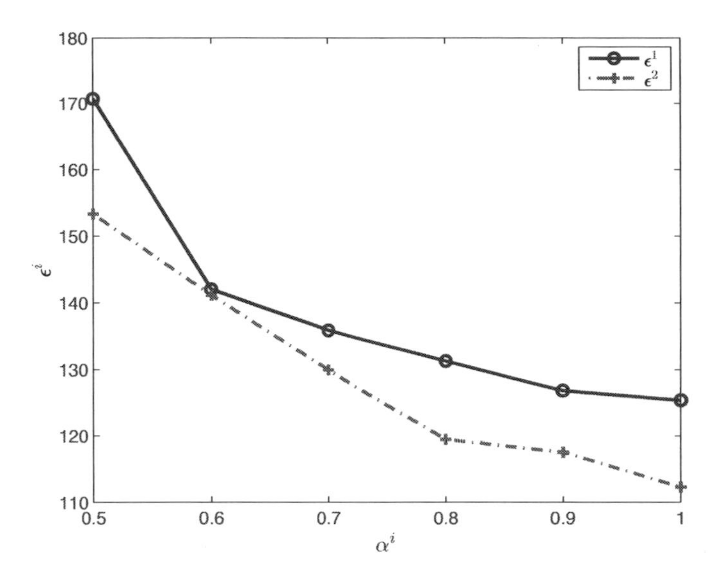

Figure 9.6 The relationship of α^i and upper bound of ϵ^i.

9.5 Conclusion

In this chapter, the multitasking optimal control problem has been addressed for a class of delta-domain NCSs with network-induced constraints and disturbances. In the presence of the disturbances, the so-called ϵ-NE has been proposed to describe the control results. The delta-domain multitasking optimal control strategies have been given such that the ϵ-NE can be achieved, and, for the obtained controllers, an upper bound for the ϵ-level has been provided explicitly. Finally, some simulations have been provided to demonstrate the feasibility of the proposed control scheme.

Part IV
Coupled Design of CPS under Attacks

Chapter 10
Coupled Design of IDS and CPS under DoS Attacks

10.1 Introduction

Recent years have witnessed the migration from proprietary standards for communications towards open international standards for modern critical infrastructures. Arming the control systems with modern information technology leads to low maintenance costs and increased system flexibility [183]. However, the growing trend of integrating the control system with new information technologies exposes it to cyber attacks [41, 94]. Among various types of attacks, the DoS attack, which usually leads to congestion in the communication channel, has been listed as the most financially expensive security incident [21]. It has been reported in [62] and [37] that a number of critical infrastructures were compromised by DoS attacks. To circumvent these newly emerged problems, the concept of resilient control has been recently proposed in [115], which emphasizes controller design in an adversary cyber environment. Recently, a number of literatures have addressed the security of control systems with respect to various types of attacks. The coupled design between the cyber and physical parts of the system leads to a system-level resilient design of the CPS.

In this chapter, a game-in-game structure for the coupled design of RCS is proposed, where the outcome of the inner game affects the solution of the outer game. Meanwhile, The analytical solutions for the cyber layer NE configuration strategy are provided. Finite-horizon H_∞ control is employed for Linear Time Variant (LTV) system model with packets dropout under the Closed-Loop Perfect State (CLPS) information pattern and Closed-Loop 1-Step Delay (CLD) information pattern, respectively.

The rest of the chapter is organized as follows. In Section 10.2, the coupled design problems for RCS are proposed based on the so-called Game-in-Game structure. In Section 10.3, the existence conditions for the unique NE in the security game and physical game are provided, respectively. The coupled design problems are then addressed. In Section 10.4, the proposed methodology is used to the voltage regulation of the Uninterrupted Power System (UPS). The conclusion is drawn in Section 10.5.

10.2 Problem Formulation

10.2.1 Structure of RCS

The hierarchical structure can be illustrated in principle in Figure 10.1. where we can see that, within the so-called Game-in-Game framework, the H_∞ control and cyber security issue are both modeled as games. The cyber layer game is nested in the physical layer game, where the outcome of the inner game affects the solution of the outer game. This coincides with the hierarchical structure in Figure 10.1, where the competition results of the IDSs and cyber attacks will affect the underlying control system. Let us specify the hierarchical game in Figure 10.1 stage by stage. We denote the players of the hierarchical game as $\mathbf{P}_{\mathbf{a}_1}$, $\mathbf{P}_{\mathbf{a}_2}$, $\mathbf{P}_{\mathbf{b}_1}$, and $\mathbf{P}_{\mathbf{b}_2}$, where $\{\mathbf{P}_{\mathbf{a}_i}\}_{i=1,2}$ is the pair of players of $\mathcal{G}_1$ and $\{\mathbf{P}_{\mathbf{b}_i}\}_{i=1,2}$ is the pair of players of $\mathcal{G}_2$.

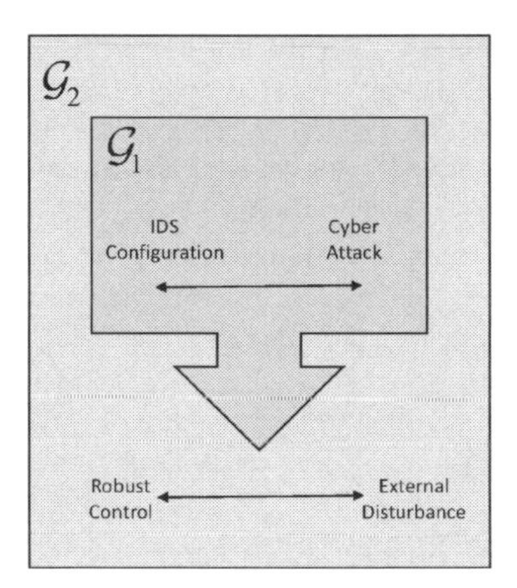

Figure 10.1 An illustration of the so-called Game-in-Game structure: The outcome of the inner game $\mathcal{G}_1$ is propagated to the outer game $\mathcal{G}_2$ and affects its solution. $\mathcal{G}_2$ can be solved by first zooming into $\mathcal{G}_1$ and then zooming out.

10.2.2 The Game

In the first stage game $\mathcal{G}_1$, $\mathbf{P}_{\mathbf{a}_1}$ represents the IDSs and determines the strategies to load defense libraries, and $\mathbf{P}_{\mathbf{a}_2}$ is the cyber attacker and determines the attacking strategies. $\mathcal{A} := \{a_1, a_2, \cdots, a_m\}$ is the set of attacks from which an attacker launches its attack. $\mathcal{L} := \{L_1, L_2, \cdots, L_n\}$ represents the defense library. $\bar{\mathcal{L}}$ denotes the set of all possible subsets of $\mathcal{L}$, with cardinality $|\bar{\mathcal{L}}| = 2^n$. Here we introduce the function $f : \bar{\mathcal{L}} \longrightarrow [0,1]$ (resp. $g : \mathcal{A} \longrightarrow [0,1]$) and use mixed strategies that are denoted as $f(F_p)$ and $g(a_q)$. Accordingly, $f(F_p)$ (resp. $g(a_q)$) can be seen as the probability of the defender (resp. attacker) choosing action F_p (resp. a_q). Thus, we have $\sum_{p=1}^{2^n} f(F_p) = 1$ (resp. $\sum_{q=1}^{m} g(a_q) = 1$). We simplify $f(F_p)$ and $g(a_q)$ as f_p and g_q henceforth. The distribution vectors over all the actions can be augmented as

$$\mathbf{f} := [f_1, \cdots, f_{2^n}]^T, \tag{10.1}$$
$$\mathbf{g} := [g_1, \cdots, g_m]^T. \tag{10.2}$$

In the second stage game $\mathcal{G}_2$, $\mathbf{P_{b_1}}$ represents H_∞ control and $\mathbf{P_{b_2}}$ is the external disturbance. Under DoS attacks, the actuator may fail to receive certain packets from the controller. The LPV model of the control system is

$$\begin{aligned} x_{k+1} &= A_k x_k + B_k u_k^a + D_k w_k + C_k, \\ u_k^a &= \alpha_k u_k, \end{aligned} \tag{10.3}$$

where A_k, B_k, C_k, and D_k, $k \in \mathbf{K} := \{0, 1, \ldots, N-1\}$ are time-varying matrices with appropriate dimensions, N is the finite horizon, $x_k \in \mathbb{R}^{\mathbf{m}}$ represents the state variable, $u_k \in \mathbb{R}^{\mathbf{n_1}}$ is the control command sent out by the controller, and $u_k^a \in \mathbb{R}^{\mathbf{n_1}}$ is the control input received by the actuator after DoS attacks, $w_k \in \mathbb{R}^{\mathbf{n_2}}$ denotes the external disturbance. α_k is a stochastic variable describing the packets dropout at time instant k, which is modeled as an independently and identically distributed (i.i.d.) Bernoulli process with distribution as

$$\mathbb{P}\{\alpha_k = 1\} = \alpha, \qquad \mathbb{P}\{\alpha_k = 0\} = 1 - \alpha, \tag{10.4}$$

for all $k \in \mathbf{K}$. In our model, these successful probabilities are dependent on the competition results of the IDSs and DoS attacks. We denote the Packet Delivery Rate (PDR) as $\alpha(\mathbf{f}, \mathbf{g})$ and it is the IDS's goal to maximize it. Correspondingly, the DoS attacker will try its best to maximize its benefit from attacks, which is denoted as $\beta(\mathbf{f}, \mathbf{g})$.

In this chapter, we only consider the case where every packet is in Transmission Control Protocol (TCP) and is acknowledged. The CLPS and CLD information patterns are considered, respectively. The information set for the CLPS is denoted as $\mathcal{I}_0^1 = \{x_0\}$, $\mathcal{I}_k^1 = \{x_1, x_2, \ldots, x_k, \alpha_0, \alpha_1, \ldots, \alpha_{k-1}\}$, and that of the CLD is $\mathcal{I}_0^2 = \{x_0\}$, $\mathcal{I}_k^2 = \{x_1, x_2, \ldots, x_{k-1}, \alpha_0, \alpha_1, \ldots, \alpha_{k-1}\}$. Note that for $\mathcal{G}_2$, we only consider the case where the information set available to both players is the same, and the asymmetric information case is beyond the scope of this chapter. We introduce $\mathcal{M}$ and $\mathcal{N}$ which are the sets of the control and disturbance strategies. If the strategy μ_k (resp. ν_k) maps the information vector into the action u_k (resp. w_k), i.e., $u_k = \mu_k(\mathcal{I}_k^i)$ and $w_k = \nu_k(\mathcal{I}_k^i)$ for

$i \in \{1,2\}$, it is called admissible. The admissible strategies μ_k and ν_k can be augmented over all the discrete-time indexes

$$\mu = \{\mu_0, \ldots, \mu_k, \ldots \mu_{N-1}\}, \quad \nu = \{\nu_0, \ldots, \nu_k, \ldots \nu_{N-1}\}. \tag{10.5}$$

The associated quadratic performance index is given as

$$J(\mu, \nu, \mathbf{f}, \mathbf{g}) = \mathbb{E}_{\alpha_k}\{\|x_N\|^2_{Q_N} + \sum_{k=0}^{N-1} \{\|x_k\|^2_{Q_k} + \alpha_k \|u_k\|^2_{R_k} - \gamma^2 \|w_k\|^2\}\}, \tag{10.6}$$

where $\|x\|^2_S = x^T S x$, and the scalar γ represents the disturbance attenuation level. We assume that $Q_N > 0$, $Q_k > 0$ and $R_k > 0$ for all $k \in \mathbf{K}$. $\mathbb{E}_{\alpha_k}\{\cdot\}$ is simplified as $\mathbb{E}\{\cdot\}$ and $J(\mu, \nu, \mathbf{f}, \mathbf{g})$ is simplified as $J(\cdot)$ henceforth.

10.2.3 Design Objective

The optimal strategies in both cyber and physical layer take the form of NE strategies. At NE, a player cannot improve its outcome by altering its strategy unilaterally while other players play the NE strategies. Thus, any deviation from the NE by the adversary (physical disturbance or cyber attacks) will increase its own cost and the benefit of the defender (H_∞ control or IDSs). The definition for the NE configuration strategy is shown as follows.

Definition 10.1. The pair $(\mathbf{f}^*, \mathbf{g}^*)$ is the NE configuration strategy for $\mathcal{G}_1$ if

$$\alpha(\mathbf{f}, \mathbf{g}^*) \le \alpha(\mathbf{f}^*, \mathbf{g}^*), \tag{10.7}$$

$$\beta(\mathbf{f}^*, \mathbf{g}) \le \beta(\mathbf{f}^*, \mathbf{g}^*), \tag{10.8}$$

where we use a shorthand notation of α^* (resp. β^*) in place of $\alpha(\mathbf{f}^*, \mathbf{g}^*)$ (resp. $\beta(\mathbf{f}^*, \mathbf{g}^*)$) henceforth.

As mentioned before, games $\mathcal{G}_1$ and $\mathcal{G}_2$ are played sequentially and PDR $\alpha(\mathbf{f}, \mathbf{g})$ will have a definite value once the inner game $\mathcal{G}_1$ is solved. Based on the rationality of the players, the estimated PDR will take the value α^*. The NE control strategy of $\mathcal{G}_2$ is defined as follows.

Definition 10.2. For the pair of fixed cyber strategies $(\mathbf{f}^*, \mathbf{g}^*)$, the pair $(\mu^*, \nu^*) \in \mathcal{M} \times \mathcal{N}$ constitutes the NE control strategy for $\mathcal{G}_2$ if, for all $(\mu, \nu) \in \mathcal{M} \times \mathcal{N}$,

$$J(\mu^*, \nu, \mathbf{f}^*, \mathbf{g}^*) \le J(\mu^*, \nu^*, \mathbf{f}^*, \mathbf{g}^*) \le J(\mu, \nu^*, \mathbf{f}^*, \mathbf{g}^*). \tag{10.9}$$

The performance index $J(\mu^*, \nu^*, \mathbf{f}^*, \mathbf{g}^*)$ is denoted as J^* henceforth.

Remark 10.3. Note that $\mathcal{G}_2$ is actually a zero-sum game. The NE strategies (μ^*, ν^*) are also called Saddle Point Equilibrium (SPE) solution. The quadratic performance index J^* is called SPE value. $J(\mu, \nu^*, \mathbf{f}^*, \mathbf{g}^*)$ is called upper value and $J(\mu^*, \nu, \mathbf{f}^*, \mathbf{g}^*)$ is called lower value.

By [14], The optimal attenuation level γ^* is defined as

$$\gamma^* := \inf\{\gamma : J(\mu, \nu^*, \mathbf{f}^*, \mathbf{g}^*) \text{ and } J(\mu^*, \nu, \mathbf{f}^*, \mathbf{g}^*) \text{ is bounded}\}.$$

Note that $\mathbf{f}^*$ and μ^* are implementable on the defense side. We define the pair of strategies $(\mathbf{f}^*, \mu^*)$ as the joint defense strategies. The coupled design problems that we are going to address in this chapter are formulated as follows.

Problem 10.4. Design the joint defense strategies $(\mathbf{f}^*, \mu^*)$ such that (10.7), (10.8) and (10.9) hold simultaneously.

We define the situation that IDSs fail to protect the control system, if and only if the optimal attenuation level γ^* is above the control system's acceptable attenuation level $\hat{\gamma}$.

Problem 10.5. Find out the sufficient conditions for $(\mathbf{f}, \mathbf{g})$ such that (1) IDSs fail to protect the control system, e.g., $\gamma^* > \hat{\gamma}$. (2) IDSs are able to protect the control system, e.g., $\gamma^* \leq \hat{\gamma}$.

10.3 Main Content

In this section, we provide answers to *Problems* 10.4 and 10.5, respectively. Note that the solutions to *Problems* 10.4 and 10.5 are all dependent on the NE configuration strategy and NE control strategy. We will first provide analytical NE solutions to $\mathcal{G}_1$ and $\mathcal{G}_2$ before the coupled design problems are addressed.

10.3.1 NE Configuration Strategy for $\mathcal{G}_1$

The security game $\mathcal{G}_1$ is modeled as a two-person nonzero-sum finite game. The reward table for the defender (resp. attacker) is $\mathbf{M}_\alpha$ (resp. $\mathbf{M}_\beta$).

$$\mathbf{M}_\alpha := \begin{array}{|c|c|c|c|c|}\hline & a_1 & a_2 & \cdots & a_m \\ \hline F_1 & \alpha_{11} & \alpha_{12} & \cdots & \alpha_{1m} \\ \hline F_2 & \alpha_{21} & \alpha_{22} & \cdots & \alpha_{2m} \\ \hline \vdots & \vdots & \vdots & \ddots & \vdots \\ \hline F_{2^n} & \alpha_{2^n 1} & \alpha_{2^n 2} & \cdots & \alpha_{2^n m} \\ \hline \end{array}$$

$$\mathbf{M}_\beta := \begin{array}{|c|c|c|c|c|}\hline & a_1 & a_2 & \cdots & a_m \\ \hline F_1 & \beta_{11} & \beta_{12} & \cdots & \beta_{1m} \\ \hline F_2 & \beta_{21} & \beta_{22} & \cdots & \beta_{2m} \\ \hline \vdots & \vdots & \vdots & \ddots & \vdots \\ \hline F_{2^n} & \beta_{2^n 1} & \beta_{2^n 2} & \cdots & \beta_{2^n m} \\ \hline \end{array}$$

The entry α_{pq} (resp. β_{pq}) $p = 1, \ldots, 2^n$, $q = 1, \ldots, m$ denotes the outcome of the defender (resp. attacker) corresponding to a particular pair of decisions made by the defender and attacker. Specifically, α_{pq} (resp. β_{pq}) is the gain of the defender (resp. attacker) if actions F_p and a_q are taken. In the sequel, the existence and uniqueness of NE in the security game $\mathcal{G}_1$ is established.

Theorem 10.6. *Let us denote*

$$\boldsymbol{R}_\beta = \begin{bmatrix} \beta_{11} - \beta_{12} & \beta_{21} - \beta_{22} & \cdots & \beta_{2^n 1} - \beta_{2^n 2} \\ \vdots & \vdots & \ddots & \vdots \\ \beta_{11} - \beta_{1m} & \beta_{21} - \beta_{2m} & \cdots & \beta_{2^n 1} - \beta_{2^n m} \\ 1 & 1 & \cdots & 1 \end{bmatrix},$$

$$\boldsymbol{R}_\alpha = \begin{bmatrix} \alpha_{11} - \alpha_{21} & \alpha_{12} - \alpha_{22} & \cdots & \alpha_{1m} - \alpha_{2m} \\ \vdots & \vdots & \ddots & \vdots \\ \alpha_{11} - \alpha_{2^n 1} & \alpha_{21} - \alpha_{2^n 2} & \cdots & \alpha_{2^n 1} - \alpha_{2^n m} \\ 1 & 1 & \cdots & 1 \end{bmatrix},$$

$$\boldsymbol{v}_\beta = \underbrace{\begin{bmatrix} 0 \, 0 \cdots 0 \, 1 \end{bmatrix}^T}_{2^n}, \quad \boldsymbol{v}_\alpha = \underbrace{\begin{bmatrix} 0 \, 0 \cdots 0 \, 1 \end{bmatrix}^T}_{m}.$$

There exists a unique NE in the security game $\mathcal{G}_1$, if the following equations

$$\boldsymbol{R}_\beta \boldsymbol{f}^* = \boldsymbol{v}_\beta, \quad \boldsymbol{R}_\alpha \boldsymbol{g}^* = \boldsymbol{v}_\alpha \tag{10.10}$$

with $2^n = m$ and invertible matrices $\boldsymbol{R}_\beta$, $\boldsymbol{R}_\alpha$ admit solutions with the property $f_p, g_q > 0$, $\forall p \in \{1, \ldots, 2^n\}, q \in \{1, \ldots, m\}$. The NE strategies $\boldsymbol{f}^*$ and $\boldsymbol{g}^*$ are hence given by

$$\boldsymbol{f}^* = \boldsymbol{R}_\beta^{-1} \boldsymbol{v}_\beta, \quad \boldsymbol{g}^* = \boldsymbol{R}_\alpha^{-1} \boldsymbol{v}_\alpha. \tag{10.11}$$

Proof. From [14], $P_{\mathbf{a}1}$'s mixed strategy $\mathbf{f}^*$ is such that $P_{\mathbf{a}2}$'s expected costs for all the actions are equal and vice versa. The following equations can be obtained

$$
\begin{cases}
f_1^*\beta_{11} + f_2^*\beta_{21} + \ldots + f_{2^n}^*\beta_{2^n 1} \\
= f_1^*\beta_{12} + f_2^*\beta_{22} + \ldots + f_{2^n}^*\beta_{2^n 2} \\
= \ldots = f_1^*\beta_{1m} + f_2^*\beta_{2m} + \ldots + f_{2^n}^*\beta_{2^n m}, \\
g_1^*\alpha_{11} + g_2^*\alpha_{12} + \ldots + g_m^*\alpha_{1m} \\
= g_1^*\alpha_{21} + g_2^*\alpha_{22} + \ldots + g_m^*\alpha_{2m} \\
= \ldots = g_1^*\alpha_{2^n 1} + g_2^*\alpha_{2^n 2} + \ldots + g_m^*\alpha_{2^n m}, \\
\quad \sum_{p=1}^{2^n} f_p = 1, \quad \sum_{q=1}^{m} g_q = 1.
\end{cases} \tag{10.12}
$$

We can rewrite (12) as $\mathbf{R}_\beta \mathbf{f}^* = \mathbf{v}_\beta$ and $\mathbf{R}_\alpha \mathbf{g}^* = \mathbf{v}_\alpha$. The proof is completed.

Remark 10.7. Note that (10.11) is a counter-intuitive feature of NE solution in mixed strategies. While computing the mixed strategy, each player only focuses on its opponent's cost function, rather than optimizes its own cost function. Similar results can also be found in [2].

10.3.2 NE Control Strategy for $\mathcal{G}_2$

Once the first stage game $\mathcal{G}_1$ is solved, the NE strategies $(\mathbf{f}^*, \mathbf{g}^*)$ can be obtained from (10.11), which leads to a fixed α^*. Now we are in a position to find the SPE strategies for game $\mathcal{G}_2$. We first provide the following lemma, which will be used in the sequel.

Lemma 10.8. *For $\boldsymbol{A} \in \mathbb{R}^{m_1 \times m_1}$, $\boldsymbol{b} \in \mathbb{R}^{1 \times m_1}$ and $\boldsymbol{c} \in \mathbb{R}^{1 \times 1}$,*

i if $\mathcal{F}(\boldsymbol{x}) = \boldsymbol{x}^T \boldsymbol{A}\boldsymbol{x} + \boldsymbol{b}\boldsymbol{x} + \boldsymbol{c} \leq 0$ for all non-zero vector $\boldsymbol{x} \in \mathbb{R}^{m_1 \times 1}$, we have $\boldsymbol{A} \leq 0$.

ii if $\mathcal{F}(\boldsymbol{x}) = \boldsymbol{x}^T \boldsymbol{A}\boldsymbol{x} + \boldsymbol{b}\boldsymbol{x} + \boldsymbol{c} \geq 0$ for all non-zero vector $\boldsymbol{x} \in \mathbb{R}^{m_1 \times 1}$, we have $\boldsymbol{A} \geq 0$.

Proof. For symmetrical matrix $\mathbf{A}$, we can rewrite $\mathcal{F}(\mathbf{x})$ as $\mathcal{F}(\mathbf{x}) = (\mathbf{x} - 1/2\mathbf{A}^{-1}\mathbf{b})^T \mathbf{A}(\mathbf{x} - 1/2\mathbf{A}^{-1}\mathbf{b}) + \mathbf{c} - 1/4\mathbf{b}^T\mathbf{A}^{-1}\mathbf{b}$. For statement (i), if $\mathbf{A} > 0$, we can find an $\mathbf{x}$ such that $\mathcal{F}(\mathbf{x}) > 0$, since $\mathbf{A}$, $\mathbf{b}$ and $\mathbf{c}$ are given constants. Statement (ii) can be proved in the same way. The proof is completed.

In what follows, we provide the conditions such that the SPE solution (10.9) can be found.

Preliminary Notation for Theorem 10.9 **and Theorem** 10.10. Let Z_k, $k \in \mathbf{K}$ be a non-negative matrix generated recursively from

$$\begin{aligned} Z_k &= Q_k + P_{u_k}^T(\alpha^* R_k + \hat{\alpha}^* B_k^T Z_{k+1} B_k) P_{u_k} - \gamma^2 P_{w_k}^T P_{w_k} \\ &\quad + H_k^T Z_{k+1} H_k, & (10.13) \\ \varsigma_k &= F_k(\varsigma_{k+1} + 2Z_{k+1}\beta_k) + 2(\alpha^* P_{u_k}^T R_k C_{u_k} \\ &\quad -\gamma^2 P_{w_k}^T C_{w_k}), & (10.14) \\ n_k &= n_{k+1} + \varsigma_{k+1}^T \beta_k + \beta_k^T Z_{k+1} \beta_k + \alpha^* C_{u_k}^T R_k C_{u_k} \\ &\quad -\gamma^2 C_{w_k}^T C_{w_k}, & (10.15) \\ Z_N &= Q_N,\ n_N = 0,\ \varsigma_N = 0. \end{aligned}$$

where

$$\hat{\alpha}^* = \alpha^*(1-\alpha^*), \tag{10.16a}$$
$$H_k = A_k - \alpha^* B_k P_{u_k} + D_k P_{w_k}, \tag{10.16b}$$
$$\Sigma_k = R_k + B_k^T Z_{k+1} B_k, \tag{10.16c}$$
$$\beta_k = C_k + D_k C_{w_k} - \alpha^* B_k C_{u_k}, \tag{10.16d}$$
$$F_k = A_k^T - \alpha^* P_{u_k}^T B_k^T + P_{w_k}^T D_k^T, \tag{10.16c}$$
$$P_{u_k} = \Lambda_{u_k}^{-1} \mathrm{T}_{u_k} Z_{k+1} A_k, \tag{10.16f}$$
$$P_{w_k} = \Lambda_{w_k}^{-1} \mathrm{T}_{w_k} Z_{k+1} A_k, \tag{10.16g}$$
$$C_{u_k} = \frac{1}{2}\Lambda_{u_k}^{-1} \mathrm{T}_{u_k}(\varsigma_{k+1} + 2Z_{k+1} C_k), \tag{10.16h}$$
$$C_{w_k} = \frac{1}{2}\Lambda_{w_k}^{-1} \mathrm{T}_{w_k}(\varsigma_{k+1} + 2Z_{k+1} C_k), \tag{10.16i}$$
$$\Lambda_{u_k} = R_k + \mathrm{T}_{u(\alpha)_k} Z_{k+1} B_k, \tag{10.16j}$$
$$\Lambda_{w_k} = \gamma^2 I - \mathrm{T}_{w_k} Z_{k+1} D_k, \tag{10.16k}$$
$$\mathrm{T}_{u(\alpha)_k} = B_k^T(I + \alpha^* Z_{k+1} D_k \Pi_k^{-1} D_k^T), \tag{10.16l}$$
$$\mathrm{T}_{u_k} = B_k^T(I + Z_{k+1} D_k \Pi_k^{-1} D_k^T), \tag{10.16m}$$
$$\mathrm{T}_{w_k} = D_k^T(I - \alpha^* Z_{k+1} B_k \Sigma_k^{-1} B_k^T). \tag{10.16n}$$

Theorem 10.9. *Consider the two-person zero-sum affine-quadratic dynamic game $\mathcal{G}_2$ with information pattern $\mathcal{I}_k^1$, and with fixed N, $\gamma > 0$ and $\alpha^* \in (0, 1]$. Then,*

i There exists a unique feedback SPE solution if and only if

$$\Pi_k = \gamma^2 I - D_k^T Z_{k+1} D_k > 0,\ k \in \mathbf{K}, \tag{10.17}$$

where the nonincreasing sequence $Z_k \geq 0$ is generated by the Riccati recursions 10.13.

ii Under condition (10.17), the matrices Λ_{u_k} and Λ_{w_k} are invertible, and the unique feedback saddle-point strategies (μ^*, ν^*) are given by

$$u_k = \mu^*(\mathcal{I}_k^1) = -P_{u_k}x_k - C_{u_k},$$
$$w_k = \nu^*(\mathcal{I}_k^1) = P_{w_k}x_k + C_{w_k}. \tag{10.18}$$

iii The saddle point value is given by

$$J^* = V_0(x_0) = x_0^T Z_0 x_0 + \varsigma_0^T x_0 + n_0. \tag{10.19}$$

Proof. To prove (i), (ii), and (iii), we need to employ dynamic programming in [14]. The cost-to-go function $V_k(x_k)$ turns out to be affine quadratic and the SPE strategies are affine in state. These facts can be verified by induction method. Starting with $k = N$, the following can be written for $k = N - 1$,

$$V_N(x_N) = x_N^T Q_N x_N, \tag{10.20}$$
$$\begin{aligned} V_{N-1}(x_{N-1}) = \min_{u_{N-1}} \max_{w_{N-1}} \mathbb{E}\{&\|x_{N-1}\|^2_{Q_{N-1}} + \alpha_{N-1}\|u_{N-1}\|^2_{R_{N-1}} \\ &-\gamma^2\|w_{N-1}\|^2 + (A_{N-1}x_{N-1} + \alpha_{N-1}B_{N-1}u_{N-1} \\ &+D_{N-1}w_{N-1} + C_{N-1})^T Q_N (A_{N-1}x_{N-1} \\ &+\alpha_{N-1}B_{N-1}u_{N-1} + D_{N-1}w_{N-1} + C_{N-1})\}. \end{aligned} \tag{10.21}$$

Note that the functional is strictly convex in u_{N-1} and strictly concave in w_{N-1}, since we have $\Pi_{N-1} = \gamma^2 I - D_{N-1}^T Q_N D_{N-1} > 0$ and $\Sigma_{N-1} = R_{N-1} + B_{N-1}^T Q_N B_{N-1} > 0$. The first order necessary and sufficient conditions for convexity and concavity yield the following equations

$$\begin{aligned} &(R_{N-1} + B_{N-1}^T Q_N B_{N-1})\mu^*(x_{N-1}) \\ &= -B_{N-1}^T Q_N (A_{N-1}x_{N-1} + D_{N-1}\nu^*(x_{N-1}) + C_{N-1}), \end{aligned} \tag{10.22}$$
$$\begin{aligned} &(\gamma^2 I - D_{N-1}^T Q_N D_{N-1})\nu^*(x_{N-1}) \\ &= D_{N-1}^T Q_N (A_{N-1}x_{N-1} + \alpha^* B_{N-1}\mu^*(x_{N-1}) + C_{N-1}), \end{aligned} \tag{10.23}$$

which suggest that the SPE strategies $(\mu_{N-1}^*, \nu_{N-1}^*)$ have to be affine in x_{N-1}. Thus, we denote $u_{N-1} = -P_{u_{N-1}}x_{N-1} - C_{u_{N-1}}$ and $w_{N-1} = P_{w_{N-1}}x_{N-1} + C_{w_{N-1}}$, and by substituting them into (10.22) and (10.23), we arrive at (10.16f), (10.16g), (10.16h) and (10.16i) at stage $N-1$. Also, the substitution of (10.22) and (10.23) back into (10.21) leads to $x_{N-1}^T Z_N x_{N-1} + \varsigma_{N-1}^T x_{N-1} + n_{N-1}$, which implies that the cost-to-go function has an affine quadratic structure with $k = N - 1$. Since $V_{N-1}(x_{N-1})$ is an affine-quadratic function, we may proceed similarly and obtain the SPE solution $(\mu_{N-2}^*, \nu_{N-2}^*)$ and the value function $V_{N-2}(x_{N-2})$. This procedure can be carried out recursively for all $k \leq N - 2$ and we can obtain (ii) and finally the saddle point value (iii) under condition (i).

For statement (i), we still need to prove the necessity part and nonnegative definiteness of Z_k. Let us suppose that, for some $\bar{k} \in \mathbf{K}$, (10.17) has a negative value, then the maximizer has the choice such that the corresponding static game becomes unbounded. The non-negative definiteness of the nonincreasing sequence Z_k, $k \in \mathbf{K}$ in (i) can be proven by the fact that, for $k = N-1$, we have

$$\begin{aligned}
V_{N-1}(x_{N-1}) &= x_{N-1}^T Z_{N-1} x_{N-1} + \varsigma_{N-1}^T x_{N-1} + n_{N-1} \\
&= \max_{w_{N-1}} \min_{u_{N-1}} \mathbb{E}\{\|x_{N-1}\|_{Q_{N-1}}^2 + \alpha_{N-1}\|u_{N-1}\|_{R_{N-1}}^2 \\
&\quad -\gamma^2 \|w_{N-1}\|^2 + x_N^T Z_N x_N\} \\
&= \max_{w_{N-1}} \mathbb{E}\{\|x_{N-1}\|_{Q_{N-1}}^2 + \alpha_{N-1}\left\|P_{u_{N-1}} x_{N-1} + C_{N-1}\right\|_{R_{N-1}}^2 \\
&\quad -\gamma^2 \|w_{N-1}\|^2 + x_N^T Z_N x_N\} \\
&\geq \mathbb{E}\{\|x_{N-1}\|_{Q_{N-1}}^2 + \alpha_{N-1}\left\|P_{u_{N-1}} x_{N-1} + C_{N-1}\right\|_{R_{N-1}}^2 \\
&\quad + x_N^T Z_N x_N \,|w_{N-1} = 0\} > 0.
\end{aligned} \tag{10.24}$$

Using Lemma 10.8, we have $Z_{N-1} \geq 0$. Furthermore, it is easy to obtain $Z_{N-1} - Z_N \geq 0$ from (10.24). The same procedure can be repeated for all $k \leq N-2$.

The uniqueness of saddle point strategies (μ_k^*, ν_k^*) in (ii) is guaranteed by the positive definiteness of Π_k and Σ_k. Substituting (10.16l) into (10.16j), we have

$$\begin{aligned}
\Lambda_{u_k} &= R_k + \mathrm{T}_{u(\alpha)_k} Z_{k+1} B_k \\
&= \Sigma_k + \alpha^* B_k^T Z_{k+1} D_k \Pi_k^{-1} D_k^T Z_{k+1} B_k > 0.
\end{aligned} \tag{10.25}$$

In the same vein, the following is obtained

$$\begin{aligned}
\Lambda_{w_k} &= \gamma^2 I - \mathrm{T}_{w_k} Z_{k+1} D_k \\
&= \Pi_k + \alpha^* D_k^T Z_{k+1} B_k \Sigma_k^{-1} B_k^T Z_{k+1} D_k > 0.
\end{aligned} \tag{10.26}$$

Hence, Λ_{u_k} and Λ_{w_k} are invertible for all $k \in \mathbf{K}$, and, as a result, (10.16f) and (10.16h) (resp. (10.16g) and (10.16i)) lead to the unique P_{u_k} and C_{u_k} (resp. P_{w_k} and C_{w_k}). The proof is completed.

We also consider the CLD information pattern, where the information available to both players is one time unit delayed. It is very practical to model the interactions between the controller and disturbance as a zero-sum game under CLD information pattern since there always exists deterministic propagation delay in the communication channel.

Theorem 10.10. *Consider the two-person zero-sum affine-quadratic dynamic game $\mathcal{G}_2$ with information pattern $\mathcal{I}_k^2$, and with fixed N, $\gamma > 0$ and $\alpha^* \in (0,1]$. Then,*

i There exists a unique noise insensitive SPE solution if and only if

$$\Upsilon_k = \gamma^2 I - \bar{D}_k^T \tilde{Z}_{k+1} \bar{D}_k > 0, \;\; \forall k \in \boldsymbol{K}, \tag{10.27}$$

where

$$\tilde{S}_{k+1} = \tilde{Z}_{k+1}(I + \bar{D}_k(\gamma^2 I - \bar{D}_k^T \tilde{Z}_{k+1}\bar{D}_k)^{-1}\bar{D}_k^T \tilde{Z}_{k+1}), \tag{10.28}$$

$$\begin{aligned}
\tilde{Z}_k = & \begin{bmatrix} Q_k & 0 \\ 0 & \alpha^* P_{u_k}^T R_k P_{u_k} \end{bmatrix} \\
& + \begin{bmatrix} A_k^T & A_k^T \\ 0 & P_{w_k}^T D_k^T \end{bmatrix} \tilde{S}_{k+1} \begin{bmatrix} A_k & 0 \\ A_k & D_k P_{w_k} \end{bmatrix} \\
& -\alpha^* \begin{bmatrix} A_k^T & A_k^T \\ 0 & P_{w_k}^T D_k^T \end{bmatrix} \tilde{S}_{k+1} \begin{bmatrix} 0 & B_k P_{u_k} \\ 0 & B_k P_{u_k} \end{bmatrix} \\
& +\alpha^* \begin{bmatrix} 0 & 0 \\ P_{u_k}^T B_k^T & P_{u_k}^T B_k^T \end{bmatrix} \tilde{S}_{k+1} \begin{bmatrix} 0 & B_k P_{u_k} \\ 0 & B_k P_{u_k} \end{bmatrix} \\
& -\alpha^* \begin{bmatrix} 0 & 0 \\ P_{u_k}^T B_k^T & P_{u_k}^T B_k^T \end{bmatrix} \tilde{S}_{k+1} \begin{bmatrix} A_k & 0 \\ A_k & D_k P_{w_k} \end{bmatrix},
\end{aligned} \tag{10.29}$$

$$\bar{D}_k = \begin{bmatrix} D_k & 0 \end{bmatrix}^T, \quad \tilde{Z}_N = diag\{Q_N, 0\}.$$

ii *Under condition (10.27), the unique feedback saddle-point strategies* (μ^*, ν^*) *are given by*

$$\begin{aligned}
u_k &= \mu^*(\mathcal{I}_k^2) = -P_{u_k}\xi_k - C_{u_k}, \\
w_k &= \nu^*(\mathcal{I}_k^2) = P_{w_k}\xi_k + C_{w_k},
\end{aligned} \tag{10.30}$$

where ξ_k *is generated by the compensator*

$$\begin{aligned}
\xi_{k+1} = & A_k x_k + \alpha_k B_k(-P_{u_k}\xi_k - C_{u_k}) \\
& + D_k(P_{w_k}\xi_k + C_{w_k}) + C_k,
\end{aligned} \tag{10.31}$$

and the saddle point value is again given by (10.19).

Proof. It follows from the interchangeability of SPE that $(\mu^*(\mathcal{I}_k^2), \nu^*(\mathcal{I}_k^2))$ is a particular representation of the SPE strategies (10.18), and the uniqueness and noise insensitivity properties follow from the dynamic programming argument. For the existence of the SPE, we have the following augmented dynamical system by adding compensator (10.31) to the original system (10.3).

$$\zeta_{k+1} = \bar{A}_k \zeta_k + \bar{D}_k w_k + \bar{C}_k \tag{10.32}$$

$$\bar{A}_k = \begin{bmatrix} A_k & -\alpha_k B_k P_{u_k} \\ A_k & -\alpha_k B_k P_{u_k} + D_k P_{w_k} \end{bmatrix}, \quad \bar{D}_k = \begin{bmatrix} D_k \\ 0 \end{bmatrix},$$

$$\bar{C}_k = \begin{bmatrix} C_k - \alpha_k B_k C_{u_k} \\ C_k - \alpha_k B_k C_{u_k} + D_k C_{w_k} \end{bmatrix}, \quad \zeta_k = \begin{bmatrix} x_k & \xi_k \end{bmatrix}^T.$$

Paralleling the proof of Theorem (10.9), we arrive at

$$
\begin{aligned}
V_k(\zeta_k) &= \min_{u_k} \max_{w_k} \mathbb{E}\{\zeta_k^T \tilde{Z}_k \zeta_k + \theta_k^T \zeta_k + \vartheta_k.\} \\
&= \min_{u_k} \max_{w_k} \mathbb{E}\{x_k^T Q_k x_k + \alpha_k \xi_k^T P_{u_k}^T R_k P_{u_k} \xi_k + \alpha_k C_{u_k}^T R_k P_{u_k} \xi_k \\
&\quad + \alpha_k \xi_k^T P_{u_k}^T R_k C_{u_k} + \alpha_k C_{u_k}^T R_k C_{u_k} - \gamma^2 w_k^T w_k \\
&\quad + (\bar{A}_k \zeta_k + \bar{D}_k w_k + \bar{C}_k)^T \tilde{Z}_{k+1} (\bar{A}_k \zeta_k + \bar{D}_k w_k + \bar{C}_k) \\
&\quad + \theta_{k+1}^T (\bar{A}_k \zeta_k + \bar{D}_k w_k + \bar{C}_k) + \vartheta_{k+1}\}.
\end{aligned} \tag{10.33}
$$

If condition (10.27) holds in statement (i), the functional $V_k(\zeta_k)$ will be strictly concave in w_k at every stage k. The necessity part in statement (i) follows from the fact that, if (10.27) is negative for some $\bar{k} \in \mathbf{K}$, the maximizer has the choice such that the game becomes unbounded. The first order necessary and sufficient condition for strict concavity yields

$$
w_k = (\gamma^2 I - \bar{D}_k^T \tilde{Z}_{k+1} \bar{D}_k)^{-1} \bar{D}_k^T (\tilde{Z}_{k+1} \bar{A}_k \zeta_k + \tilde{Z}_{k+1} \bar{C}_k + \theta_{k+1}). \tag{10.34}
$$

By substituting w_k back into $V_k(\zeta_k)$, and by requiring it to be satisfied for all x_k $(k \in \mathbf{K})$, we arrive at (10.30). The proof is completed.

For the H_∞ control, the minimal disturbance attenuation level γ^* is always desired as long as the SPE solution exists. Thus, we introduce the optimal disturbance attenuation level $\gamma_c(\tilde{\alpha}, N) = \inf\{\gamma : \gamma^2 I - D_k^T Z_{k+1}(\tilde{\alpha}, \gamma) D_k > 0,\ k \in \mathbf{K}\}$ for CLPS information pattern and $\tilde{\gamma}_c(\tilde{\alpha}, N) = \inf\{\gamma : \gamma^2 I - \bar{D}_k^T \tilde{Z}_{k+1}(\tilde{\alpha}, \gamma) \bar{D}_k > 0,\ k \in \mathbf{K}\}$ for CLD information pattern.

Remark 10.11. It has been proven in [14] that the feedback SPE solution under CLPS information pattern requires the least stringent existence condition. Thus, condition (10.27) implies condition (10.17) and we have $\gamma_c(\tilde{\alpha}, N) < \tilde{\gamma}_c(\tilde{\alpha}, N)$. This suggests that the propagation delay in the communication channel also has a negative effect on the disturbance attenuation level, which will be shown later via simulation.

Remark 10.12. According to [13], the SPE value $J(\mu^*, \nu^*, \mathbf{f}^*, \mathbf{g}^*)$ is also the upper bound of $J(\mu^*, w_{[1,N]}, \mathbf{f}^*, \mathbf{g}^*)$, where $w_{[1,N]} = \{w_1, \ldots, w_N\}$ is a l_2 sequence.

10.3.3 Coupled Design Problem

For *Problem 10.4*, the joint defense strategies $(\mathbf{f}^*, \mu^*)$ are obtained by applying Theorem 10.6 and Theorem 10.9 (resp. Theorem 10.10) sequentially under CLPS (resp. CLD) information pattern. This coincides with the practical situation since the DoS attacker has to sneak through the cyber defense facility first (where $\mathcal{G}_1$ is solved) before it can actually compromise the physical control system (where $\mathcal{G}_2$ is solved). Specifically, the joint defense strategies $(\mathbf{f}^*, \mu^*)$ can be found by the following algorithm.

Algorithm 1. Find the joint defense strategies $(\mathbf{f}^*, \mu^*)$.
Step 1: Apply Theorem 10.6 to get the NE configuration strategy $(\mathbf{f}^*, \mathbf{g}^*)$.
Step 2: Calculate the estimated PDR α^* by $\alpha^* = \mathbb{E}_{\mathbf{f}^*,\mathbf{g}^*}\{\mathbf{M}_\alpha\} = \mathbf{f}^{*T}\mathbf{M}_\alpha \mathbf{g}^*$.
Step 3: Apply the estimated PDR α^* of Theorem 10.9 (resp. Theorem 10.10) to get the NE control strategy (μ^*, ν^*) with the CLPS (resp. CLD) information pattern.

In order to solve *Problem 10.5*, we introduce the following instrumental lemma that shows the optimal attenuation level $\gamma_c(\alpha, N)$ is monotonously decreasing with respect to α for given N.

Lemma 10.13. *For given N and $\alpha_1 \leq \alpha_2$, we have $\gamma_c(\alpha_2, N) \leq \gamma_c(\alpha_1, N)$.*

Proof. It follows from the Lemma 10.13 in [100] that the cost-to-go function $V_k(x_k)$ is a non-increasing function of α. That is to say, if $\alpha_1 \leq \alpha_2$, we have

$$\begin{aligned} & x_k^T(Z_k(\alpha_2, \tilde{\gamma}) - Z_k(\alpha_1, \tilde{\gamma}))x_k + (\varsigma_k^T(\alpha_2, \tilde{\gamma}) - \varsigma_k^T(\alpha_2, \tilde{\gamma}))x_k \\ & + (n_k(\alpha_2, \tilde{\gamma}) - n_k(\alpha_2, \tilde{\gamma})) \leq 0, \end{aligned} \tag{10.35}$$

for all x_k. By using Lemma 10.8, we have $Z_k(\alpha_2, \tilde{\gamma}) \leq Z_k(\alpha_1, \tilde{\gamma})$. Thus, the constraints in $\gamma_c(\tilde{\alpha}, N) = \inf\{\gamma : \gamma^2 I - D_k^T Z_{k+1}(\tilde{\alpha}, \gamma) D_k > 0,\ k \in \mathbf{K}\}$ is less restrictive for α_2 than α_1, which leads to the conclusion of $\gamma_c(\alpha_2, N) \leq \gamma_c(\alpha_1, N)$. The proof is completed.

To this end, we are ready to present the following proposition which provides an answer to *Problem 10.5*. Let us start with providing the sufficient conditions such that IDSs fail to protect the control system.

Proposition 10.14. *If $\boldsymbol{P}_{\mathbf{a}_2}$ adopts the attack strategy $\boldsymbol{g}^*$ and $\gamma_c(\alpha(\boldsymbol{f}^*, \boldsymbol{g}^*), N) > \hat{\gamma}$, we have $\gamma_c(\alpha(\boldsymbol{f}, \boldsymbol{g}^*), N) > \hat{\gamma}$ for all $0 \leq \boldsymbol{f} \leq 1$.*

Proof. Let us first introduce the set $\Omega(\mathbf{f}) := \{\alpha(\mathbf{f}, \mathbf{g}^*) : 0 < \alpha(\mathbf{f}, \mathbf{g}^*) \leq 1\}$. This set can be seen as the collection of all the possible results of $\mathcal{G}_1$ if the attacker is rational. It follows from (10.8) that $\max_{\mathbf{f}\in[0,1]} \Omega(\mathbf{f}) = \alpha^*$. According to Lemma 10.13, we have $\gamma_c(\alpha(\mathbf{f}, \mathbf{g}^*), N) \geq \gamma_c(\alpha(\mathbf{f}^*, \mathbf{g}^*), N)$. The proof is completed.

The sufficient conditions that IDSs are able to protect the control system are shown as follows.

Proposition 10.15. *If $\mathcal{G}_1$ is a zero-sum game, e.g., $\boldsymbol{M}_\alpha = -\boldsymbol{M}_\beta$, if $\gamma_c(\alpha(\boldsymbol{f}^*, \boldsymbol{g}^*), N) \leq \hat{\gamma}$, we have $\gamma_c(\alpha(\boldsymbol{f}^*, \boldsymbol{g}), N) \leq \hat{\gamma}$ for all $0 \leq \boldsymbol{g} \leq 1$.*

Proof. If $\mathcal{G}_1$ is a zero-sum game, the inequality 10.8 can be rewritten as $\alpha(\mathbf{f}, \mathbf{g}^*) \leq \alpha(\mathbf{f}^*, \mathbf{g}^*) \leq \alpha(\mathbf{f}^*, \mathbf{g})$. Let us first introduce the set $\Phi(\mathbf{g}) := \{\alpha(\mathbf{f}^*, \mathbf{g}) : 0 < \alpha(\mathbf{f}^*, \mathbf{g}) \leq 1\}$ and we have $\min_{\mathbf{g}\in[0,1]} \Phi(\mathbf{g}) = \alpha^*$. According to Lemma 10.13, we have $\gamma_c(\alpha(\mathbf{f}^*, \mathbf{g}), N) \leq \gamma_c(\alpha(\mathbf{f}^*, \mathbf{g}^*), N)$. The proof is completed.

Remark 10.16. Proposition 10.14 suggests that, if $\gamma_c(\alpha^*, N) > \hat{\gamma}$ and the attacker adopts $\mathbf{g}^*$ as the attacking strategy, the IDSs will fail to protect the control system no matter what defense strategies are used. Proposition 10.15 suggests that, if $\mathcal{G}_1$ is a zero-sum game and $\gamma_c(\alpha^*, N) \leq \hat{\gamma}$, the IDSs are able to protect the control system by adopting the configuration strategy $\mathbf{f}^*$ regardless of the attack strategy $\mathbf{g}$.

Remark 10.17. From [100] and [64], we can see that the PDR is always a prescribed value and does not need to be designed. Therefore, the classical H_∞ control is only dependent on the strategies of the physical layer players. In this chapter, we exploit the "causes" of the packets dropout under the DoS attack. The strategies of the cyber layer players are considered since they also exert influence on the disturbance attenuation level of the H_∞ control. The proposed method is more comprehensive and practical for CPSs.

10.4 Numerical Simulation

In this section, the resilient control problem is investigated associated with UPS, which is used to provide uninterrupted, high quality, and reliable power for critical systems, such as life supporting system, emergency system, and data storage system [140]. In practice, the UPS can be subject to load variation, disturbance signals, bounded control signals, and DoS attacks. Thus, resilience is vital for UPS such that the voltage is maintained at a desired setting under DoS attacks in the cyber layer and load disturbances in the physical layer.

10.4.1 Dynamic Model

Table 10.1 Parameters of UPS

Symbol	Meaning
$L_f = 1mH$	Inductance
$R_{Lf} = 15m\Omega$	Resistance
$C_f = 300\mu F$	Capacitor
$Y_o = 1S$	Admissible load admittance
$\mathrm{T}_s = 0.0008s$	Sampling interval

The dynamic model of the UPS is from [140], whose parameters are shown in Table 10.1. The matrices and vectors are given by

$$A = \begin{bmatrix} 1 - \mathtt{T}_s R_{L_f}/L_f & \mathtt{T}_s/L_f & 0 \\ -\mathtt{T}_s/C_f & 1 - \mathtt{T}_s Y_o/C_f & 0 \\ 0 & \mathtt{T}_s & 1 \end{bmatrix}, \quad B = \begin{bmatrix} \mathtt{T}_s/L_f \\ 0 \\ 0 \end{bmatrix},$$

$$D = \begin{bmatrix} 0 & \mathtt{T}_s/C_f & 0 \\ -\mathtt{T}_s/L_f & \mathtt{T}_s Y_o/C_f & 0 \end{bmatrix}^T, C = \begin{bmatrix} 0\,0\,0 \end{bmatrix}^T,$$

$$x_k = \begin{bmatrix} x_{1k}\; x_{2k}\; x_{3k} \end{bmatrix}^T, \quad w_k = \begin{bmatrix} i_k^o\; v_k^{rms} \end{bmatrix}^T, \tag{10.36}$$

where $\begin{bmatrix} x_{1k}\; x_{2k} \end{bmatrix}^T = \begin{bmatrix} i_k^{L_f}\; v_k^{rms} - v_k^{orms} \end{bmatrix}^T$, $x_{3k+1} = \mathtt{T}_s x_{2k} + x_{3k}$, v_k^{orms} is the Root Mean Square (RMS) value of capacitor voltage, v_k^{rms} represents the RMS value of the sinusoidal reference, $i_k^{L_f}$ is the inductor current, i_k^o denotes the output disturbance. We assign the disturbance $w_k = \begin{bmatrix} 0\; sin(0.5k) \end{bmatrix}^T$. The associated quadratic performance index is

$$\mathbb{E}\{\|x_{201}\|_{20}^2 + \sum_{k=0}^{200} \{\|x_k\|_{20}^2 + \alpha_k \|u_k\|^2 - \gamma^2 \|w_k\|^2\}\}.$$

Consider a two-person game with one party being the defender IDSs $\mathbf{P_{a_1}}$ and the other being the DoS attacker $\mathbf{P_{a_2}}$. The defender has two possible actions for each play, i.e., either to load library l_1 (F_1) or load library l_2 (F_2), while the attacker has two possible actions either to choose attack a_1 or a_2. Suppose that l_1 can successfully detect a_1, and l_2 can successfully detect a_2. The reward table for the defender and attacker are given by

$\mathbf{M}_\alpha =$

	a_1	a_2
F_1	0.9	0.77
F_2	0.8	0.85

$\mathbf{M}_\beta =$

	a_1	a_2
F_1	0.5	2
F_2	1.5	1

(10.37)

10.4.2 Simulation Results

Firstly, we calculate the joint strategies $(\mathbf{f}^*, \mu^*)$. Using Theorem 10.9, we have $\mathbf{f}^* = \begin{bmatrix} 0.25\; 0.75 \end{bmatrix}^T$ and $\mathbf{g}^* = \begin{bmatrix} 0.4444\; 0.5556 \end{bmatrix}^T$. The estimated PDR can be obtained by $\alpha^* = \mathbb{E}_{\mathbf{f}^*, \mathbf{g}^*}\{\mathbf{M}_\alpha\} = \mathbf{f}^{*T} \mathbf{M}_\alpha \mathbf{g}^* = 0.8278$. We employ the estimated PDR α^* and set $\gamma = 100$. The initial values are chosen to be $x_0 = \begin{bmatrix} 1\;1\;0 \end{bmatrix}^T$. The control results under different information patterns that curves with CLPS information pattern and CLD information pattern are shown in Figure 10.2. The minimal attenuation level under different information patterns are $\gamma_c(0.8278, 201) = 37.4$ and $\tilde{\gamma}_c(0.8278, 201) = 100.1$, which verifies the fact that the propagation delay has a negative effect on the optimal disturbance attenuation level.

For the assessment of the security system, if the acceptable attenuation level is $\hat{\gamma} = 36$, it follows from Proposition 10.14 that the security system with

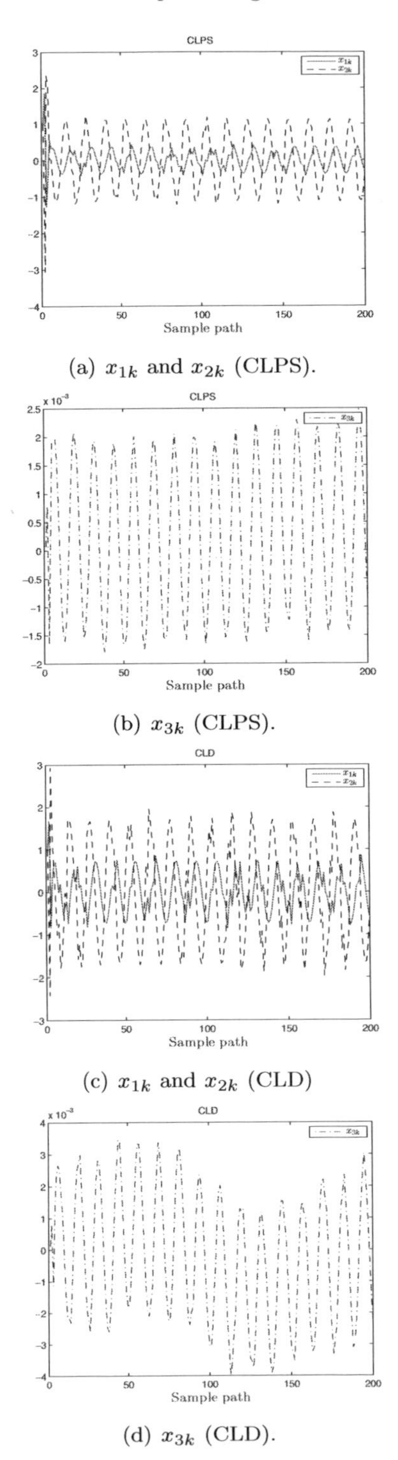

(a) x_{1k} and x_{2k} (CLPS).

(b) x_{3k} (CLPS).

(c) x_{1k} and x_{2k} (CLD)

(d) x_{3k} (CLD).

Figure 10.2 The resultant curves with CLPS information pattern and CLD information pattern.

the reward table 10.37 can not protect the control system since $\gamma_c(\alpha^*, 201) = 37.4 > 36$. If the ability of the security system is enhanced and

$$\mathbf{M}_\alpha = -\mathbf{M}_\beta = \begin{array}{|c|c|c|}\hline & a_1 & a_2 \\ \hline F_1 & 0.98 & 0.92 \\ \hline F_2 & 0.92 & 0.98 \\ \hline \end{array}. \tag{10.38}$$

Again, by applying Theorem 10.9, we have $\mathbf{f}_e^* = \begin{bmatrix} 0.5 \; 0.5 \end{bmatrix}^T$ and $\mathbf{g}_e^* = \begin{bmatrix} 0.5 \; 0.5 \end{bmatrix}^T$. According to Proposition 10.15, the security system with the reward table 10.38 is able to protect the control system since $\gamma_c(\alpha^*, 201) = 35.8 < 36$. That is to say, if we adopt the configuration $\mathbf{f}_e^* = \begin{bmatrix} 0.5 \; 0.5 \end{bmatrix}^T$, the optimal attenuation level γ_c will not exceed the acceptable attenuation level $\hat{\gamma}$ regardless of the attacking strategy $\mathbf{g}$.

10.5 Conclusion and Future Work

CPSs in many critical infrastructures are exposed to public networks and face challenges from malicious adversaries in the cyber world. In this chapter, we employ resilient control such that the control system maintains an acceptable level of operation under cyber attacks and physical disturbances. The so-called Game-in-Game theoretical framework has been proposed to co-design the IDS configuration and H_∞ control. The existence conditions for the unique NE solutions in both cyber game and physical game have been provided. We have developed the joint defense strategies and evaluated the cyber security system. The effectiveness of the proposed method has been verified by a numerical example.

It is worth mentioning that some future works remain to be done and more problems can be addressed within the Game-in-Game framework. For example, the intelligent DoS attackers can be considered which can modify their strategies according to the defender's strategies. The Sensor-to-Actuator (S-C) packets dropout can also be considered as well as the C-A packets dropout in this chapter. We can also exploit the RCS where each control packet is transmitted using the User Datagram Protocol (UDP) type network [64], which is a more complicated case than the TCP case.

Chapter 11
Attack-Tolerant Control for Nonlinear NCSs

11.1 Introduction

As we all known, networks technologies have been widely applied in NCSs [48, 145, 93]. NCSs are not a class of isolated systems anymore by transmitting control and measurement signals over public networks. Not only are mobility and interoperability increased, but also cost in maintenance and installation is reduced in NCSs [75, 181, 149]. However, NCSs are much vulnerable to various malicious attacks [106, 105, 184], such as data injection attacks [98], replay attacks [188], DoS attacks [21], and so on. It is worth emphasizing that NCSs have a very high demand for real time, however, actuators or controllers do not receive signals on time due to DoS attacks [128, 171, 175]. Therefore, DoS attacks which influence availability of signals cause serious harm on NCSs. Thus a filtering algorithm is designed to monitor malicious attacks firstly. Then combined with the designed filtering algorithm, a control algorithm is designed to deal with malicious attacks. In security research of NCSs, there are two main methods of combining filtering algorithms with control algorithms. One is based on the method of Fault Detection and Isolation (FDI). The FDI which determines whether alarm is mainly based on comparison of a given threshold value and the difference between a measured value and a estimated value [9]. The other is based on the method of IDS. The IDS is placed in transmission networks to filter malicious attacks before NCSs are attacked; thus integrity and availability of signals are ensured in NCSs [170, 193]. Note that both algebraic graph methods and game theory are used to solve security problems on NCSs [193, 143]. Moreover, the game theory which are a class of useful methods on modeling security issues in NCSs is well-established in [8]. Nevertheless, not only do DoS attacks cause damage, but also actuator saturation has an effect on NCSs.

In this chapter, an optimal control problem is investigated for NCSs with actuator saturation under DoS attacks. In the attack model, game theory is proposed to describe interaction between the IDS and DoS attacks, and to get the best delivery package rate. Moreover, conditions of existence and uniqueness are presented on NE strategies. In the control model, an iterative ADP algorithm is applied to solve the optimal control problem with actuator saturation. Furthermore, convergence of the iterative ADP algorithm is proved to ensure stability of NCSs. BP neural networks are used to realize the iterative ADP algorithm. Finally, a numerical example is provided to illustrate the effectiveness of the proposed design techniques.

The rest of this chapter is organized as follows: In Section 11.2, a model of the NCS subject to DoS attacks is established and the IDS is deployed in transmission networks to resist the DoS attacks. Moreover, design objectives are given about optimal strategies of defense and attack according to game theory. Then an iterative ADP algorithm is derived and convergence of the iterative ADP algorithm is obtained in Section 11.3. Furthermore, BP neural networks are used to realize the iterative ADP algorithm in Section 11.4. Finally, in Section 11.5, a numerical simulation is given to demonstrate validity and applicability of the proposed approach. Section 11.6 concludes this chapter.

11.2 Problem Statement and Preliminaries

In this chapter, actuator saturation is considered in a NCS where transmission networks between controller and actuator are attacked by DoS attacks; however, feedback networks are transmitted by reliable networks as in [174]. The framework of the NCS with actuator saturation under DoS attacks is shown in Figure 11.1.

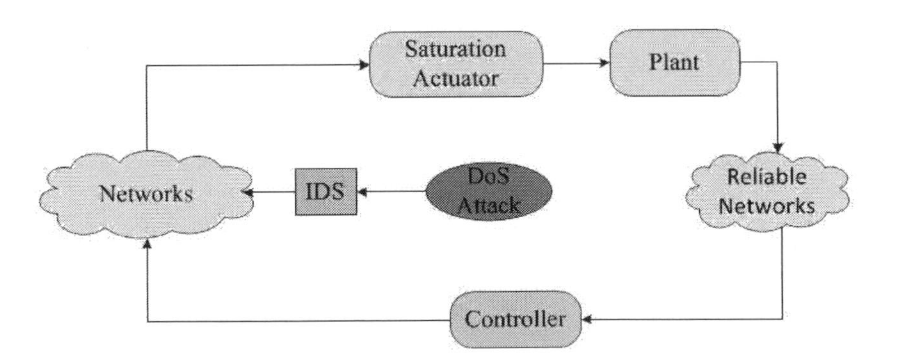

Figure 11.1 The NCS under DoS attacks.

11.2.1 Control Model

Consider the discrete-time nonlinear system with actuator saturation as follows

$$x(k+1) = f(x(k)) + g(x(k))\text{sat}(u(k)). \tag{11.1}$$

where $x(k) \in \mathbf{R}^n$ is state vector, $u(k) \in \mathbf{R}^m$ is control vector, $f(x(k))$ and $g(x(k))$ are two differentiable functions, k is discrete-time variable. The discrete-time nonlinear system (1) is assumed to be controllable and satisfy Lipschitz stability on $\Omega \in \mathbf{R}^n$. Moreover, sat: $\mathbf{R}^m \to \mathbf{R}^m$, and

$$\text{sat}(u(k)) = [\text{sat}(u_1(k)), \text{sat}(u_2(k)), \cdots, \text{sat}(u_m(k))]^T$$

with

$$\text{sat}(u_i(k)) = \text{sgn}(u_i(k)) \min\{\bar{u}_i, |u_i(k)|\},\ i \in 1, 2, \cdots, m.$$

where $\bar{u}_i$ is the saturation boundary. When DoS attacks are considered, the discrete-time nonlinear system (11.1) is formulated as follows

$$x(k+1) = f(x(k)) + \alpha_k g(x(k))\text{sat}(u(k)). \tag{11.2}$$

where α_k is a random distribution variable according to Bernoulli distribution, i.e., if no packets dropout occurs, $\alpha_k = 1$, otherwise, $\alpha_k = 0$. Therefore, the random variable α_k indicates effect of DoS attacks. Let

$$P\{\alpha_k = 1\} = \alpha,$$
$$P\{\alpha_k = 0\} = 1 - \alpha.$$

where α is a constant with $\alpha \in [0, 1]$. Thereby, α is a delivery package rate that is dependent on the game result between IDS and DoS attacks in transmission networks.

In this chapter, an optimal state feedback controller design scheme is mainly investigated for the discrete-time nonlinear system (11.2). Let $v(k) = \text{sat}(u(k))$ be the saturation control input. The optimal state feedback controller not only makes the discrete-time nonlinear system (11.2) stable, but also minimizes the following performance function

$$J = \mathbb{E}\left\{ x^T(N)Q_N x(N) + \sum_{k=0}^{N-1} \left(x^T(k)Q_k x(k) + \alpha_k W(v(k)) \right) \right\}. \tag{11.3}$$

where N is finite time length. $Q_N > 0$, $Q_k > 0$ and

$$W(v(k)) = 2\int_0^{v(k)} \phi^{-T}(M^{-1}s)MR_k ds$$

in which $s \in \mathbf{R}^m$, $\phi \in \mathbf{R}^m$, $R_k > 0$, $M = \text{diag}\{\bar{u}\}$, $\phi(\cdot)$ is a monotone bounded increasing odd function whose first derivative is bounded. Therefore, tanh$(\cdot)$ is selected to represent function $\phi(\cdot)$ in this chapter. Hence, the control vector $v(k)$ is guaranteed within range of actuator saturation by the performance function (11.3).

According to Bellman principle of optimality [174], an HJB equation is obtained as follows

$$V^*(x(k)) = \min \mathbb{E}\left\{x^T(k)Q_k x(k) + \alpha_k W(v(k)) + V^*(x(k+1))\right\}. \quad (11.4)$$

Therefore, the optimal control problem is transformed into solving the HJB equation (11.4) to find an optimal state feedback controller.

11.2.2 Attack Model

The attack model is described by a game model in which there are a DoS attacker and an IDS. Note that DoS attacks destroy packets transmission; however, the IDS is placed in transmission networks to resist the DoS attacks. Once the DoS attacks occur, the IDS triggers an alarm and takes active response measures. In the attack model, the DoS attacker is one player to determine attack strategies. Similarly, the IDS is the other player in the game model to determine defense strategies. An attack set is denoted by $\mathcal{A} := \{a_1, a_2, \cdots, a_m\}$ and the DoS attacker selects attack behaviors from the attack set $\mathcal{A}$ to destroy the NCS (11.2). Equally, $\mathcal{D} := \{d_1, d_2, \cdots, d_n\}$ is a defense set and the IDS chooses defensive behaviors from defense set $\mathcal{D}$ to resist the DoS attacks in the NCS (11.2). An example is given in Figure 11.2 to illustrate the performance of IDS defensive behaviors under different DoS attacks. In Figure 11.2, the IDS contains two defensive behaviors d_1 and d_2, where two blue boxes denote defensive behavior d_1, two green boxes denote defensive behavior d_2, the DoS attacker contains two attack behaviors a_1 and a_2, where two black circles denote one kind of DoS attack a_1, two purple circles denote the other kind of DoS attack a_2, two red crosses denote the DoS attack is defended successfully, two yellow hexagons denote the DoS attack is defended unsuccessfully. That is, attack behaviors a_1 and a_2 are detected by defensive behaviors d_1 and d_2, respectively.

Denote $f := \mathcal{D} \to [0, 1]$ and $g := \mathcal{A} \to [0, 1]$. Then a mixed strategy $g(a_q)$ is a probability of the DoS attacker choosing attack behavior a_p. A mixed strategy $f(d_p)$ is a probability of the IDS choosing defensive behavior d_p. Moreover, there exist the following equalities

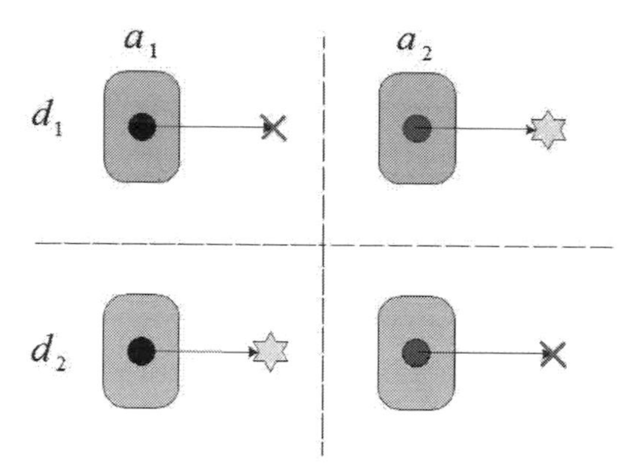

Figure 11.2 The performance of IDS defensive behaviors under different DoS attacks.

$$\sum_{p=1}^{n} f(d_p) = 1, \; p \in [1, 2, \cdots, n],$$
$$\sum_{q=1}^{m} g(a_q) = 1, \; q \in [1, 2, \cdots, m].$$

For simplicity, $f(d_p)$ and $g(a_q)$ are represented as f_p and g_q, respectively. Let $\mathbf{f}$ and $\mathbf{g}$ be

$$\mathbf{f} := [f_1, f_2, \cdots, f_n]^T,$$
$$\mathbf{g} := [g_1, g_2, \cdots, g_m]^T.$$

Denote $\alpha(\mathbf{f}, \mathbf{g})$ and $\beta(\mathbf{f}, \mathbf{g})$, which mean that both the delivery package rate α and an attack benefit β are determined by $\mathbf{f}$ and $\mathbf{g}$. The IDS is to maximize the delivery package rate $\alpha(\mathbf{f}, \mathbf{g})$; however, the DoS attacker is to maximize the attack benefit $\beta(\mathbf{f}, \mathbf{g})$. Under the framework of the game model, there exists an optimal allocation strategy in the form of an NE strategy. Any game players in the NE point do not get any benefit by altering its strategy unilaterally. Thus, a definition for the NE strategy is shown in the following.

Definition 11.1. If the following inequalities

$$\alpha(\mathbf{f}, \mathbf{g}^*) \leq \alpha(\mathbf{f}^*, \mathbf{g}^*),$$
$$\beta(\mathbf{f}^*, \mathbf{g}) \leq \beta(\mathbf{f}^*, \mathbf{g}^*).$$

set up simultaneously, strategy set $(\mathbf{f}^*, \mathbf{g}^*)$ is called the NE strategy of the game model.

For simplicity, α^* and β^* are used to take the place of $\alpha(\mathbf{f}^*, \mathbf{g}^*)$ and $\beta(\mathbf{f}^*, \mathbf{g}^*)$, respectively. The payoff matrix for the defender is denoted as $\mathbf{N}_\alpha$.

Similarly, the payoff matrix for the attacker is denoted as $\mathbf{N}_\beta$. Then there exist

$$\mathbf{N}_\alpha := \begin{array}{|c|c|c|c|c|}\hline & a_1 & a_2 & \cdots & a_m \\ \hline d_1 & \alpha_{11} & \alpha_{12} & \cdots & \alpha_{1m} \\ \hline d_2 & \alpha_{21} & \alpha_{22} & \cdots & \alpha_{2m} \\ \hline \vdots & \vdots & \vdots & \ddots & \vdots \\ \hline d_n & \alpha_{n1} & \alpha_{n2} & \cdots & \alpha_{nm} \\ \hline \end{array}, \tag{11.5}$$

$$\mathbf{N}_\beta := \begin{array}{|c|c|c|c|c|}\hline & a_1 & a_2 & \cdots & a_m \\ \hline d_1 & \beta_{11} & \beta_{12} & \cdots & \beta_{1m} \\ \hline d_2 & \beta_{21} & \beta_{22} & \cdots & \beta_{2m} \\ \hline \vdots & \vdots & \vdots & \ddots & \vdots \\ \hline d_n & \beta_{n1} & \beta_{n2} & \cdots & \beta_{nm} \\ \hline \end{array}. \tag{11.6}$$

In the payoff matrices (11.5) and (11.6), the element α_{pq} with $p \in \{1, 2, \cdots, n\}, q \in \{1, 2, \cdots, m\}$ is used to express benefit of the defender when the IDS uses defensive behavior d_p and the DoS attacker uses attack behavior a_q. Similarly, the element β_{pq} with $p \in \{1, 2, \cdots, n\}, q \in \{1, 2, \cdots, m\}$ is used to express benefit of the attacker when the IDS uses defensive behavior d_p and the DoS attacker uses attack behavior a_q. Then both existence and uniqueness of the optimal allocation strategy are given in the following lemma.

Lemma 11.2. *[170] Denote matrices $\mathbf{R}_\beta, \mathbf{R}_\alpha, \mathbf{v}_\beta$, and $\mathbf{v}_\alpha$ as follows*

$$\mathbf{R}_\beta = \begin{bmatrix} \beta_{11} - \beta_{12} & \beta_{21} - \beta_{22} & \cdots & \beta_{n1} - \beta_{n2} \\ \vdots & \vdots & \ddots & \vdots \\ \beta_{11} - \beta_{1m} & \beta_{21} - \beta_{2m} & \cdots & \beta_{n1} - \beta_{nm} \\ 1 & 1 & \cdots & 1 \end{bmatrix},$$

$$\mathbf{R}_\alpha = \begin{bmatrix} \alpha_{11} - \alpha_{12} & \alpha_{21} - \alpha_{22} & \cdots & \alpha_{n1} - \alpha_{n2} \\ \vdots & \vdots & \ddots & \vdots \\ \alpha_{11} - \alpha_{1m} & \alpha_{21} - \alpha_{2m} & \cdots & \alpha_{n1} - \alpha_{nm} \\ 1 & 1 & \cdots & 1 \end{bmatrix},$$

$$\mathbf{v}_\beta = \underbrace{\begin{bmatrix} 0 \, 0 \cdots 0 \, 1 \end{bmatrix}^T}_{n}, \quad \mathbf{v}_\alpha = \underbrace{\begin{bmatrix} 0 \, 0 \cdots 0 \, 1 \end{bmatrix}^T}_{m}.$$

If matrices $\mathbf{R}_\beta$ and $\mathbf{R}_\alpha$ are invertible, meanwhile, the following equalities

$$\mathbf{R}_\beta \mathbf{f}^* = \mathbf{v}_\beta, \ \mathbf{R}_\alpha \mathbf{g}^* = \mathbf{v}_\alpha, \ n = m$$

have solutions with $f_p, g_q > 0, \forall p \in \{1, 2, \cdots, n\}, \forall q \in \{1, 2, \cdots, m\}$*, then there exists a unique NE solution. The NE strategy* $(\mathbf{f}^*, \mathbf{g}^*)$ *is given by*

$$\mathbf{f}^* = \mathbf{R}_\beta^{-1}\mathbf{v}_\beta, \ \mathbf{g}^* = \mathbf{R}_\alpha^{-1}\mathbf{v}_\alpha.$$

11.3 Iterative ADP Algorithm

11.3.1 Formula Derived for Iterative ADP Algorithm

By Lemma 11.2, the optimal delivery package rate is obtained by $\alpha^* = \mathbf{f}^{*\mathbf{T}}\mathbf{N}_\alpha\mathbf{g}^*$ for the discrete-time nonlinear system (11.2). That is, $\mathbb{E}\{\alpha_k\} = \alpha^*$. By solving the following equality

$$\frac{\partial V^*(x(k))}{\partial v(k)} = \frac{\partial \mathbb{E}\left(x^T(k)Q_k x(k) + \alpha_k W(v(k)) + V^*(x(k+1))\right)}{\partial v(k)} = 0\,, \quad (11.7)$$

the optimal state feedback controller $v^*(k)$ is obtained as follows

$$v^*(k) = M \tanh\left(-\frac{1}{2}(MR_k)^{-1}g^T(x(k))\frac{\partial V^*(x(k+1))}{\partial x(k+1)}\right). \quad (11.8)$$

The optimal control policy $v^*(k)$ is dependent on the optimal performance index function $V^*(x(k))$, where $V^*(x(k))$ is a solution of the HJB equation (11.4). However, it is difficult to solve the HJB equation (11.4) directly. Therefore, an iterative ADP algorithm is proposed to solve the optimal control policy (11.8). For each state $x(k)$, the performance index function $V(x(k))$ and control policy $v(k)$ are updated by the iterative ADP algorithm. The iterative ADP algorithm starts with $i = 0$ and the initial performance index function is $V_0(x(k)) = 0$. Hence, the following control policy for $i = 0$ is calculated as

$$v_0(x(k)) = \arg\min_{v(k)} \mathbb{E}\left\{x^T(k)Q_k x(k) + \alpha_k W(v(k)) + V_0(x(k+1))\right\}. \quad (11.9)$$

Then performance index function is updated as follows

$$V_1(x(k)) = \mathbb{E}\left\{x^T(k)Q_k x(k) + \alpha_k W(v_0(x(k))) + V_0(x(k+1))\right\}. \quad (11.10)$$

Therefore, the iterative equalities are given as follows

$$v_i(x(k)) = \arg\min_{v(k)} \mathbb{E}\left\{x^T(k)Q_k x(k) + \alpha_k W(v(k)) + V_i(x(k+1))\right\}, \quad (11.11)$$

$$V_{i+1}(x(k)) = \mathbb{E}\left\{x^T(k)Q_k x(k) + \alpha_k W(v_i(x(k))) + V_i(x(k+1))\right\}, \quad (11.12)$$

for $i = 1, 2, \cdots, +\infty$.

11.3.2 Properties of Iterative ADP Algorithm

In this subsection, convergences of $v(k)$ and $V(x(k))$ are obtained. For a given state $x(k)$, the iterative ADP algorithm improves the control policy (11.11) and performance index function (11.12) in each iteration. There exist $v_i(k) \to v^*(k)$ and $V_{i+1}(x(k)) \to V^*(x(k))$ when $i \to \infty$.

The boundedness and monotonicity of $V_{i+1}(x(k))$ are obtained by the following theorem.

Theorem 11.3. *Let the discrete-time nonlinear system (11.2) be controllable, then the performance index $V_{i+1}(x(k))$ is nondecreasing and there is an upper bound H with $0 \leq V_{i+1}(x(k)) \leq H, \forall i \in \mathbf{N}$.*

Proof. Denote a new performance index function $Z_{i+1}(x(k))$ as follows

$$Z_{i+1}(x(k)) = \mathbb{E}\left\{x^T(k)Q_k x(k) + \alpha_k W(v_{i+1}(k)) + Z_i(x(k+1))\right\},$$

where $Z_0(x(k)) = V_0(x(k)) = 0$.

Thereby, $Z_i(x(k)) \leq V_{i+1}(x(k))$ is obtained by mathematical induction. Firstly, the following inequality

$$V_1(x(k)) - Z_0(x(k)) = \mathbb{E}\left\{x^T(k)Q_k x(k) + \alpha_k W(v_0(k))\right\} \geq 0$$

is established when $i = 0$. Therefore, $V_1(x(k)) \geq Z_0(x(k))$ holds. Secondly, assume that $V_i(x(k)) \geq Z_{i-1}(x(k))$ establishes at $i-1$ for arbitrary $x(k)$. Lastly, the following equalities

$$Z_i(x(k)) = \mathbb{E}\left\{x^T(k)Q_k x(k) + \alpha_k W(v_i(k)) + Z_{i-1}(x(k+1))\right\},$$
$$V_{i+1}(x(k)) = \mathbb{E}\left\{x^T(k)Q_k x(k) + \alpha_k W(v_i(k)) + V_i(x(k+1))\right\}$$

set up for each i. Moreover, the following equality is obtained as

$$V_{i+1}(x(k)) - Z_i(x(k)) = V_i(x(k+1)) - Z_{i-1}(x(k+1)) \geq 0.$$

That is, the following inequality

$$Z_i(x(k)) \leq V_{i+1}(x(k))$$

is established. Hence, $Z_i(x(k)) \leq V_{i+1}(x(k))$ is obtained for $\forall i \in \mathbf{N}$ by mathematical induction.

According to the principle of optimality, an optimal solution is less than a feasible solution. Therefore, there exists

$$V_i(x(k)) \leq Z_i(x(k)) \leq V_{i+1}(x(k)).$$

The monotonicity of the performance index function $V_{i+1}(x(k))$ is obtained.

Then let $\nu_i(k)$ be an admissible control policy and $\Lambda_{i+1}(x(k))$ be a performance index function obtained by $\nu_i(k)$ with $\Lambda_0(x(k)) = V_0(x(k)) = 0$. It is easily gotten that $V_{i+1}(x(k)) \geq 0$. $\Lambda_{i+1}(x(k))$ is updated as follows

$$\Lambda_{i+1}(x(k)) = \mathbb{E}\left\{x^T(k)Q_k x(k) + \alpha_k W(\nu_i(k)) + \Lambda_i(x(k+1))\right\}. \quad (11.13)$$

It follows from equation (11.13) that

$$\begin{aligned}&\Lambda_i(x(k+1))\\ &= \mathbb{E}\left\{x^T(k+1)Q_k x(k+1) + \alpha_{k+1} W(\nu_{i-1}(k+1)) + \Lambda_{i-1}(x(k+2))\right\}.\end{aligned}$$

Therefore, it is obtained that

$$\begin{aligned}&\Lambda_{i+1}(x(k))\\ &= \mathbb{E}\left\{x^T(k)Q_k x(k) + \alpha_k W(\nu_i(k)) + \Lambda_i(x(k+1))\right\}\\ &= \mathbb{E}\left\{x^T(k)Q_k x(k) + \alpha_k W(\nu_i(k))\right.\\ &\quad \left.+x^T(k+1)Q_k x(k+1) + \alpha_{k+1} W(\nu_{i-1}(k+1)) + \Lambda_{i-1}(x(k+2))\right\}\end{aligned}$$

$$\begin{aligned}&= \mathbb{E}\left\{x^T(k)Q_k x(k) + \alpha_k W(\nu_i(k))\right.\\ &\quad +x^T(k+1)Q_k x(k+1) + \alpha_{k+1} W(\nu_{i-1}(k+1))\\ &\quad \vdots\\ &= \mathbb{E}\left\{x^T(k)Q_k x(k) + \alpha_k W(\nu_i(k))\right.\\ &\quad +x^T(k+1)Q_k x(k+1) + \alpha_{k+1} W(\nu_{i-1}(k+1))\\ &\quad +x^T(k+2)Q_k x(k+2) + \alpha_{k+2} W(\nu_{i-2}(k+2))\\ &\quad +\cdots\\ &\quad \left.+x^T(k+i)Q_k x(k+i) + \alpha_{k+i} W(\nu_0(k+i)) + \Lambda_0(x(k+i+1))\right\},\end{aligned}$$

where $\Lambda_0(x(k+i+1)) = 0$.

Let $U_i(x(k)) = \mathbb{E}\left\{x^T(k)Q_k x(k) + \alpha_k W(\nu_i(k))\right\}$, then equation (11.14) is rewritten as

$$\begin{aligned}&\Lambda_{i+1}(x(k))\\ &= \sum_{j=0}^{i} U_{i-j}(x(k+j))\\ &= \sum_{j=0}^{i} \mathbb{E}\left\{x^T(k+j)Q_k x(k+j) + \alpha_{k+j} W(\nu_{i-j}(k+j))\right\}\\ &\leq \lim_{i\to\infty} \sum_{j=0}^{i} \mathbb{E}\left\{x^T(k+j)Q_k x(k+j) + \alpha_{k+j} W(\nu_{i-j}(k+j))\right\}.\end{aligned}$$

Since $\nu_i(k)$ is an admissible control policy, there exists $x(k) \to 0$ when $k \to \infty$. Therefore, the upper bound H is established with

$$\Lambda_{i+1}(x(k)) \leq \lim_{i\to\infty} \sum_{j=0}^{i} U_{i-j}(x(k+j)) \leq H, \forall i \in \mathbf{N}.$$

According to the principle of optimality, an optimal solution is less than a feasible solution. Hence, it is gotten that

$$0 \leq V_{i+1}(x(k)) \leq \Lambda_{i+1}(x(k)) \leq H, \forall i \in \mathbf{N}.$$

The boundedness of the performance index function $V_{i+1}(x(k))$ is obtained. The proof is completed.

The performance index function $V_{i+1}(x(k))$ is a nondecreasing and bounded function by Theorem 11.3. Therefore, there exists $V_{i+1}(x(k)) \to V^*(x(k))$ with increase in number of iteration. When the optimal performance index function $V^*(x(k))$ is obtained, the optimal control policy $v^*(k)$ is gotten by equation (11.8).

11.4 Realization of Iterative ADP Algorithm by Neural Networks

In the discrete-time nonlinear system (11.2), both the performance index function $V(x(k))$ and the control policy $v(k)$ are nonlinear. Furthermore, equations (11.11) and (11.12) are expressed inexplicit in the iterative ADP algorithm. Hence, BP neural networks are used to approximate $V(x(k))$ and $v(k)$.

BP neural networks are a kind of multilayer feedforward neural networks trained by an error back propagation algorithm. The structure of BP neural networks is shown in Figure 11.3. In Figure 11.3, the BP neural networks are composed of the input layer, the hidden layer, and the output layer and numbers of the input layer neurons, the hidden layer neurons, and the output layer neurons are denoted by m, n, and p, respectively. Meanwhile, the weight matrix between the input layer and the hidden layer is denoted by W_{ih}. The weight matrix between the hidden layer and the output layer is denoted by W_{ho}.

The calculation process of BP neural networks is composed of two parts which are a forward calculation and an error back propagation. The input layer neurons receive input information from outside and pass the information to the hidden layer neurons. In the hidden layer, the tangent sigmoid transfer function $tansig(\cdot)$ represented by $\sigma(\cdot)$ is chosen as an excitation function and used to process information. The hidden layer is designed for a single hidden layer structure in this chapter. Then the hidden layer transfers information to

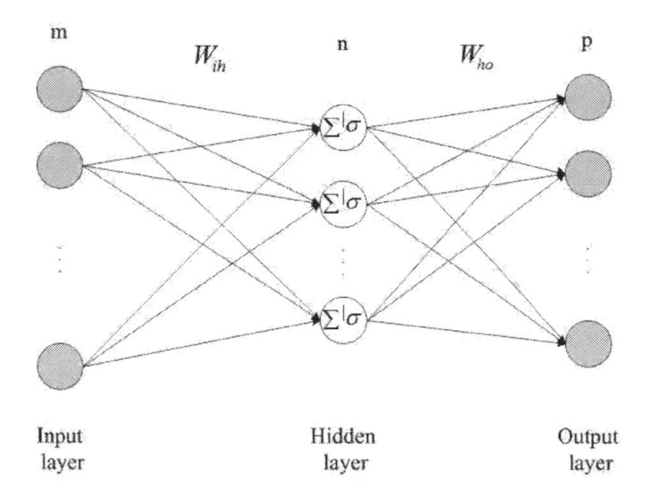

Figure 11.3 The structure of BP neural networks.

the output layer neurons. The forward calculation is completed after information is processed by the output layer to outside. When an actual output does not agree with an expected output, the error back propagation is launched. Moreover, errors of the output layer and weights of each layer are updated by a gradient descent method. The calculation process continues until the errors of output layer are reduced to an acceptable level or pre-set learning times.

A structure of the iterative ADP algorithm by BP neural networks is shown in Figure 11.4 In Figure 11.4, the action neural networks and critic neural networks share the same structure of BP neural networks. The action neural networks are used to approximate the control policy $v(k)$ and the critic neural networks are used to approximate the performance index function $V(x(k))$.

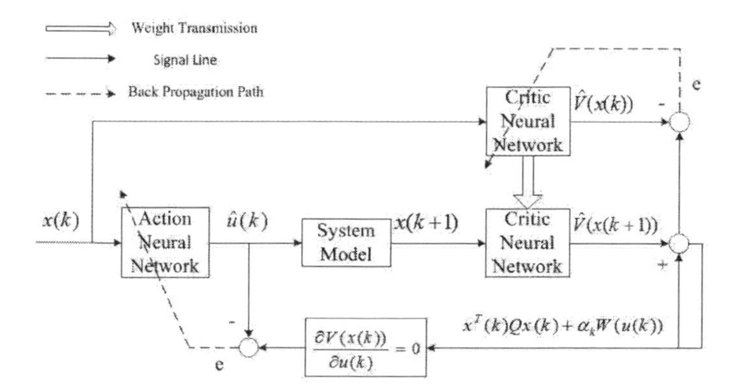

Figure 11.4 A structure of the iterative ADP algorithm by neural networks.

Meanwhile, solid lines indicate forward calculation and dashed lines indicate error back propagation.

In action networks, state $x(k)$ is given to produce output $\hat{u}(k)$. The following expression

$$\hat{v}_i(x(k)) = W_{ho}^T \sigma \left(W_{ih}^T x(k) \right).$$

is used to approximate the control policy $v(k)$. The output target of action networks is solved according to the principle of optimality shown as follows

$$v_i(k) = M \tanh \left(-\frac{1}{2} (M R_k)^{-1} g^T(x(k)) \frac{\partial V^*(x(k+1))}{\partial x(k+1)} \right).$$

Therefore, the output error of action networks is taken as

$$e_a = \hat{v}_i(x(k)) - v_i(x(k)).$$

To minimize error e_a in the action networks, there exists the following equality

$$E_a = \frac{1}{2} e_a^T e_a.$$

The weights in the action networks are updated by a gradient descent method as follows

$$\begin{aligned} w_a(k) &= w_a(k) + \Delta w_a(k), \\ \Delta w_a(k) &= l_a(-\frac{\partial E_a}{\partial w_a}), \\ \frac{\partial E_a}{\partial w_a} &= \frac{\partial E_a}{\partial V(x(k))} \frac{\partial V(x(k))}{\partial w_a}, \end{aligned}$$

where l_a is the learning rate of the action networks.

In critic networks, state $x(k)$ is given to produce function $\hat{V}(x(k))$. The following expression

$$\hat{V}_i(x(k)) = W_{ho}^T \sigma \left(W_{ih}^T x(k) \right).$$

is used to approximate the performance index function $V(x(k))$. The objective function of critic networks is written as

$$V_{i+1}(x(k)) = \mathbb{E} \left\{ x^T(k) Q_k x(k) + \alpha_k W(v_i(k)) + V_i(x(k+1)) \right\}.$$

Hence, the error function of critic networks is taken as

$$e_c = \hat{V}_{i+1}(x(k)) - V_{i+1}(x(k)).$$

To minimize error e_c in the critic networks, there exists the following equality

$$E_c = \frac{1}{2}e_c^2.$$

Similarly, the weights in the critic networks are updated by a gradient descent method as follows

$$\begin{aligned}
w_c(k) &= w_c(k) + \Delta w_c(k), \\
\Delta w_c(k) &= l_c(-\frac{\partial E_c}{\partial w_c}), \\
\frac{\partial E_c}{\partial w_c} &= \frac{\partial E_c}{\partial V(x(k))}\frac{\partial V(x(k))}{\partial w_c},
\end{aligned}$$

where l_c is the learning rate of the critic networks.

The iterative ADP algorithm by BP neural networks is given as follows

Algorithm The iterative ADP algorithm based on BP neural networks

Step 1: Give the initial state $x(0)$ and the calculation accuracy ε.
Step 2: Set the initial iteration $i = 0$, $V_0(x(k)) = 0$.
Step 3: Calculate the $v_0(k)$ by (11.9) and the $V_1(x(k))$ by (11.10).
Step 4: Calculate the $v_i(k)$ by (11.11) and the $V_{i+1}(x(k))$ by (11.12).
Step 5: If $|V_{i+1}(x(k)) - V_i(x(k))| \le \varepsilon$, perform Step 6. Otherwise, $i = i + 1$, perform Step 4.
Step 6: Stop.

11.5 Numerical Example

Consider the following discrete-time nonlinear system with actuator saturation under DoS attacks as

$$x(k+1) = f(x(k)) + \alpha_k g(x(k))v(k), \tag{11.14}$$

where

$$f(x(k)) = \begin{bmatrix} x_1(k)\exp(x_2^3(k))x_2(k) \\ x_2^3(k)x_1(k) \end{bmatrix}, \quad g(x(k)) = \begin{bmatrix} 0.5 & 0 \\ 0 & 0.5 \end{bmatrix}.$$

The following performance index function is selected as

$$\begin{aligned}
&V^*(x(k)) \\
&= \min \mathbb{E}\left\{ x^T(k)Q_k x(k) + 2\int_0^{v(k)} \tanh^{-T}(M^{-1}s)MR_k ds + V^*(x(k+1)) \right\},
\end{aligned}$$

where

$$M = \begin{bmatrix} 0.5 & 0 \\ 0 & 0.5 \end{bmatrix}, \quad R_k = Q_k = I_{2\times 2}.$$

Hence, the optimal networked controller (11.8) is gotten in this example. Consider a two-person game in which one player is a defender $\mathbf{P_{a1}}$ and the other is an attacker $\mathbf{P_{a2}}$. The defender $\mathbf{P_{a1}}$ can select two kinds of defensive strategies, i.e., d_1 and d_2. There exist two kinds of attack strategies a_1 and a_2 to be chosen by the attacker $\mathbf{P_{a2}}$. A defensive behavior only identifies and resists against a particular attack behavior. That is, d_1 and d_2 can defense a_1 and a_2 successfully, respectively. Payoff matrices are given as follows

$\mathbf{N}_\alpha :=$

	a_1	a_2
d_1	0.92	0.75
d_2	0.72	0.88

, $\mathbf{N}_\beta :=$

	a_1	a_2
d_1	0.9	0.77
d_2	0.8	0.85

.

According to Lemma 11.2, the parameter in the NE strategy for the defender is given as follows

$$\mathbf{f}^* = \begin{bmatrix} 0.278 \; 0.722 \end{bmatrix}^T,$$

while the parameter in the NE strategy for the attacker is taken as follows

$$\mathbf{g}^* = \begin{bmatrix} 0.394 \; 0.606 \end{bmatrix}^T.$$

Therefore, the delivery package rate $\alpha^* = \mathbf{f}^{*\mathbf{T}}\mathbf{N}_\alpha \mathbf{g}^* = 0.817$ is gotten. Three-layer BP neural networks are chosen so that the action networks and the critic networks are with structures of 2-8-2 and 2-8-1, respectively. Choose $k = 50$, $\varepsilon = 10^{-6}$, and the learning rate $l_a = l_c = 0.01$. The initial state is shown as follows

$$x(0) = \begin{bmatrix} 1 \; -1 \end{bmatrix}^T.$$

The action networks and the critic networks are trained for 1000 iteration times in each step to guarantee neural networks training error is less than 10^{-6}. The iteration error in BP neural networks is shown in Figure 11.5. In Figure 11.5, it is shown that the iteration error tends to zero with an increase in the number of iterations. Then the control input trajectories of the discrete-time nonlinear system (11.14) are shown in Figure 11.6. In Figure 11.6, it is gotten that the control input is limited to boundary of actuator saturation. The state trajectories are shown in Figure 11.7. In Figure 11.7, it is obtained that the state tends to zero and the discrete-time nonlinear system (11.14) achieves stability.

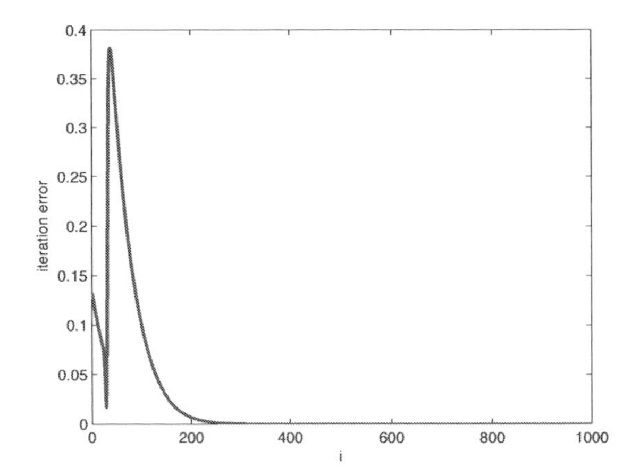

Figure 11.5 Iteration error.

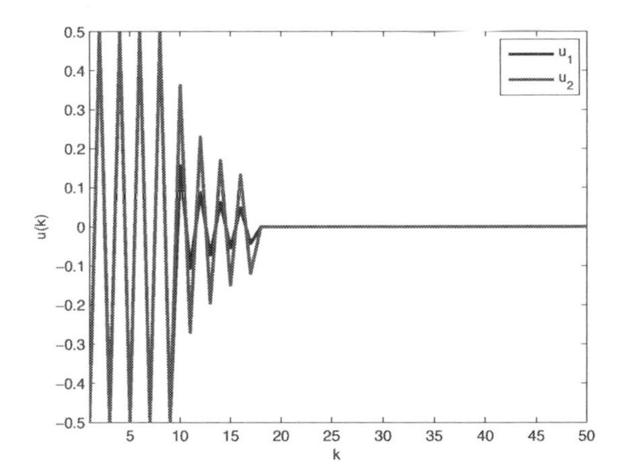

Figure 11.6 Control input trajectories.

11.6 Conclusion

In this chapter, the optimal control problem has been considered in the NCS with actuator saturation under DoS attacks. In network layer, the optimal defense strategy and the optimal attack strategy have been designed by the game theoretic approach. In control layer, the optimal control problem with actuator saturation has been analyzed. The iterative ADP algorithm has been proposed to deal with actuator saturation. Then BP neural networks have been used to realize the iterative ADP algorithm. Validity of the proposed method has been verified by a numerical simulation.

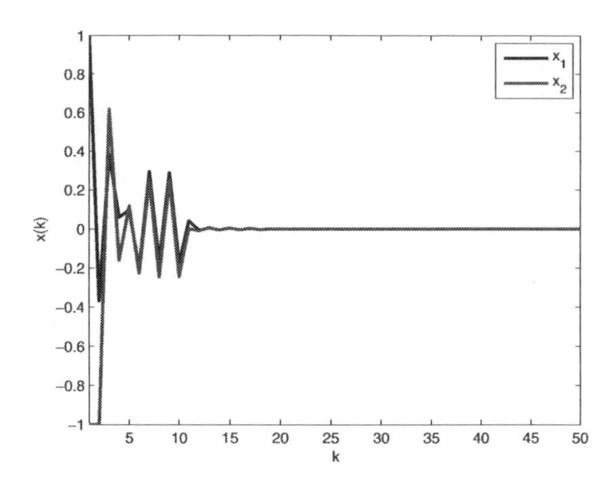

Figure 11.7 State trajectories.

Chapter 12
Coupled Design of CPS under DoS Attacks

12.1 Introduction

The past decades have witnessed the migration from proprietary standards for communications towards open international standards for modern critical infrastructures. However, it is difficult to update software and hardware applications for legacy control systems. Malicious attackers, on the other hand, can easily launch an attack since the amount of knowledge needed to successfully execute an attack is decreasing. As a consequence, many incidents related to damages of cyber attacks on Industrious Control Systems (ICSs) have already been reported in [70]. In [40], the air traffic control system tower at Worcester Regional Airport (MA) USA was shut down by a hacker. In [129], it has been reported that the power grid in the U.S. was penetrated by cyber spies and some key infrastructure was compromised by the intrusion. It is also reported in [129] that the Siemens SCADA systems have been attacked by the computer worm, Stuxnet. ICSs are widely used in the electric, water, oil, and gas industries and they are critical to the operation of U.S. infrastructures. The aforementioned attacks have incurred environment and financial losses. The information technologies employed in ICSs are vastly vulnerable and have a direct effect on the physical component of the system. Hence it is essential to take into account cyber security when designing ICSs.

In this chapter, we use dynamical systems to capture the physical layer of the system, and focus on IDSs at the cyber layer of the system for defense against malicious behaviors. IDSs are often used to detect and raise alarm for cyber attacks such as the DoS attack, which can cause delays and congestion in the communication channel. For an ICS equipped with IDSs, the integrated design involves both IDS configuration and controller design. In [191] and [192], the authors have addressed this issue by proposing a coupled optimality criteria for designing resilient control systems. The cyber state and controlled plant are modeled as a coupled continuous Markov process and the controllers are designed via an iterative method. In this work, we consider a specific cyber defense mechanism, DoS attack, and study its consequence on ICSs.

The rest of the chapter is organized as follows: In Section 12.2.2, we present a unifying framework and the hybrid system model for the resilient control system. This model describes the controlled plant under DoS attack. In Section 12.2.3, we present the methodologies of finding the optimal defense stra-tegy in the cyber layer. In Section 12.2.4, the H_∞ optimal control design in the physical layer is given. In Section 12.2.5, we propose an algorithm for the co-design of optimal defense mechanism and control strategy. In Section 12.3, the method proposed is applied to the control of a power system and the results are corroborated by numerical simulations. In Section 12.4, conclusions are drawn and directions of future work are presented.

12.2 Resilient and H_∞ Optimal Control

In this section, we consider the problem in which the adversary launches DoS attacks to the networked control system. A hybrid discrete-time dynamical system model is established consisting of IDSs at the cyber layer and the underlying dynamical system. Figure 12.1 illustrates the interplay between cyber and physical layers of the system. A Markov chain is used to capture the dynamics of the cyber state, while the physical layer dynamics are captured by a discrete-time dynamical system with Sensor-to-Controller (S-C) and Controller-to-Actuator (C-A) delays. An adversary intends to launch a DoS attack to jam the communication channels in the networked control system and cause congestion in the communication networks. The configuration policies against the attacker at the cyber layer affect the control system through S-C and C-A delays. In addition, the control system performance under best-effort controller at the physical layer needs to be taken into account when designing a configuration policy. The resilient control design involves the co-design of the cyber configuration policy as well as the optimal controller for the physical dynamical system.

12.2.1 Attacks on ICS

According to [70], the attacks on the ICS can be summarized in Figure 12.1 A_3 and A_5 represent deception attacks, where the false information $\tilde{y} \neq y$ and $\tilde{u} \neq u$ are sent from sensors and controllers. A_1 and A_4 represent direct attacks against the actuators or the plant. A_2 is the DoS attack, where the controller is prevented from receiving sensor measurements and the actuator from receiving control signals. Note that DoS attacks are most commonly used by the adversary that involve jamming the communication channels, compromising

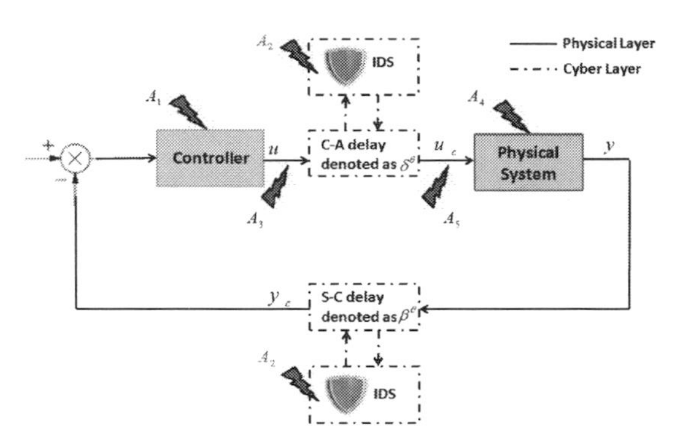

Figure 12.1 The control system can be subject to many attacks. They can be on the controller, the plant, and the communication networks. The IDS can be made for defending the networked control system.

devices, and attacking the routing protocols, etc. We will restrict our attention to DoS attacks in this chapter, leaving the deception attacks for future works. Since IDS is often designed to detect unauthorized uses of networks [190], we will use IDS configuration as the defense mechanism.

12.2.2 System Framework

In this subsection, the system framework is provided of resilient control. In Figure 12.2, we can see the cyber layer and the control layer is interconnected through a defence and attack mechanism and the control performance. The cyber layer is modeled as a competitive Markov jump process and the control layer is modeled as a stochastic model in which the stochastic variable is distributed according to the Bernouli distribution. The controlled plant is described by the model as follows.

$$\begin{cases} x_{k+1} = Ax_k + B_2 u_{c,k} + B_1 \omega_k, \\ z_k = Dx_k, \end{cases} \tag{12.1}$$

where $x_k \in \mathbb{R}^n$ and $u_{c,k} \in \mathbb{R}^m$ are the state variable and the control signal received by the actuator, ω_k is the disturbance belonging to $l_2[0,\infty)$. A, B_1, B_2, and D are matrices with appropriate dimensions. The measurement with randomly varying communication delays is described by

$$\begin{cases} y_k = Cx_k, \\ y_{c,k} = (1-\delta^\theta) y_k + \delta^\theta y_{k-1}, \end{cases} \tag{12.2}$$

$\theta \in \Theta := \{\theta_1, \theta_2, \ldots, \theta_s\}$ is the failure state in the cyber layer of the system. The stochastic variable δ^θ is distributed according to the Bernouli distribution

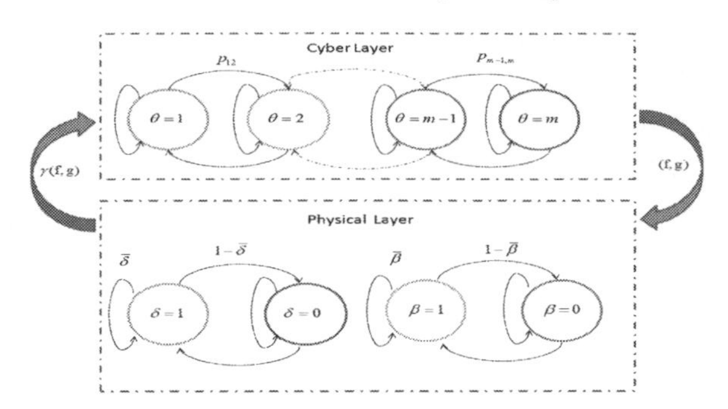

Figure 12.2 The system framework contains two decision problems. The decision problem at the cyber layer a competitive Markov decision problem, capturing the interactions between an attacker and a defender whose actions $(\mathbf{f}, \mathbf{g})$ affect the cyber state θ and the transition probabilities. The decision problem at the physical layer is to design an optimal controller of a dynamical system S-C delay δ^θ and C-A delay β^θ for achieving control performance γ. The two decision problems are interleaved and coupled.

given by

$$\begin{aligned} \bar{\delta}^\theta &:= \Pr\{\delta^\theta = 1\} = E\{\delta^\theta\}, \\ \Pr\{\delta^\theta = 0\} &= 1 - E\{\delta^\theta\} = 1 - \bar{\delta}^\theta, \end{aligned} \tag{12.3}$$

In this chapter, we propose an observer-based control strategy described by

$$\begin{cases} \hat{x}_{k+1} = A\hat{x}_k + B_2 u_{c,k} + L^\theta(y_{c,k} - \bar{y}_{c,k}), \\ \quad \bar{y}_{c,k} = (1 - \bar{\delta}^\theta)C\hat{x} + \bar{\delta}^\theta C\hat{x}_{k-1}, \end{cases} \tag{12.4}$$

$$\begin{cases} \quad u_k = K^\theta \hat{x}_k, \\ u_{c,k} = (1 - \beta^\theta)u_k + \beta^\theta u_{k-1}, \end{cases} \tag{12.5}$$

where $y_{c,k} \in \mathbb{R}^p$ is the measured output, $y_k \in \mathbb{R}^p$ is the actual output, and $u_k \in \mathbb{R}^m$ is the control signal generated by the controller. $K \in \mathbb{R}^{m\times n}$ and $L \in \mathbb{R}^{n\times p}$ denote the controller gains and observer gains that are to be designed. The stochastic variable β^θ, mutually independent of δ^θ, is also a Bernoulli distributed white sequence with expected value $\bar{\beta}^\theta$. Note that the S-C delay is described by the situation that $\delta^\theta = 1$ and the C-A delay is described by $\beta^\theta = 1$. In the sequel, the optimal strategy designed in different layers will be shown and the coupled design for the holistic hybrid model is presented in the end.

12.2.3 Optimal Defense Mechanism

Since IDS is designed to detect unauthorized uses of systems and networks, we use it to defend the networked control system. In practice, IDS is deployed at different levels to monitor the traffic of applications and networks, that is, the IDS is configured with different security enforcement. In this chapter, we use the IDS configuration to represent the cyber defense strategy.

An attacker launches its attack from his attack space $\mathcal{A} := \{a_1, a_2, \cdots, a_M\}$. The set $\mathcal{L} := \{L_1, L_2, \cdots, L_N\}$ denote the defence library and $\bar{\mathcal{L}}$ denote the set of all possible sets of $\mathcal{L}$, with cardinality $|\bar{\mathcal{L}}| = 2^N$. Let $F_i \in \bar{\mathcal{L}}, i \in \{1, 2, \cdots, 2^N\}$ be a configuration set of all libraries. As shown in Figure 12.3, we need different configurations of libraries to detect different attacks. Stationary mixed strategy is used in which $f(\theta, F_i)$ and $g(\theta, a_j)$ are the prob-

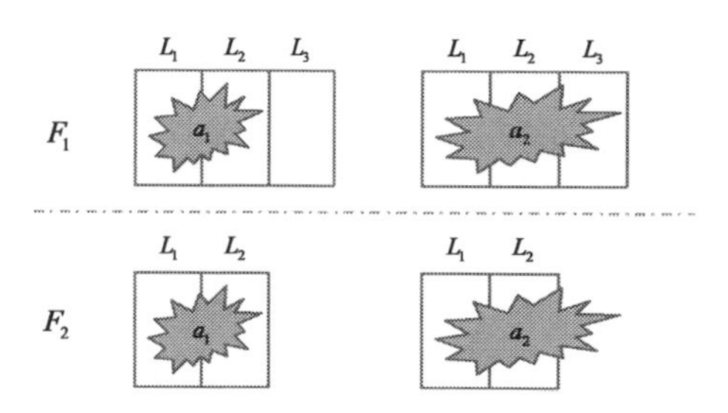

Figure 12.3 An example to illustrate the necessity of different IDS configurations: Library configurations F_1 and F_2 are used to detect different attacks a_1 and a_2; F_2 outperforms F_1 for detecting a_1 since F_1 uses more libraries than F_2 does and degrades the system performance. However, F_1 can detect a_2 better than F_2 does since a_2 is not fully detectable by F_2.

abilities of the detector and attacker choosing actions $F_i \in \bar{\mathcal{L}}$ and $a_j \in \mathcal{A}$, respectively. We denote $f(\theta, F_i)$ and $g(\theta, a_j)$ as $f_i(\theta)$ and $g_j(\theta)$. Note that $f_i(\theta)$ and $g_j(\theta))$ are functions of the random jumping process $\{\theta_n\}$. $\{\theta_n\}$ is a finite state discrete Markov jump process, that is, θ takes discrete values in a given finite set $\Theta := \{\theta_1, \theta_2, \cdots, \theta_s\}$. Functions $f_i : \Theta \longrightarrow [0, 1], i \in \{1, 2, \cdots, 2^N\}$ and $g_j : \Theta \longrightarrow [0, 1], j \in \{1, 2, \cdots, M\}$, need to satisfy $\sum_{i=1}^{2^N} f_i(\theta) = 1$ and $\sum_{j=1}^{M} g_j(\theta) = 1$. The cyber system switches among different states and the transition probabilities $\mathbb{P}(\theta'(n+1) \,|\theta(n), a_j, F_i),\ \theta'(n+1), \theta(n) \in \Theta$ are dependent on the defense mechanism and attack strategy and

$$\sum_{\theta' \in \Theta} \mathbb{P}(\theta' \,|\theta(n), F_i, a_j) = 1.$$

Function $r : \Theta \times \mathcal{A} \times \bar{\mathcal{L}} \longrightarrow \mathbb{R}$ defines the cost of the possible action pair (F_i, a_j) for a certain cyber state θ. The defender can be seen as the minimizer who minimizes the cost function $r(\theta, F_i, a_j)$, and the attacker can be seen as the maximizer who maximizes the cost function, Assuming that the game between the attacker and the defender is zero-sum, we have the relation

$$\begin{aligned} r(\theta, F_i, a_j) &= r^a(\theta, F_i, a_j) \\ &= -r^l(\theta, F_i, a_j), \end{aligned} \tag{12.6}$$

We augment the distribution vectors over all the cyber states Θ and have

$$\begin{aligned}
\mathbf{f}(\theta) &:= [f_1(\theta), \cdots, f_{2^N}(\theta)]^T, \\
\mathbf{g}(\theta) &:= [g_1(\theta), \cdots, g_M(\theta)]^T, \\
\mathbf{F}_s &:= [\mathbf{f}(\theta_1), \cdots, \mathbf{f}(\theta_s)]^T \in \mathbf{F}, \\
\mathbf{G}_s &:= [\mathbf{g}(\theta_1), \cdots, \mathbf{g}(\theta_s)]^T \in \mathbf{G}, \\
\theta \in \Theta &:= \{\theta_1, \theta_2, \cdots \theta_s\}.
\end{aligned}$$

$v_\beta : \Theta \times \mathbf{F} \times \mathbf{G} \longrightarrow \mathbb{R}$ is the β-discounted payoff if

$$v_\beta = \sum\nolimits_{n=0}^{\infty} \beta^n E^{\mathbf{f}(\theta),\mathbf{g}(\theta)} r(\theta, F_i, a_j),$$

where $\beta \in (0,1)$ is a discounted factor.

Remark 12.1. Note that the distribution of stochastic variable δ^θ and β^θ, which reflect the Quality-of-Service (QoS) of the communication network, is actually dependent on the states and attack and defence mechanism in the cyber layer. Let us define four mappings $H_1 : \Theta \times \mathbf{F} \times \mathbf{G} \to \mathbb{R}$, $W_1 : \Theta \times \mathbf{F} \times \mathbf{G} \to \mathbb{R}$, $H_2 : \Theta \times \bar{\mathcal{L}} \times \mathcal{A} \to \mathbb{R}$ and $W_2 : \Theta \times \bar{\mathcal{L}} \times \mathcal{A} \to \mathbb{R}$. $\bar{\delta}^\theta$ and $\bar{\beta}^\theta$can be seen as the result of these mappings;

$$\begin{aligned}
\bar{\delta}^\theta &:= H_1(\theta, \mathbf{f}(\theta), \mathbf{g}(\theta)) = \ \mathbf{f}(\theta)^T H(\theta) \ \mathbf{g}(\theta), \\
\bar{\beta}^\theta &:= W_1(\theta, \mathbf{f}(\theta), \mathbf{g}(\theta)) = \ \mathbf{f}(\theta)^T W(\theta) \ \mathbf{g}(\theta),
\end{aligned}$$

where $[H(\theta)]_{ij} = H_2(\theta, F_i, a_j)$ and $[W(\theta)]_{ij} = W_2(\theta, F_i, a_j)$.

The following definition will present the characterization of the optimal defence mechanism.

Definition 12.2. (Saddle-Point Equilibrium) Let

$$\mathbf{v}_\beta = [v_\beta(\theta_1), \cdots, v_\beta(\theta_s)]^T,$$

A pair $(\mathbf{F}_s^*, \mathbf{G}_s^*)$ is a saddle-point of the β-discounted game if

$$\mathbf{v}_\beta(\mathbf{F}_s^*, \mathbf{G}_s) \le \mathbf{v}_\beta(\mathbf{F}_s^*, \mathbf{G}_s^*) \le \mathbf{v}_\beta(\mathbf{F}_s, \mathbf{G}_s^*), \tag{12.7}$$

in which $\mathbf{F}_s^*, \mathbf{G}_s^*$ are the saddle-point equilibrium defense and attack strategies.

The algorithm to find the optimal defence mechanism $\boldsymbol{F}_s^*$ iteratively is presented in the following theorem.

Theorem 12.3. *The discounted zeros-sum, stochastic game possesses a value* $\boldsymbol{v}_\beta$ *for* $\forall F_i \in \bar{\mathcal{L}},\ \ a_j \in \mathcal{A}$ *that is the unique solution of equations*

$$v_\beta^{N+1}(\theta) = \boldsymbol{val}\{R(\theta)\}, \tag{12.8}$$

$$[R(\theta)]_{ij} = r(\theta, F_i, a_j) + \beta \sum_{\theta' \in \Theta} \mathbb{P}(\theta' \,|\theta, F_i, a_j) v_\beta^N(\theta')]. \tag{12.9}$$

where ***val*** *is a function that yields the game value of a zero-sum matrix game. Then, we can obtain the saddle-point equilibrium strategies to achieve the value with*

$$(\boldsymbol{f}^*(\theta), \boldsymbol{g}^*(\theta)) \in \boldsymbol{arg\ val}\{R(\theta)\},$$

where ***arg val*** *yields the mixed strategies that yield the value of the game.*

12.2.4 H_∞ Optimal Control

The dynamic system model described by (1)-(4) is a hybrid discrete model that has been investigated earlier in [156]. We extend the idea and propose the H_∞ index for the discrete hybrid model which is the expectation over $\mathbf{f}(\theta)$ and $\mathbf{g}(\theta)$ for a given θ. If the initial condition is zero, the H_∞ index γ_θ satisfies

$$E^{\ \mathbf{f}(\theta),\ \mathbf{g}(\theta)} \left\{ \sum\nolimits_{k=0}^{\infty} \{\|z_k\|\} \right\} < \gamma_\theta^2 \sum\nolimits_{k=0}^{\infty} \{\|\omega_k\|^2\} \tag{12.10}$$

for all $\theta \in \Theta$. Before presenting the main theorem of this subsection, we make the following assumptions:

Assumption 12.1 *The dwell time of the hybrid model is long enough, allowing the transient effects to dissipate.*

This assumption is reasonable since the time scale on which cyber events happen is often on the order of days. In contrast, physical systems evolve on the time scale of seconds, which means $k = \varepsilon n$. Under this assumption, the global stability of the controlled system is determined by the stability of each subsystem. The theorem below shows how to convert the conditions satisfying the H_∞ index (9) into the Linear Matrix Inequalities (LMIs), which makes easy convenient to calculate by computational tools.

Theorem 12.4. *Given scalars $\gamma_\theta > 0$ and the strategy pair $(\boldsymbol{f}(\theta),\ \boldsymbol{g}(\theta))$ or all $\theta \in \Theta$. The hybrid model described by (1)-(5) is exponentially mean-square stable and the H_∞-norm constraint (9) is achieved for all nonzero ω_k if there exist positive definite matrices $P_{11}^\theta \in \mathbb{R}^{m\times m}$, $P_{22}^\theta \in \mathbb{R}^{(n-m)\times(n-m)}$, $S_1^\theta \in \mathbb{R}^{n\times n}$ and $P_2^\theta \in \mathbb{R}^{n\times n}$ and $S_2^\theta \in \mathbb{R}^{n\times n}$, and real matrices $M^\theta \in \mathbb{R}^{m\times n}$, $N^\theta \in \mathbb{R}^{n\times p}$ such that the following LMIs hold, where*

$$P_1^\theta := U_1^T P_{11}^\theta U_1 + U_2^T P_{22}^\theta U_2,$$

and $U_1 \in \mathcal{R}^{m\times n}$ *and* $U_2 \in \mathcal{R}^{(n-m)\times n}$ *satisfies*

$$\begin{bmatrix} U_1 \\ U_2 \end{bmatrix} B_2 V = \begin{bmatrix} \Sigma \\ 0 \end{bmatrix}, \quad \Sigma = diag\{\sigma_1, \sigma_2, \cdots, \sigma_m\},$$

and $\sigma_i (i = 1, 2, \cdots, m)$ *are eignvalues of* B_2. *The controller gain and observer gain are given by:*

$$K^\theta = V\Sigma^{-1} {P_{11}^\theta}^{-1} \Sigma V^T M^\theta, \quad L^\theta = {S_1^\theta}^{-1} N^\theta. \tag{12.11}$$

$$\Pi^\theta = \begin{bmatrix} \Pi_{11}^\theta & * \\ \Pi_{21}^\theta & \Pi_{22}^\theta \end{bmatrix} < 0, \tag{12.12}$$

where

$$\Pi_{11}^\theta = \begin{bmatrix} P_2^\theta - P_1^\theta & * & * & * & * \\ 0 & S_2^\theta - S_1^\theta & * & * & * \\ 0 & 0 & -P_2^\theta & * & * \\ 0 & 0 & 0 & -S_2^\theta & * \\ 0 & 0 & 0 & 0 & -\gamma_\theta^2 I \end{bmatrix},$$

$$\Pi_{22}^\theta = \begin{bmatrix} -P_1^\theta & * & * & * & * \\ 0 & -S_1^\theta & * & * & * \\ 0 & 0 & -P_1^\theta & * & * \\ 0 & 0 & 0 & -S_1^\theta & * \\ 0 & 0 & 0 & 0 & -I \end{bmatrix},$$

$$\Pi_{21}^\theta = \begin{bmatrix} \Pi_{21}^\theta(1,1) & \Pi_{21}^\theta(1,2) \\ \Pi_{21}^\theta(2,1) & \Pi_{21}^\theta(2,2) \end{bmatrix},$$

$$\Pi_{21}^\theta(1,1) = \begin{bmatrix} P_1^\theta A + (1-\bar{\beta}^\theta) B_2 M^\theta & -(1-\bar{\beta}^\theta) B_2 M^\theta \\ 0 & S_1^\theta A - (1-\bar{\delta}^\theta) N^\theta C \end{bmatrix},$$

$$\Pi_{21}^\theta(1,2) = \begin{bmatrix} \bar{\beta}^\theta B_2 M^\theta & -\bar{\beta}^\theta B_2 M^\theta & P_1^\theta B_1 \\ 0 & -\bar{\delta}^\theta N^\theta C & S_1^\theta B_1 \end{bmatrix},$$

$$\Pi_{21}^\theta(2,1) = \begin{bmatrix} \alpha_1^\theta B_2 M^\theta & -\alpha_1^\theta B_2 M^\theta \\ \alpha_2^\theta N^\theta C & 0 \\ D & 0 \end{bmatrix},$$

$$\Pi_{21}^\theta(2,2) = \begin{bmatrix} -\alpha_1^\theta B_2 M^\theta & \alpha_1^\theta B_2 M^\theta & 0 \\ -\alpha_2^\theta N^\theta C & 0 & 0 \\ 0 & 0 & 0 \end{bmatrix},$$

$$\alpha_1^\theta = [(1-\bar{\beta}^\theta \bar{\beta}^\theta]^{1/2},$$
$$\alpha_2^\theta = [(1-\bar{\delta}^\theta \bar{\delta}^\theta]^{1/2}.$$

Proof. The proof follows the steps described in [156] and hence is omitted here due to page limitations.

Note that the LMIs in Theorem 12.4 leads to the convex optimization problem as follows.

$$\hat{\gamma}_\theta := \min_{\substack{P_{11}>0, P_{22}>0, P_2>0 \\ S_1>0, S_2>0, M, M}} \gamma_\theta \qquad (12.13)$$
$$\text{subject to} \quad (11)$$

Since γ_θ is influenced by the cyber state and strategy, it is actually dependent on the three-tuple $(\theta,\ \mathbf{f}(\theta),\ \mathbf{g}(\theta))$. Let us define two mappings $C_1 : \Theta \times \mathbf{F} \times \mathbf{G} \to \mathcal{R}$ and $C_2 : \Theta \times \bar{\mathcal{L}} \times \mathcal{A} \to \mathcal{R}$. $\hat{\gamma}_\theta$ can be seen as the value of the mapping:

$$\hat{\gamma}_\theta = C_1(\theta, \mathbf{f}(\theta), \mathbf{g}(\theta)) = \ \mathbf{f}(\theta)^T C(\theta)\ \mathbf{g}(\theta),$$

where $[C(\theta)]_{ij} = C_2(\theta, F_i, a_j)$. Since the design in both physical and cyber layer have been specified, the co-design procedure will be discussed in the next section.

Definition 12.5. A two person zero-sum game $\mathbf{R}(\theta) = [r_{ij}(f(\theta, F_i), g(\theta, a_j))]_{F_i \in \bar{\mathcal{L}},\ a_j \in \mathcal{A}}$ at a particular state θ is said to have a value $V^*(\theta)$

$$\begin{aligned} V^*(\theta) &= \max_{\mathbf{f}(\theta)} \min_{\mathbf{g}(\theta)} \mathbf{f}^T(\mathbf{s})\mathbf{R}(\mathbf{s})\mathbf{g}(\mathbf{s}) \\ &= \min_{\mathbf{g}(\theta)} \max_{\mathbf{f}(\theta)} \mathbf{f}^T(\theta)\mathbf{R}(\theta)\mathbf{g}(\theta) \end{aligned} \qquad (12.14)$$

Problem 12.6. Find a controller such that

- Find the game value of the β-discounted game $\mathbf{v}_\beta$.
- Find the corresponding saddle point $(\mathbf{f}^*, \mathbf{g}^*)$.

Let us denote the defense mechanism and attack strategy to be $f_i(\theta)$ and $g_j(\theta)$, respectively. Note that the distribution of stochastic variable δ and β, which reflect the Quality-of-Service (QoS) of the communication network, is influenced by the actions of the attacker and the defender on the cyber layer. We denote $\delta,\ \beta,\ \bar{\delta},\ \bar{\beta}$ as $\delta(f_i(\theta), g_j(\theta))$, $\beta(f_i(\theta), g_j(\theta))$, $\bar{\delta}(f_i(\theta), g_j(\theta))$, and $\bar{\beta}(f_i(\theta), g_j(\theta))$ henceforth. The subproblem for the control layer is proposed.

Problem 12.7. Find a controller such that

- The closed-loop system is exponentially mean-square stable
- Under the zero-initial condition, the controlled output z satisfies $\sum_{k=0}^{\infty} E^{f_i(\theta), g_j(\theta)}\{\|z_k\|\} < \gamma^2 \sum_{k=0}^{\infty} E\{\|\omega_k\|^2\}$

where γ is the noise attenuation level and is a function of $f_i(\theta)$ and $g_j(\theta)$. Denote γ to be $V(f_i(\theta), g_j(\theta))$ henceforth.

12.2.5 Coupled Design

In this subsection, we provide a cross layer design based on the previous results to demonstrate how to design the resilient controller using a holistic view. The main problem we will address in this chapter is formulated as below:

Problem 12.8. The resilient control of the cyber-physical system against the DoS attack is to find a set of control and observer gains K^θ and L^θ in (12.4) and (12.5) satisfying H_∞ optimal performance $\hat{\gamma}_\theta$, and the optimal cyber policy $\mathbf{F}_s^*$ and $\mathbf{G}_s^*$.

The coupled design means that, on one hand, cyber defence mechanism takes into account the H_∞ index, in which $r(\theta, F_i, a_j) = C_2(\theta, F_i, a_j)$. On the other hand, the H_∞ optimal controller is designed with $\bar{\delta}^\theta = \mathbf{f}^*(\theta)^T H(\theta)\mathbf{g}^*(\theta)$ and $\bar{\beta}^\theta = \mathbf{f}^*(\theta)^T W(\theta)\mathbf{g}^*(\theta)$. We propose the following algorithm for coupled design.

Algorithm 10 Algorithm for Coupled Design

Given: $H_2(\theta, F_i, a_j)$ and $W_2(\theta, F_i, a_j)$ for all $\theta \in \Theta, F_i \in \bar{\mathcal{L}}, a_j \in \mathcal{A}$,
Output: K^θ and L^θ for all $\theta \in \Theta$; $\mathbf{F}_s^*$ and $\mathbf{G}_s^*$.

1. **Initialization**:
2. Initialize v_β^0 and $\beta = 0.5$.
3. **Iterative update**:
4. **while** $(\boldsymbol{v}_\beta^{h+1} - \mathbf{v}_\beta^h > [\varepsilon, \varepsilon, \cdots, \varepsilon]^T)$ **do**
5. Solve the convex optimization problem (12) and we can obtain $C_2(\theta, F_i, a_j)$. We establish the equation $r(\theta, F_i, a_j) = C_2(\theta, F_i, a_j)$.
6. Calculate the cost matrix $R(\theta)$ using (8)
7. Find $v_\beta^{N+1}(\theta)$ using the following LMIs

$$\begin{aligned} \text{(LMG)} \quad v_\beta^{N+1}(\theta) &= \max_{\tilde{y}} \tilde{y}' 1_m \\ \textbf{s.t} \quad R^T(\theta)\, \tilde{y} &\leq 1_n \\ \tilde{y} &\geq 0 \end{aligned}$$

8. **end while**
9. Obtain $\mathbf{F}_s^*$ using $\mathbf{f}^*(\theta) = \tilde{y} v_\beta(\theta)$ and solve the dual problem of (MG), which can be found in [16] to get $\mathbf{G}_s^*$ and $\mathbf{g}^*(\theta)$
10. Use Theorem 12.4 to obtain the controller gain and the observer gain for all $\theta \in \Theta$ with

$$K^\theta = V\Sigma^{-1} {P_{11}^\theta}^{-1} \Sigma V^T M^\theta, \quad L^\theta = {S_1^\theta}^{-1} N^\theta.$$

The following problem is the cross layer design problem we will address in this note and it is composed of problem 12.6 and problem 12.7.

Problem 12.9. The coupling design contains both the decision process and control process.

- Given $\bar{\delta}(f_i(\theta), g_j(\theta))$ and $\bar{\beta}(f_i(\theta), g_j(\theta))$ for all $\theta \in \Theta, i \in \bar{\mathcal{L}}, j \in \mathcal{A}$, find the saddle point $(\mathbf{f}^*, \mathbf{g}^*)$ of the β-discounted game $\mathbf{v}_\beta$, where $r(\theta(k), F_i, a_j)$ denote the H_∞ performance index of the underlying control system.
- The closed-loop system is exponentially mean-square stable
- Under the zero-initial condition, develop the controller such that the controlled output z satisfies
  $\sum_{k=0}^{\infty} E^{\mathbf{f}^*(\theta(k)=i),\mathbf{g}^*(\theta(k)=i)}\{\|z_k\|\} < \gamma^2 \sum_{k=0}^{\infty} E\{\|\omega_k\|^2\}$
  $i = 1, 2, \cdots, n$.

Note that the proposed algorithm above involves a value iteration method for computing the stationary mixed saddle-point equilibrium for the stochastic game, in which a linear program (LMG) is solved at each step. "LMG" stands for linear program for matrix games. The mixed Nash equilibrium of a matrix game is computed by solving a linear program (LMG). The algorithm also invokes the computational tools for solving a set of LMIs for obtaining H_∞ robust controller in the form of (12.4) and (12.5) that achieve optimal control system performances.

12.3 Numerical Simulation

In this section, we investigate the resilient control problem of UPS. UPS usually provides uninterrupted, high quality, and reliable power for vital loads, such as life supporting system, data storage systems, or emergency equipment systems. Thus, the resilience and robustness of the UPS is essential. We perform an integrated design of the optimal defense mechanism for IDS and the optimal control strategy for PWM inverter such that the output AC voltage can maintain the desired setting under the influence of malicious attacks. The discrete-time model at half-load operating point can be found in [156].

$$A = \begin{bmatrix} 0.9226 & -0.6330 & 0 \\ 1.0 & 0 & 0 \\ 0 & 1.0 & 0 \end{bmatrix},$$

$$B_1 = \begin{bmatrix} 0.5 \\ 0 \\ 0.2 \end{bmatrix}, \quad B_2 = \begin{bmatrix} 1 \\ 0 \\ 0 \end{bmatrix},$$

$$D = \begin{bmatrix} 0.1 \; 0 \; 0 \end{bmatrix},$$

$$C = \begin{bmatrix} 23.738 \; 20.287 \; 0 \end{bmatrix}.$$

For the cyber layer, two states are considered: a normal state θ_1 and a compromised state θ_2. We use library l_1 to detect a_1 and use l_2 to detect a_2. Suppose that the system can only load one library each time. We provide the following tables with elements to be the action pairs($[H(\theta)]_{ij}$,$[W(\theta)]_{ij}$). At θ_1, we have

	a_1	a_2
F_1	(0.01,0.01)	(0.05,0.05)
F_2	(0.03,0.03)	(0.01,0.01)

and the transition probabilities are

	a_1	a_2
F_1	(1,0)	(0,1)
F_2	(0,1)	(1,0)

At state θ_2, we have

	a_1	a_2
F_1	(0.06,0.06)	(0.1,0.1)
F_2	(0.08,0.08)	(0.06,0.06)

and the transition probabilities are the same as in θ_1. Then, using Theorem 12.3, we have the cost/reward table for state θ_1,

	a_1	a_2
F_1	0.0994	0.1641
F_2	0.1232	0.0994

and for state θ_2

	a_1	a_2
F_1	0.1961	0.8084
F_2	0.3148	0.1961

respectively. Using Algorithm 9, we obtain the game values at states θ_1 and θ_2 to be $\mathbf{v}_{0.5} = [0.3370 \;\; 0.5299]^T$ The optimal mixed strategies are $\mathbf{f}^*(\theta_1) = [0.4273\ 0.5726]^T$, $\mathbf{f}^*(\theta_2) = [0.2329\ 0.7671]^T$, $\mathbf{g}^*(\theta_1) = [0.5726\ 0.4273]^T$, and $\mathbf{g}^*(\theta_2) = [0.7671\ 0.2329]^T$. In Figure 12.4-12.7, we show the cyber states and physical system performance when an attacker launches an attack a_1. Figure 12.4 shows cyber state of the system under the saddle-point configuration policy. Figure 12.5 shows the steady-state performance of the dynamical system under co-designed controller when it switches between two cyber states. Figure 12.6 shows the H control result under different cyber states, and Figure 12.7 shows the performance under an H robust controller without considering cyber-layer of the system. Comparing Figure 12.7 with Figure 12.5, we can see that the H_∞ performance in Figure 12.5 is much better than the one in Figure 12.7, and the system in Figure 12.7 is more vulnerable to attacks and moves to the compromised state more frequently.

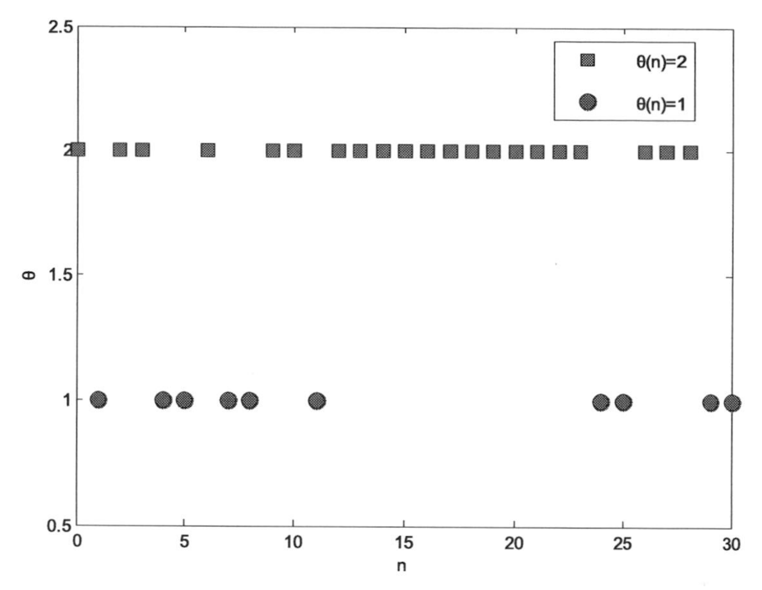

Figure 12.4 shows the evolution of the cyber state over time.

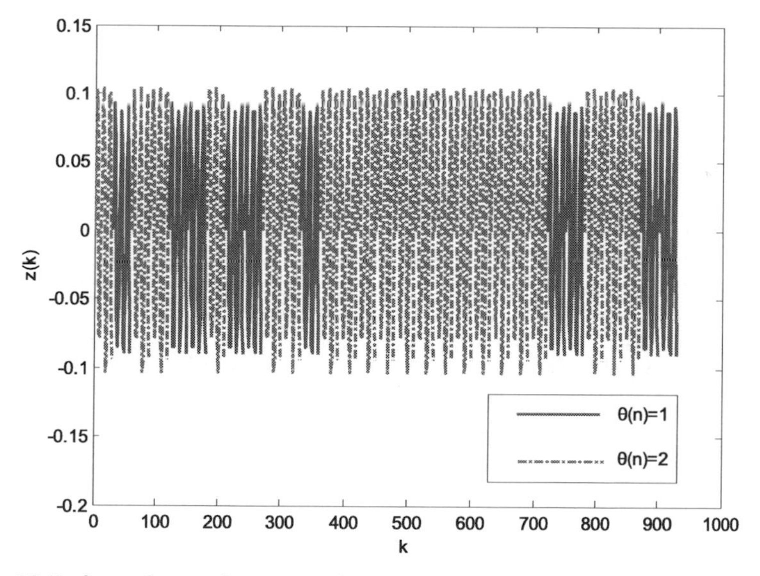

Figure 12.5 shows the steady-state performance under resilient control.

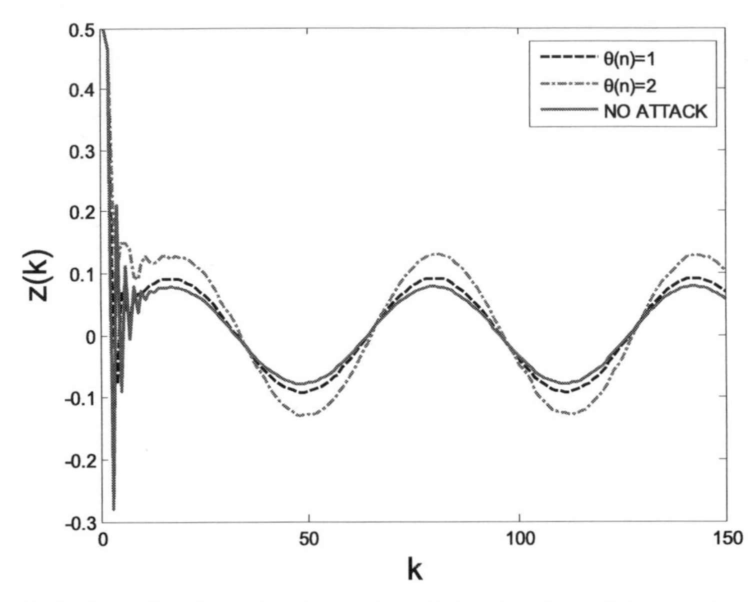

Figure 12.6 shows the H_∞ optimal control applied to the physical dynamical system for each cyber state.

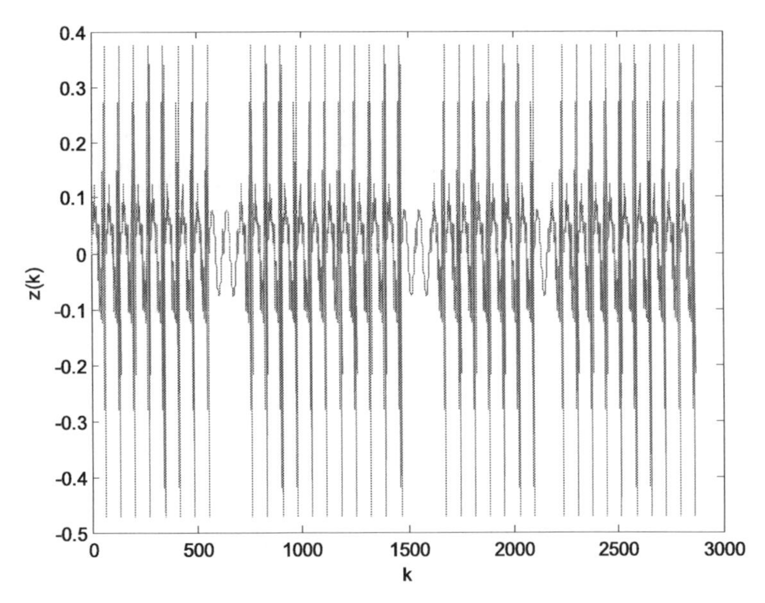

Figure 12.7 shows the performance under an H_∞ robust controller without considering cyber-layer of the system.

12.4 Conclusion

Industrial control systems in many critical infrastructures are subject to malicious cyber attacks. The goal of resilient control systems is to protect the system from such attacks and maintain an acceptable level of operation in the face of cyber attacks and uncertainties. In this chapter, we have proposed a methodology to design the IDS configuration policy at the cyber-layer and the controller for the physical layer dynamical system. We have used a co-design algorithm based on value iterations and linear matrix inequalities to compute the H_∞ optimal control and the cyber security policy. Using numerical examples, we have shown that the design methodology yields a controller that outperforms the H_∞ controller without taking cyber defense into account. This chapter has focused on the denial-of-service attacks and their impact on the cyber security policies and performance of the dynamical system. As future work, we can consider different cyber attack models and study more sophisticated defense strategies. In addition, H_∞ control problem can be also viewed as a game problem between disturbances and controller. By adopting a game-theoretic perspective, we will employ the concepts and tools from our recent initiative on multi-resolution and multi-layer games.

Chapter 13
Attack-Tolerant Control under DoS Attacks

13.1 Introduction

Nowadays, arming automation systems with modern IT results in a number of benefits, such as low maintenance and installation costs, and increased system interoperability. However, on the other hand, the exposure to public networks suggests that the control network is not simple and isolated anymore and can be easily compromised by malicious attacks; see, e.g., [63], [11], and [194]. In this chapter, the term 'resilience', standing for minimizing the impact of the adversary, is captured in both the physical layer and the cyber layer. In the physical layer, the adversary usually takes the form of external disturbance, and resilience is actually the robustness of the system to such disturbances. Thus, we introduce H_∞ optimal control to resist external disturbance in the physical layer. In the cyber layer, we restrict the adversary to be DoS attacks, and resilience can be seen as defending against such attacks. We use IDSs to detect the anomaly in the cyber layer, such as overlong time delays. Then, IDSs raise alarms so that malicious attacks can be removed automatically. Thus, IDSs are used to defend against malicious behavior in the cyber layer. In a nutshell, resilient control suggests a coupled design of H_∞ optimal control and IDSs configuration for CPSs, and such a layered design can be seen in Figure 13.1

Contributions of this chapter are summarized as follows: firstly, a hybrid model for RCS is proposed, in which the cyber security part is modeled as a stochastic Markov game and evolves on a different time scale from that of the physical control part. Two types of discrete operators, the delta operator and forward shift operator, are used in the hybrid model according to their respective numerical advantages. Secondly, the joint optimality is proposed to capture the interdependence of the physical and cyber parts of the hybrid model. And last but not least, a coupled design methodology is proposed to achieve the goal of joint optimality. The algorithms of the coupled design methodology are unified into the framework of ILMIs.

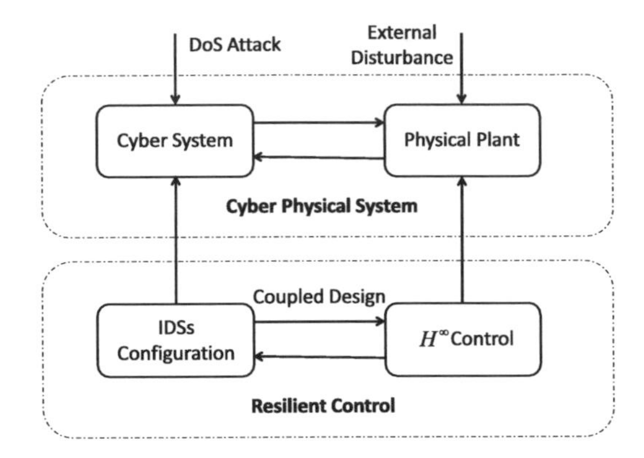

Figure 13.1 Layered Design of Resilient Control: H_∞ control is designed to provide robustness to the physical plant against external disturbances, while IDSs are configured to provide security to the cyber system. The coupled design between the cyber and physical parts of the system leads to a system-level resilient design of the cyber-physical control system.

In Section 13.2, some preliminary results on delta operators are presented. In Section 13.3, we present the hybrid model for RCS. We also provide the designing goal for RCS in terms of joint optimality. In Section 13.4, the conditions of the optimal defense policy and H_∞ optimal control policy are given. Then, a coupled design methodology is proposed by exploiting the relations between the cyber layer and the physical layer. In Section 13.5, we apply the proposed method to UPS and use numerical simulations to demonstrate the effectiveness of our approach and results.

13.2 Preliminaries

In this section, we present some preliminary results on delta operator which will be used henceforth. Let us first recall the q operator, also known as the forward shift operator:

$$qx_k := x_{k+1}. \tag{13.1}$$

The delta operator is defined as $\delta := (q-1)/\mathrm{T}_s$, and can be further expressed as:

$$\delta x_t = \begin{cases} \frac{dx_t}{dt}, & \mathrm{T}_s = 0, \\ \frac{x_{t+\mathrm{T}_s} - x_t}{\mathrm{T}_s}, & \mathrm{T}_s \neq 0, \end{cases} \tag{13.2}$$

where $\mathtt{T}_s$ is a sampling period. It follows from 13.2 that $\delta x_k \to \dot{x}_t$ as $\mathtt{T}_s \to 0$. When $\mathtt{T}_s = 1$, we have $\delta x_k = x_{k+1} - x_k$, which is equivalent to the conventional forward shift operator. Also, the convergence process of $\delta x_k \to \dot{x}_t$ is monotonous. Thus, if the sampling period $\mathtt{T}_s$ is more than one, the numerical advantage of delta operator will be inferior to the shift operator; if $0 < \mathtt{T}_s \ll 1$, the delta operator will outperform shift operator. To sum it up, the conventional forward shift operator q takes the advantage in the fast time scale, and the δ operator has better numerical properties in the slow time scale. Thus, we use q operator to model the fast changing variable, and use δ operator to model the slow changing variable. The definition of the Lyapunov function in the delta domain is defined as follows:

Definition 13.1. [162](Delta Domain Lyapunov Function) Let $V(x_k)$ be a Lyapunov function in delta domain. A delta operator system is stochastic asymptotically stable, if the following conditions hold:

i $V(x_k) \geq 0$, with equality if and only if $x_k = 0$,
ii $\delta V(x_k) = \mathbb{E}[V(x_{k+\mathtt{T}_s}) - V(x_k)]/\mathtt{T}_s < 0$.

Before ending this section, we present the following lemma, which will be used to prove our main results in the sequel.

Lemma 13.2. *[162] For any time functions x_k and y_k, there exists*

$$\delta(x_k y_k) = \delta x_k y_k + x_k \delta y_k + T_s \delta x_k \delta y_k,$$

where T_s is a sampling period.

13.3 Problem Statement

The structure of RCS considered in this chapter is shown in Figure 13.2 We can see from Figure 13.2 that a bank of alternative candidate controllers are built offline and switches with the outcomes of IDSs. The switching is orchestrated by a specially designed logic that uses the measurements that are the sensor information on which the cyber states locate. The purpose of this logic is to decide which candidate controller should be put in the feedback loop at each time instant n. IDSs raise alarms and govern the switching dynamics of the hybrid system. The hybrid model architecture is shown in Figure 13.2. Throughout this chapter, the following assumptions are made.

Assumption 13.1 *Cyber events happen on the order of days, while the underlying physical system evolves on the time scale of seconds. Thus, we have $k = \varepsilon n, \ \ 0 < \varepsilon \ll 1$.*

Assumption 13.2 *The dwell time of the switched control system is large enough and the transient effects have dissipated before next switching.*

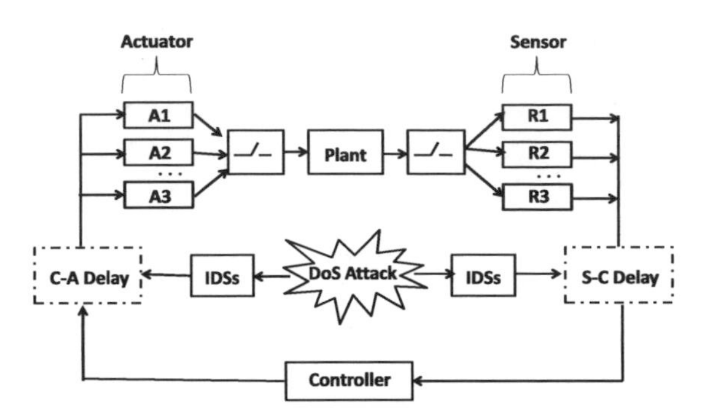

Figure 13.2 Structure of the hybrid model:IDSs are optimally configured to protect the communication channels of control systems from DoS attacks. The observer and controller switch between multiple configurations according to cyber states.

Assumption 13.3 *We assume that the attacker and the defender have completely opposite goals. In particular, for all $i \in \Theta$, $a_q \in \mathcal{A}$ and $F_p \in \bar{\mathcal{L}}$, we have*

$$\begin{aligned} r(i, F_p, a_q) &:= r^a(i, F_p, a_q) \\ &= -r^d(i, F_p, a_q). \end{aligned} \tag{13.3}$$

In a sense, only function $r(i, F_p, a_q)$ is needed, which the attacker (resp. defender) tries to maximize (resp. minimize).

These assumptions are reasonable and also employed in [191]. The state space representation of the model is described by a linear stochastic delta operator model shown as follows:

$$\begin{cases} \delta x_k = Ax_k + B_2 u_{c,k} + B_1 \omega_k, \\ y_k = Cx_k, \\ z_k = Dx_k, \end{cases} \tag{13.4}$$

where $x_k \in \mathbb{R}^n$ is the physical state, $z_k \in \mathbb{R}^r$ is the controlled output, $u_{c,k} \in \mathbb{R}^m$ is the signal received by the actuator, $y_k \in \mathbb{R}^p$ is the actual output, $\omega_k \in \mathbb{R}^q$ is the disturbance input belonging to $l_2[0, \infty)$, A, B_1, B_2, C, and D are matrices with appropriate dimensions. The S-C delay and C-A delay are shown as follows:

$$\text{S-C Delay}: \quad y_{c,k} = (1 - \varsigma^{\theta_n}) y_k + \varsigma^{\theta_n} y_{k-1}, \tag{13.5}$$

$$\text{C-A Delay}: \quad u_{c,k} = (1 - \beta^{\theta_n}) u_k + \beta^{\theta_n} u_{k-1}. \tag{13.6}$$

where $y_{c,k} \in \mathbb{R}^p$ is the measured output, $u_k \in \mathbb{R}^m$ is the control command. ς^{θ_n} (resp. β^{θ_n}) represents the S-C (resp. C-A) delay. $\theta_n : \mathbb{R}_+ \to \Theta := \{1, 2, \cdots, s\}$ is the switching signal here, which excites a particular cyber state over any given time interval. The transition law of the θ_n at n dependents on the strategies of the attacker and the defender. We introduce the stationary mixed strategy [190] here which is denoted as $f(i, F_p)$ and $g(i, a_q)$. $f(i, F_p)$ (resp. $g(i, a_q)$) is the probability of the defender (resp. attacker) choosing action $F_p \in \bar{\mathcal{L}}$ (resp. $a_q \in \mathcal{A}$) at stage $\theta_n = i$. Function $f : \Theta \times \bar{\mathcal{L}} \longrightarrow [0, 1]$ (resp. $g : \Theta \times \mathcal{A} \longrightarrow [0, 1]$) needs to satisfy $\sum_{p=1}^{2^N} f(i, F_p) = 1$ (resp. $\sum_{q=1}^{M} g(i, a_q) = 1$). Thus, the transition probability from state i at time θ_n to state j at time $\theta_{n+\Delta}$, $i, j \in \Theta$, is

$$Pr(\theta_{n+\Delta} = j|\theta_n = i) = \begin{cases} \lambda_{ij}(\mathbf{f}(i), \mathbf{g}(i)) \\ 1 + \lambda_{ij}(\mathbf{f}(i), \mathbf{g}(i)) \end{cases} \tag{13.7}$$

where

$$\begin{aligned} \mathbf{f}(i) &:= [f(i, F_1), \cdots, f(i, F_{2^N}), \forall i \in \Theta, \\ \mathbf{g}(i) &:= [g(i, a_1), \cdots, g(i, a_M)]^T, \forall i \in \Theta, \end{aligned}$$

$\Delta > 0$ is on the same time scale of n, $\lambda_{ij}(\mathbf{f}(i), \mathbf{g}(i))$ is the average transition rate in terms of $\bar{\lambda}_{ij}(F_p, a_q)$, $\forall i, j \in \Theta$, which means

$$\lambda_{ij}(\mathbf{f}(i), \mathbf{g}(i)) = \sum_{q=1}^{M} \sum_{p=1}^{2^N} f(i, F_p) g(i, a_q) \bar{\lambda}_{ij}(F_p, a_q) \tag{13.8}$$

Note that the S-C delay and C-A delay are both stochastic variables distributed according to a Bernouli distribution:

$$\begin{aligned} &\Pr\{\varsigma^{\theta_n} = 1\} = \mathbb{E}\{\varsigma^{\theta_n}\} = \bar{\varsigma}^{\theta_n}, \\ &\Pr\{\varsigma^{\theta_n} = 0\} = 1 - \mathbb{E}\{\varsigma^{\theta_n}\} = 1 - \bar{\varsigma}^{\theta_n}, \\ &\Pr\{\beta^{\theta_n} = 1\} = \mathbb{E}\{\beta^{\theta_n}\} = \bar{\beta}^{\theta_n}, \\ &\Pr\{\beta^{\theta_n} = 0\} = 1 - \mathbb{E}\{\beta^{\theta_n}\} = 1 - \bar{\beta}^{\theta_n}, \end{aligned} \tag{13.9}$$

in which $\Pr\{\cdot\}$ represent the occurrence probability of the event $'\cdot'$. An observer-based controller is used which is shown as follows:

$$\text{Controller} : u_k = K^{\theta_n} \hat{x}_k, \tag{13.10}$$

$$\text{Observer} : \begin{cases} \hat{x}_{k+1} = A\hat{x}_k + B_2 u_{c,k} + L^{\theta_n}(y_{c,k} - \bar{y}_{c,k}) \\ \bar{y}_{c,k} = (1 - \bar{\varsigma}^{\theta_n}) C\hat{x} + \bar{\varsigma}^{\theta_n} C\hat{x}_{k-1}, \end{cases} \tag{13.11}$$

where $\hat{x}_k$ is the estimation state, $K^{\theta_n} \in \mathbb{R}^{m \times n}$ is the control gain, and $L^{\theta_n} \in \mathbb{R}^{n \times p}$ is the observer gain. They both are to be designed. We define the estimation error to be:

$$e_k := x_k - \hat{x}_k. \tag{13.12}$$

Substituting (13.5), (13.6), (13.10) and (13.11) into (13.4) and (13.12), we have the following closed-loop model:

$$\begin{aligned}
\delta x_k &= [A + (1 - \bar{\beta}^{\theta_n}) B_2 K^{\theta_n}] x_k - (1 - \bar{\beta}^{\theta_n}) B_2 K^{\theta_n} e_k \\
&\quad + \bar{\beta}^{\theta_n} B_2 K^{\theta_n} x_{k-1} - \bar{\beta}^{\theta_n} B_2 K^{\theta_n} e_{k-1} + B_1 \omega_k \\
&\quad - (\beta^{\theta_n} - \bar{\beta}^{\theta_n}) B_2 K^{\theta_n} x_k + (\beta^{\theta_n} - \bar{\beta}^{\theta_n}) B_2 K^{\theta_n} e_k \\
&\quad + (\beta^{\theta_n} - \bar{\beta}^{\theta_n}) B_2 K^{\theta_n} x_{k-1} \\
&\quad - (\beta^{\theta_n} - \bar{\beta}^{\theta_n}) B_2 K^{\theta_n} e_{k-1}, \\
\delta e_k &= [A - (1 - \bar{\varsigma}^{\theta_n}) L^{\theta_n} C] e_k - \bar{\varsigma}^{\theta_n} L^{\theta_n} C e_{k-1} \\
&\quad + (\varsigma^{\theta_n} - \bar{\varsigma}^{\theta_n}) L^{\theta_n} C x_k - (\varsigma^{\theta_n} - \bar{\varsigma}^{\theta_n}) L^{\theta_n} C x_{k-1} \\
&\quad + B_1 \omega_k.
\end{aligned} \tag{13.13}$$

The definition of H_∞ optimal index in the delta domain which will be used later is introduced as follows:

Definition 13.3. (Delta Domain H_∞ index) The closed-loop model (13.13) is said to satisfy delta domain H_∞ index over any open interval on which a switching signal $\theta_n = i \in \Theta$ is constant if we have:

$$S_0^\infty \mathbb{E}_{\varsigma^i, \beta^i}\{z_k^T z_k\} dk < \gamma_i^2 S_0^\infty \mathbb{E}_{\varsigma^i, \beta^i}\{\omega_k^T \omega_k\} dk, \tag{13.14}$$

where γ_i is the noise attenuation level and

$$S_0^\infty f_k dk = \begin{cases} \int_0^\infty f_t dt & \text{continuous time,} \\ \mathrm{T}_s \sum_{n=0}^\infty f_{nT_s} & \text{discrete time.} \end{cases}$$

We denote

$$\gamma_i := \sup_\omega \frac{S_0^\infty \mathbb{E}_{\varsigma^i, \beta^i}\{z_k^T z_k\} dk}{S_0^\infty \mathbb{E}_{\varsigma^i, \beta^i}\{\omega_k^T \omega_k\} dk}.$$

The H_∞ optimal index is defined as

$$\begin{aligned}
& J_I^i(\varsigma^i, \beta^i) \\
& := \inf_u \gamma_i = \inf_u \sup_\omega \frac{S_0^\infty \mathbb{E}_{\varsigma^i, \beta^i}\{z_k^T z_k\} dk}{S_0^\infty \mathbb{E}_{\varsigma^i, \beta^i}\{\omega_k^T \omega_k\} dk},
\end{aligned} \tag{13.15}$$

and $\mathbf{J}_I := \left[J_I^1, J_I^2, \cdots, J_I^s\right]^T$.

Note that time delays ς^i and β^i are determined by actions of the defender and attacker and cyber states, namely the tuple $\{F_p, a_q, i\}$ Let us define mapping $H : \Theta \times \bar{\mathcal{L}} \times \mathcal{A} \to \mathbb{R}$ (resp. $W : \Theta \times \bar{\mathcal{L}} \times \mathcal{A} \to \mathbb{R}$) to reflect the interdependence. The scalar ς^i (resp. β^i) can be seen as the result of the mapping $H : \Theta \times \bar{\mathcal{L}} \times \mathcal{A} \to \mathbb{R}$ (resp. $W : \Theta \times \bar{\mathcal{L}} \times \mathcal{A} \to \mathbb{R}$), which is denoted as $H(i, F_p, a_q) := \bar{\varsigma}^i$ (resp. $W(i, F_p, a_q) := \bar{\beta}^i$). With a little abuse of notation, $J_I^i(\varsigma^i, \beta^i)$ can also be denoted as $J_I^i(F_p, a_q)$.

We incorporate the H_∞ index $\mathbf{J}_I$ into the Markov game, since each action in the cyber layer is chosen considering the underlying physical control layer.

To be specific, we let

$$r(i, F_p, a_q) := J_I^i(F_p, a_q). \tag{13.16}$$

In the Markov game, the objective of the defender (resp. attacker) is to minimize (resp. maximize) the sum of discounted expected rewards:

$$J_{II}^i(\mathbf{F}_s, \mathbf{G}_s, J_I^i) := \sum_{n=0}^{\infty} \beta^n \mathbb{E}_{\mathbf{F}_s, \mathbf{G}_s}(J_I^{i,n} | \theta_0 = i) \tag{13.17}$$

where θ_0 indicates the initial time and $J_{II}^i(\mathbf{F}_s, \mathbf{G}_s, J_I^i)$ represents the value for state $\theta_0 = i$ under strategy pair $(\mathbf{F}_s, \mathbf{G}_s)$ and given J_I^i. $J_{II}^i(\mathbf{F}_s, \mathbf{G}_s, J_I^i)$ indicates that the players here optimize their strategies not only with respect to instantaneous costs but also taking into account future discounted costs. The discounted factor β satisfying $\beta \in (0, 1]$ can be thought of as goading the player into trying to win sooner rather than later.

The optimal configuration policy takes the form of saddle point equilibrium (SPE) of the Markov game. At SPE, a player cannot improve its outcome by altering its policy unilaterally while other players play the SPE strategy. Thus, any deviation from the SPE solution by the attacker will increase the benefit of the defender and cost of the attacker. The definition of the SPE configuration is presented as follows:

Definition 13.4. [191](SPE Configuration) Let

$$\mathbf{J}_{II} = \left[J_{II}^1, \cdots, J_{II}^s\right]^T,$$

The pair $(\mathbf{F}_s^*, \mathbf{G}_s^*)$ is the SPE configurations if the following establishes for all $i \in \Theta$ and $\mathbf{J}_I \in \mathbb{R}^s$:

$$J_{II}^i(\mathbf{F}_s^*, \mathbf{G}_s; \mathbf{J}_I) \leq J_{II}^i(\mathbf{F}_s^*, \mathbf{G}_s^*; \mathbf{J}_I) \leq J_{II}^i(\mathbf{F}_s, \mathbf{G}_s^*; \mathbf{J}_I). \tag{13.18}$$

For simplicity, we denote $J_{II}^i(\mathbf{F}_s^*, \mathbf{G}_s^*; \mathbf{J}_I)$ as J_{II}^{i*}, and $\mathbf{J}_{II}(\mathbf{F}_s^*, \mathbf{G}_s^*; \mathbf{J}_I)$ as $\mathbf{J}_{II}^*$ henceforth.

In the sequel, we present the joint optimal policy, based on which the designing goal of resilient control is given.

Definition 13.5. (Joint Optimal Policy) Under the hybrid model structure shown in Figure 13.2, a joint optimal policy for the resilient control is a tuple $\{\mathbf{F}_s^*, \mathbf{G}_s^*, u^*, \omega^*\}$ such that the following criterions are satisfied simultaneously for all $\theta_n = i \in \Theta$:

$$\begin{aligned}
\mathbf{A}_1 &: \ J_I^{i*}(\varsigma_*^i, \beta_*^i) := \inf_u \sup_\omega \frac{S_0^\infty \mathbb{E}_{\varsigma_*^i, \beta_*^i}\{z_k^T z_k\} dk}{S_0^\infty \mathbb{E}_{\varsigma_*^i, \beta_*^i}\{\omega_k^T \omega_k\} dk}, \\
\mathbf{A}_2 &: \ J_{II}^{i*} := \inf_{\mathbf{F}_s} \sup_{\mathbf{G}_s} J_{II}^i(\mathbf{F}_s, \mathbf{G}_s; \mathbf{J}_I),
\end{aligned}$$

where ς_*^i and β_*^i are estimated time delays caused by the adversary. Let us further define the joint optimality as:

$$\mathbf{J}_I^* := \left[J_I^{1*}, \cdots, J_I^{s*}\right]^T, \tag{13.19}$$

$$\mathbf{J}_{II}^* := \left[J_{II}^{1*}, \cdots, J_{II}^{s*}\right]^T. \tag{13.20}$$

The objective of resilient control is to find a set of control and observer gains $\{K^i\}_{i=1}^s$ and $\{L^i\}_{i=1}^s$, and optimal defense strategy $\mathbf{F}_s^*$ and $\mathbf{G}_s^*$ such that criterion $\mathbf{A}_1$ and criterion $\mathbf{A}_2$ are satisfied simultaneously, which suggests that the worst adversaries in both cyber and physical layer are both minimized.

Remark 13.6. $J_{II}^i(\mathbf{F}_s, \mathbf{G}_s, J_I^i)$ actually provides a measure of how well a controller would perform in a conceptual experiment. When a particular $J_{II}^i(\mathbf{F}_s^1, \mathbf{G}_s^1, J_I^i)$ is large, we know that the strategy pair $(\mathbf{F}_s^1, \mathbf{G}_s^1)$ may lead to poor performance in the underlying physical layer and we should avoid using it.

13.4 Optimal Strategy Design

In this section, the conditions for the H_∞ optimal control policy and optimal defense policy are derived, respectively. Then, we provide coupled design methodology for RCS.

13.4.1 Optimal Defense Policy

In this subsection, we present two iterative methods to find the saddle point of (13.17). According to the availability of the transition probability, we adopt value iterative method or Q-learning method. Before moving on, we present the following theorem to find the optimal strategy, which is a combination of Shapleys Theorem [113] and Q-learning process.

Theorem 13.7. *Given J_I^i, $\forall i \in \Theta$, the Q-functions $Q : \Theta \times \mathcal{A} \times \bar{\mathcal{L}} \to \mathbb{R}$ is defined as follows:*

$$Q(i, F_p, a_q) := J_I^i + \beta \sum_{j\in\Theta} \mathbb{P}(j\,|i, F_p, a_q) J_{II}^j. \tag{13.21}$$

and $\boldsymbol{Q}^i := [Q(i, F_p, a_q)]_{F_p\in\bar{\mathcal{L}}, a_q\in\mathcal{A}}$.

B$_1$*: if the transition probability $\mathbb{P}(j|i, F_p, a_q)$, $\forall i, j \in \Theta$ is available, $\boldsymbol{J}_{II}^*$ is obtained iteratively by*

$$J_{II}^{i,N+1} = \boldsymbol{val}\{\boldsymbol{Q}^{i,N}\}. \tag{13.22}$$

B$_2$: *if the transition probability is not available,* $\boldsymbol{J}_{II}^*$ *is obtained iteratively by*

$$Q^{N+1}(i, F_p, a_q) = (1 - \rho_N)Q^N(i, F_p, a_q) \\ +\rho_N[J_I^i + \beta J_{II}^{j,N}], \tag{13.23}$$

where ρ_N *is the step size, which satisfies the following with probability 1:*

$$\sum\nolimits_{N=0}^{\infty} \rho_N = \infty, \quad \sum\nolimits_{N=0}^{\infty} \rho_N^2 < \infty.$$

The two iterations should be run for N_{max} *iterations, where* N_{max} *is sufficiently large. Then, we have the saddle point equilibrium as:*

$$(\boldsymbol{f}^*(i), \boldsymbol{g}^*(i)) \in \boldsymbol{arg\ val}\{\boldsymbol{Q}^{i,N_{max}}\}. \tag{13.24}$$

The proof of Theorem 13.7 is shown as follows:

Proof. For Algorithm $\mathbf{B}_1$, the proof can be seen in [120]. For Algorithm $\mathbf{B}_2$, we adopt a methodology to prove the convergence of the Q-learning process for the Markov game without any additional assumptions on the structure of the equilibrium, which is quite different from the proof in [58].

We define the operator P_N as follows:

$$\{P_N\mathbf{Q}\}(i, F_p, a_q) = J_I^i + \beta\mathbf{val}\{\mathbf{Q}^j\}, \forall i, j \in \Theta. \tag{13.25}$$

Then, we have

$$\begin{aligned} &\mathbb{E}\{\{P_N\mathbf{Q}_*\}(i, F_p, a_q)\} \\ &= \sum\nolimits_{j\in\Theta} \mathbb{P}(j\,|i, F_p, a_q)[J_I^i + \beta\mathbf{val}\{\mathbf{Q}_*^j\}] \\ &= J_I^i + \beta\sum\nolimits_{j\in\Theta} \mathbb{P}(j\,|i, F_p, a_q)\mathbf{val}\{\mathbf{Q}_*^j\} \\ &= \mathbf{Q}_*(i, F_p, a_q). \end{aligned} \tag{13.26}$$

Thus, we show that condition $\mathbf{D}_1$ in Lemma 13.21 is satisfied. On the other hand, we have

$$\begin{aligned} &\|P_N\mathbf{Q} - P_N\mathbf{Q}_*\|_\infty \\ &= \beta \max_{i\in\Theta} |\mathbf{val}\{\mathbf{Q}^i\} - \mathbf{val}\{\mathbf{Q}_*^i\}|. \end{aligned} \tag{13.27}$$

From Definition (13.20), we have

$$|\mathbf{Q}^i - \mathbf{Q}_*^i| \leqslant \|\mathbf{Q}^i - \mathbf{Q}_*^i\|. \tag{13.28}$$

Using the fact that $\mathbf{val}\{A\} \geq \mathbf{val}\{B\}$ for $A \geq B$, we arrive at

$$|\mathbf{val}\{\mathbf{Q}^i\} - \mathbf{val}\{\mathbf{Q}_*^i\}| \leqslant \|\mathbf{Q}^i - \mathbf{Q}_*^i\|. \tag{13.29}$$

Substituting (13.29) into (13.27), we have

$$\begin{aligned} \|P_N\mathbf{Q} - P_N\mathbf{Q}_*\|_\infty &\leq \beta \max_{i\in\Theta} \|\mathbf{Q}^i - \mathbf{Q}^i_*\| \\ &= \beta\|\mathbf{Q} - \mathbf{Q}_*\|_\infty. \end{aligned} \tag{13.30}$$

The proof is completed.

When the transition probability $\mathbb{P}(j|i, F_p, a_q)$, $\forall i, j \in \Theta$ is available, the algorithm to find the optimal defense mechanism $\boldsymbol{F}_s^*$ iteratively is presented in the following theorem.

Theorem 13.8. *[113] The discounted zeros-sum, stochastic game possesses a value $\boldsymbol{v}_\beta^*$ for $\forall F_p \in \bar{\mathcal{L}},\ \ a_q \in \mathcal{A}$ that is the unique solution of equations*

$$v_\beta^{i\ N+1} = \boldsymbol{val}\{R^i\}, \tag{13.31}$$

$$[R^i]_{pq} = r^i(F_p, a_q) + \beta \sum_{j\in\Theta} \mathbb{P}(j\,|i, F_p, a_q) v_\beta^{j\ N}]. \tag{13.32}$$

When the discrepancy between two iterations fall below a prescribed threshold, the iteration process ends and $\boldsymbol{v}_\beta^*$ is obtained. The corresponding saddle point equilibrium strategies are

$$(\boldsymbol{f}^{i*}, \boldsymbol{g}^{i*}) \in \boldsymbol{arg\ val}\{R^i\}. \tag{13.33}$$

If the transition probability is unknown, we use Q-learning approach, which is a model-free reinforcement learning methodology. The Q-function $[Q^i]_{pq}$: $\Theta \times \mathcal{A} \times \bar{\mathcal{L}} \to \mathbb{R}$ is defined as follows:

$$[Q^i]_{pq} := r^i(F_p, a_q) + \beta \sum_{j\in\Theta} \mathbb{P}(j\,|i, F_p, a_q) v_\beta^j. \tag{13.34}$$

Now we are ready to present the following theorem to find the optimal defense mechanism without using the transition probability.

Theorem 13.9. *The value $\boldsymbol{v}_\beta^*$ can be obtained by iteratively calculating the Q-function for all $F_p \in \bar{\mathcal{L}},\ \ a_q \in \mathcal{A}$*

$$\begin{aligned} [Q^i]_{pq}^{N+1} &= (1 - \rho_n)[Q^i]_{pq}^N \\ &\quad + \rho_n[r^i(F_p, a_q) + \beta v_\beta^{j\ N}], \end{aligned} \tag{13.35}$$

where ρ_n is the step size. This algorithm should be run for N_{max} iterations, where N_{max} is sufficiently large. Then, the value $\boldsymbol{v}_\beta^*$ is obtained, and

$$(\boldsymbol{f}^{i*}, \boldsymbol{g}^{i*}) \in \boldsymbol{arg\ val}\{Q^i\} \tag{13.36}$$

Proof. The convergence proof can be found in [58].

Remark 13.10. The existence of saddle point in Theorem 13.7 is guaranteed by Theorem 13.15 in [58], which states that every n-player discounted stochastic game possesses at least one Nash equilibrium point in stationary strategies.

Remark 13.11. In engineering practice, agents often have information limitations. When the transition probability is not known, we use algorithm $\mathbf{B}_2$ in Theorem 13.7, which is called Q-learning and is actually a reinforcement learning method [58]. The Q-learning method is based on recent experience, which can be collected by monitoring the attack and defense consequence of the network. A concept related to information collection is a honeypot, which is an isolated, unprotected, and monitored component of the networked system [4]. We can deploy honeypots in various environments to evaluate the actual attacks and for research purposes. It is worth mentioning that, although Q-learning method ($\mathbf{B}_2$) requires less information than value iterative method ($\mathbf{B}_1$), its convergence speed is much slower, which can be seen in the simulation part of this chapter.

13.4.2 H_∞ Optimal Control

In this subsection, the conditions for the closed-loop system (13.13) to be both exponentially mean-square stable and satisfying the H_∞ index in the delta domain are obtained in terms of LMIs. We present the following definition and lemma which will be used in the sequel.

Definition 13.12. (Exponentially Mean Square Stable) The closed-loop system (13.13) is exponentially mean square stable if there exist constants $b_1 > 0$ and $b_2 \in (0,1)$ such that

$$\mathbb{E}\{\|\eta_k\|^2\} \leq b_1 b_2^m \mathbb{E}\{\|\eta_0\|^2\}, \tag{13.37}$$

where $\eta_k = \begin{bmatrix} x_k \; e_k \; x_{k-1} \; e_{k-1} \end{bmatrix}^T$, η_0 is the initial value and $m \in \mathbb{R}_+$.

Lemma 13.13. *Let $V(\eta_k)$ be a Lyapunov functional in the delta domain. If there exist real scalars $\mu > 0$, $v > 0$, $\psi \in (0,1)$ and $T_s \in (0,1)$ such that*

$$\mu \|\eta_k\|^2 \leq V(\eta_k) \leq v \|\eta_k\|^2, \tag{13.38}$$

and

$$\begin{aligned} \mathbb{E}\{\delta V(\eta_k)\} &= \frac{\mathbb{E}\{V(\eta_{k+1}) \,|\eta_k\} - V(\eta_k)}{T_s} \\ &\leq -\psi V(\eta_k), \end{aligned} \tag{13.39}$$

then the closed-loop system is said to be exponentially mean-square stable with $\omega_k = 0$.

Proof. The proof follows essentially from [176]. It follows from (13.39) that

$$\mathbb{E}\{V(\eta_k)\,|\eta_{k-1}\} \leq (1-\psi\mathtt{T}_s)V(\eta_{k-1}). \tag{13.40}$$

Using the property

$$\mathbb{E}\{V(\eta_k)\,|\eta_{k-2}\} = \mathbb{E}\{\mathbb{E}\{V(\eta_k)\,|\eta_{k-1}\}\,|\eta_{k-2}\}, \tag{13.41}$$

we have

$$\begin{aligned}\mathbb{E}\{V(\eta_k)\,|\eta_{k-2}\} &\leq \mathbb{E}\{(1-\psi\mathtt{T}_s)V(\eta_{k-1})\,|\eta_{k-2}\}\\ &= (1-\psi\mathtt{T}_s)E\{V(\eta_{k-1})\,|\eta_{k-2}\}.\end{aligned} \tag{13.42}$$

If we apply (13.39) again, we have

$$\mathbb{E}\{V(\eta_k)\,|\eta_{k-2}\} \leq (1-\psi\mathtt{T}_s)^2V(\eta_{k-2}). \tag{13.43}$$

Continuing doing this, we get

$$\mathbb{E}\{V(\eta_k)\,|\eta_0\} \leq (1-\psi\mathtt{T}_s)^nV(\eta_0). \tag{13.44}$$

It follows directly from (13.38) that

$$V(\eta_k) \geq \mu\left\|\eta_k\right\|^2, \qquad V(\eta_0) \leq \nu\left\|\eta_0\right\|^2. \tag{13.45}$$

Substituting (13.45) into (13.44), we have

$$\mathbb{E}\left\|\eta_k\right\|^2 \leq (1-\psi\mathtt{T}_s)^m\frac{\nu}{\mu}\left\|\eta_0\right\|^2. \tag{13.46}$$

Since the scalars ψ and $\mathtt{T}_s$ satisfy $\psi \in (0,1)$ and $\mathtt{T}_s \in (0,1)$, we have $\psi\mathtt{T}_s \in (0,1)$. According to Definition 13.12, the proof is completed.

Remark 13.14. Lemma 13.13 is actually an integration of Theorem 13.7 and Theorem 13.15 in [131].

Theorem 13.15. *Given the sampling period T_s, controller gains $\{K^i\}_{i=1}^s$ and observer gains $\{L^i\}_{i=1}^s$, if there exist matrices $\{P_1^i\}_{i=1}^s$, $\{P_2^i\}_{i=1}^s$, $\{S_1^i\}_{i=1}^s$ and $\{S_2^i\}_{i=1}^s$ satisfying the following LMIs*

$$\Pi^i = \begin{bmatrix} \Pi_{11}^i & * \\ \Pi_{21}^i & \Pi_{22}^i \end{bmatrix} < 0, \ \forall i \in \Theta, \tag{13.47}$$

where

$$\Pi_{11}^i = \begin{bmatrix} \Pi_{111}^i & * & * & * \\ \Pi_{121}^i & \Pi_{122}^i & * & * \\ \bar{\beta}^i K^{i^T} B_2^T P_1^i & 0 & -P_2^i & * \\ -\bar{\beta}^i K^{i^T} B_2^T P_1^i & -\bar{\varsigma}^i C^T L^{i^T} S_1^i & 0 & -S_2^i \end{bmatrix},$$

$$\Pi_{21}^i = \begin{bmatrix} \Pi_{211}^i & -(1-\bar{\beta}^i)B_2K^i & \bar{\beta}^i B_2 K^i & -\bar{\beta}^i B_2 K^i \\ 0 & \Pi_{212}^i & 0 & -\bar{\varsigma}^i L^i C \\ B_2 K^i & -B_2 K^i & -B_2 K^i & B_2 K^i \\ L^i C & 0 & -L^i C & 0 \end{bmatrix},$$

$$\Pi_{22}^i = -\frac{1}{\mathtt{T}_s} diag\{P_1^{i^{-1}}, S_1^{i^{-1}}, \alpha_1^{i^{-2}} P_1^{i^{-1}}, \alpha_2^{i^{-2}} S_1^{i^{-1}}\},$$

$$\Pi_{111}^i = P_2^i + A^T P_1^i + (1-\bar{\beta}^i) K^{i^T} B_2^T P_1^i + P_1^i A \\ +(1-\bar{\beta}^i) P_1^i B_2 K^i,$$

$$\Pi_{121}^i = -(1-\bar{\beta}^i) K^{i^T} B_2^T P_1^i,$$

$$\Pi_{122}^i = S_2^i + A - (1-\bar{\varsigma}^i) L^i C^T S_1^i + S_1^i A - (1-\bar{\varsigma}^i) L^i C,$$

$$\Pi_{211}^i = A + (1-\bar{\beta}^i) B_2 K^i,$$

$$\Pi_{212}^i = A - (1-\bar{\varsigma}^i) L^i C,$$

$$\alpha_1^i = [(1-\bar{\beta}^i)\bar{\beta}^i]^{1/2}, \alpha_2^i = [(1-\bar{\varsigma}^i)\bar{\varsigma}^i]^{1/2},$$

then the closed-loop system (13.13) is globally exponentially mean-square stable.

Proof. Define a Lyapunov Functional in the delta domain:

$$V(\eta_k) = V_1(x_k) + V_2(x_{k-1}) + V_3(e_k) + V_4(e_{k-1}), \tag{13.48}$$

where

$$V_1(x_k) = x_k^T P_1^i x_k,$$
$$V_2(x_{k-1}) = \mathtt{T}_s x_{k-1}^T P_2^i x_{k-1},$$
$$V_3(e_k) = e_k^T S_1^i e_k,$$
$$V_4(e_{k-1}) = \mathtt{T}_s e_{k-1}^T S_2^i e_{k-1}.$$

By invoking Lemma 13.3, we get

$$\begin{aligned}
&\mathbb{E}_{\varsigma^i,\beta^i}\{\delta V_1(x_k)\} \\
=&\ \mathbb{E}_{\varsigma^i,\beta^i}\{(\delta x_k)^T P_1^i x_k\} \\
&+\mathbb{E}_{\varsigma^i,\beta^i}\{x_k^T P_1^i \delta x_k\} \\
&+\mathbb{E}_{\varsigma^i,\beta^i}\{\mathtt{T}_s(\delta x_k)^T P_1^i \delta x_k\} \\
=&\ \{[A+(1-\bar{\beta}^i)B_2K^i]x_k-(1-\bar{\beta}^i)B_2K^i e_k \\
&+\bar{\beta}^i B_2K^i x_{k-1}-\bar{\beta}^i B_2K^i e_{k-1}\}P_1^i x_k \\
&+x_k^T P_1^i\{[A+(1-\bar{\beta}^i)B_2K^i]x_k \\
&-(1-\bar{\beta}^i)B_2K^i e_k \\
&+\bar{\beta}^i B_2K^i x_{k-1}-\bar{\beta}^i B_2K^i e_{k-1}\} \\
&+\{[A+(1-\bar{\beta}^i)B_2K^i]x_k-(1-\bar{\beta}^i)B_2K^i e_k \\
&+\bar{\beta}^i B_2K^i x_{k-1}-\bar{\beta}^i B_2K^i e_{k-1}\}^T P_1^i \\
&\{[A+(1-\bar{\beta}^i)B_2K^i]x_k-(1-\bar{\beta}^i)B_2K^i e_k \\
&+\bar{\beta}^i B_2K^i x_{k-1}-\bar{\beta}^i B_2K^i e_{k-1}\} \\
&+\mathtt{T}_s(1-\bar{\beta}^i)\bar{\beta}^i\{B_2K^i x_k-B_2K^i e_k-B_2K^i x_{k-1} \\
&+B_2K^i e_{k-1}\}^T P_1^i\{B_2K^i x_k-B_2K^i e_k \\
&-B_2K^i x_{k-1}+B_2K^i e_{k-1}\},
\end{aligned} \tag{13.49, 13.50}$$

$$\begin{aligned}
&\mathbb{E}_{\varsigma^i,\beta^i}\{\delta V_3(e_k)\} \\
=&\ \mathbb{E}_{\varsigma^i,\beta^i}\{(\delta e_k)^T S_1^i e_k\} \\
&+\mathbb{E}_{\varsigma^i,\beta^i}\{e_k^T S_1^i \delta e_k\}+\mathbb{E}_{\varsigma^i,\beta^i}\{\mathtt{T}_s(\delta e_k)^T S_1^i \delta e_k\} \\
=&\ \{[A-(1-\bar{\varsigma}^i)L^iC]e_k-\bar{\varsigma}^i L^iCe_{k-1}\}^T S_1^i x_k \\
&+x_k^T S_1^i\{[A-(1-\bar{\varsigma}^i)L^iC]e_k-\bar{\varsigma}^i L^iCe_{k-1}\} \\
&+\mathtt{T}_s\{[A-(1-\bar{\varsigma}^i)L^iC]e_k-\bar{\varsigma}^i L^iCe_{k-1} \\
&+(\varsigma^i-\bar{\varsigma}^i)L^iCx_k-(\varsigma^i-\bar{\varsigma}^i)L^iCx_{k-1}\}^T S_1^i \\
&\{[A-(1-\bar{\varsigma}^i)L^iC]e_k-\bar{\varsigma}^i L^iCe_{k-1} \\
&+(\varsigma^i-\bar{\varsigma}^i)L^iCx_k-(\varsigma^i-\bar{\varsigma}^i)L^iCx_{k-1}\}.
\end{aligned} \tag{13.51, 13.52}$$

According to Definition 13.1, we have

$$\begin{aligned}
&\mathbb{E}_{\varsigma^i,\beta^i}\{\delta V_2(x_{k-1})\} \\
=&\ \mathbb{E}_{\varsigma^i,\beta^i}\{\frac{V_2(x_k)-V_2(x_{k-1})}{\mathtt{T}_s}\} \\
=&\ x_k^T P_2^i x_k - x_{k-1}^T P_2^i x_{k-1},
\end{aligned} \tag{13.53}$$

$$\begin{aligned}
&\mathbb{E}_{\varsigma^i,\beta^i}\{\delta V_4(e_{k-1})\} \\
=&\ \mathbb{E}_{\varsigma^i,\beta^i}\{\frac{V_4(e_k)-V_4(e_{k-1})}{\mathtt{T}_s}\} \\
=&\ e_k^T S_2^i e_k - e_{k-1}^T S_2^i e_{k-1}.
\end{aligned} \tag{13.54}$$

It follows directly from (13.48) that

$$\begin{aligned}
&\mathbb{E}_{\varsigma^i,\beta^i}\{\delta V(\eta_k)\} \qquad (13.55)\\
&= \mathbb{E}_{\varsigma^i,\beta^i}\{\delta V_1(x_k)\}\\
&+\mathbb{E}_{\varsigma^i,\beta^i}\{\delta V_2(x_{k-1})\} + \mathbb{E}_{\varsigma^i,\beta^i}\{\delta V_3(e_k)\}\\
&+\mathbb{E}_{\varsigma^i,\beta^i}\{\delta V_4(e_{k-1})\}. \qquad (13.56)
\end{aligned}$$

Substitute (13.50)-(13.54) into (13.56) and apply Shur's Complement [121], we have:

$$\mathbb{E}_{\varsigma^i,\beta^i}\{\delta V(\eta_k)\} = \eta_k^T \Pi^i \eta_k. \qquad (13.57)$$

It follows from (13.47) that

$$\begin{aligned}
\mathbb{E}_{\varsigma^i,\beta^i}\{\delta V(x_k)\} &= \eta_k^T \Pi^i \eta_k \leq -\lambda_{\min}(-\Pi^i)\eta_k^T \eta_k\\
&< -\alpha^i \eta_k^T \eta_k, \qquad (13.58)
\end{aligned}$$

where

$$\begin{aligned}
&0 < \alpha^i < \min\{\lambda_{\min}(-\Pi^i), \sigma^i\},\\
&\sigma^i := \max\{\lambda_{\max}(P_1^i), \lambda_{\max}(S_1^i), \lambda_{\max}(P_2^i), \lambda_{\max}(S_2^i)\}.
\end{aligned}$$

Then we have

$$\begin{aligned}
\mathbb{E}_{\varsigma^i,\beta^i}\{\delta V(\eta_k)\} < -\alpha^i \eta_k^T \eta_k &< -\frac{\alpha^i}{\sigma^i} V(\eta_k)\\
&:= -\psi^i V(\eta_k).
\end{aligned}$$

Thus, the closed-loop subsystem (13.13) is exponentially mean square stable between switching times. The global mean square stability of (13.13) is guaranteed by Assumption 13.2 and Lemma 13.2 in [85]. The proof is completed.

Theorem 13.16. *Given scalars $\{\gamma_i\}_{i=1}^s$, if there exist symmetric matrices $\{P_1^i\}_{i=1}^s$, $\{P_2^i\}_{i=1}^s$, $\{S_1^i\}_{i=1}^s$ and $\{S_2^i\}_{i=1}^s$, and real matrices $\{K^i\}_{i=1}^s$ and $\{L^i\}_{i=1}^s$ such that the following LMIs are feasible*

$$\Theta^i = \begin{bmatrix} \Theta_{11}^i & * \\ \Theta_{21}^i & \Theta_{22}^i \end{bmatrix} < 0, \qquad (13.59)$$

$$\Theta_{11}^i = \begin{bmatrix} \Theta_{111}^i & * & * & * & * \\ \Theta_{121}^i & \Theta_{122}^i & * & * & * \\ \bar{\beta}^i K^{i^T} B_2^T P_1^i & 0 & -P_2^i & * & * \\ -\bar{\beta}^i K^{i^T} B_2^T P_1^i & \Theta_{142}^i & 0 & -S_2^i & * \\ B_1^T P_1^i & B_1^T S_1^i & 0 & 0 & -\gamma_i{}^2 I \end{bmatrix},$$

$$\Theta_{21}^i = \begin{bmatrix} \Theta_{211}^i & \Theta_{212}^i & \bar{\beta}^i B_2 K^i & -\bar{\beta}^i B_2 K^i & B_1 \\ 0 & \Theta_{222}^i & 0 & -\bar{\varsigma}^i L^i C & B_1 \\ B_2 K^i & -B_2 K^i & -B_2 K^i & B_2 K^i & 0 \\ L^i C & 0 & -L^i C & 0 & 0 \\ D & 0 & 0 & 0 & 0 \end{bmatrix},$$

$$\begin{aligned}
\Theta_{22}^i &= diag\{\Theta_{221}^i, -I\}, \\
\Theta_{221}^i &= \frac{1}{T_s} diag\{P_1^{i^{-1}}, S_1^{i^{-1}}, \alpha_1^{i^{-2}} P_1^{i^{-1}}, \alpha_2^i S_1^{i^{-1}}\}, \\
\Theta_{111}^i &= P_2^i + A^T P_1^i + P_1^i A + (1-\bar{\beta}^i) K^{i^T} B_2^T P_1^i \\
&\quad + (1-\bar{\beta}^i) P_1^i B_2 K^i, \\
\Theta_{121}^i &= -(1-\bar{\beta}^i) K^{i^T} B_2^T P_1^i, \\
\Theta_{122}^i &= S_2^i + [A - (1-\bar{\varsigma}^i) L^i C]^T S_1^i + S_1^i [A - (1-\bar{\varsigma}^i) L^i C], \\
\Theta_{142}^i &= -\bar{\varsigma}^i C^{i^T} L^{i^T} S_1^i, \\
\Theta_{211}^i &= A + (1-\bar{\beta}^i) B_2 K^i, \\
\Theta_{212}^i &= -(1-\bar{\beta}^i) B_2 K^i, \\
\Theta_{222}^i &= A - (1-\bar{\varsigma}^i) L^i C,
\end{aligned}$$

then the closed-loop system (13.13) is globally mean-square stable and achieves H_∞ constraint (13.14). Moreover, the controller and observer gain are given by

$$K^i = P^{i^{-1}} M^i,\ L^i = S^{i^{-1}} N^i. \tag{13.60}$$

Proof. It is easy to see that (13.59) implies (13.47). Thus, the global mean-square stability of (13.13) can be guaranteed by Theorem 13.7.

$$\begin{aligned}
&S_0^\infty \mathbb{E}_{\varsigma^i, \beta^i} \{ z_k^T z_k - \gamma_i{}^2 \omega_k^T \omega_k + \delta V(\eta_k) \} \\
&= S_0^\infty \begin{bmatrix} \eta_k \\ \omega_k \end{bmatrix}^T \Lambda^i \begin{bmatrix} \eta_k \\ \omega_k \end{bmatrix},
\end{aligned}$$

where

$$\Lambda^i = \begin{bmatrix} \Pi^i + \bar{D}^T\bar{D} & * \\ \Lambda_{21}^i & -\gamma_i{}^2 I + \mathtt{T}_s B_1^T P_1^i B_1 + \mathtt{T}_s B_1^T S_1^i B_1 \end{bmatrix},$$
$$\Lambda_{21}^i = \begin{bmatrix} B_1^T P_1^i \, 0 \, 0 \, 0 \end{bmatrix} + \mathtt{T}_s B_1^T P_1^i B_3^i$$
$$+\mathtt{T}_s B_1^T S_1^i B_4^i + \begin{bmatrix} 0 \; B_1^T S_1^i \; 0 \; 0 \end{bmatrix},$$
$$\bar{D} = \begin{bmatrix} D \, 0 \, 0 \, 0 \end{bmatrix},$$
$$B_3^i = \begin{bmatrix} B_{31}^i \; -(1-\bar{\beta}^i)B_2 K^i \; \bar{\beta}^i B_2 K^i \; -\bar{\beta}^i B_2 K^i \end{bmatrix},$$
$$B_{31}^i = A + (1-\bar{\beta}^i)B_2 K^i,$$
$$B_4^i = \begin{bmatrix} 0 \; A - (1-\bar{\varsigma}^i)L^i C \; 0 \; -\bar{\varsigma}^i L^i C \end{bmatrix}.$$

Applying Shur complement to (13.59), we have $\Lambda^i < 0$, which is equivalent to

$$S_0^\infty \mathbb{E}_{\varsigma^i,\beta^i}\{z_k^T z_k - \gamma_i{}^2 \omega_k^T \omega_k + \delta V(\eta_k)\}dk < 0. \tag{13.61}$$

Since the system is under zero-initial condition and is exponentially mean square stable, we have $S_0^\infty \delta V(\eta_k)dk = 0$. It is easy to see that

$$S_0^\infty \mathbb{E}_{\varsigma^i,\beta^i}\{z_k^T z_k\}dk < \gamma_i{}^2 S_0^\infty \mathbb{E}_{\varsigma^i,\beta^i}\{\omega_k^T \omega_k\}dk. \tag{13.62}$$

The proof is completed.

Theorem 13.17. *Given scalars $\{\gamma_i\}_{i=1}^s$, if there exist symmetric matrices $\{P_{11}^i\}_{i=1}^s$, $\{P_{22}^i\}_{i=1}^s$, $\{P_2^i\}_{i=1}^s$, $\{S_1^i\}_{i=1}^s$ and $\{S_2^i\}_{i=1}^s$, and real matrices $\{M^i\}_{i=1}^s$ and $\{N^i\}_{i=1}^s$ such that the following LMIs are feasible*

$$\Gamma^i = \begin{bmatrix} \Gamma_{11}^i & * \\ \Gamma_{21}^i & \Gamma_{22}^i \end{bmatrix} < 0, \tag{13.63}$$

$$\Gamma_{11}^i = \begin{bmatrix} \Gamma_{111}^i & * & * & * & * \\ \Gamma_{121}^i & \Gamma_{122}^i & * & * & * \\ \bar{\beta}^i M^{iT} B_2^T & 0 & -P_2^i & * & * \\ -\bar{\beta}^i M^{iT} B_2^T & -\bar{\varsigma}^i C^T N^{iT} & 0 & -S_2^i & * \\ B_1^T P_1^i & B_1^T S_1^i & 0 & 0 & -\gamma_i{}^2 I \end{bmatrix},$$

$$\Gamma_{21}^i = \begin{bmatrix} \Gamma_{211}^i & \Gamma_{212}^i & \bar{\beta}^i B_2 M^i & -\bar{\beta}^i B_2 M^i & P_1^i B_1 \\ 0 & \Gamma_{222}^i & 0 & -\bar{\varsigma}^i N^i C & S_1^i B_1 \\ B_2 M^i & -B_2 M^i & -B_2 M^i & B_2 M^i & 0 \\ N^i C & 0 & -N^i C & 0 & 0 \\ D & 0 & 0 & 0 & 0 \end{bmatrix},$$

$$\begin{aligned}
\Gamma_{22}^i &= diag\{\Gamma_{221}^i, -I\}, \\
\Gamma_{111}^i &= P_2^i + A^T P_1^i + P_1^i A + (1-\bar{\beta}^i) M^{i^T} B_2^T \\
&\quad +(1-\bar{\beta}^i) B_2 M^i, \\
\Gamma_{121}^i &= -(1-\bar{\beta}^i) M^{i^T} B_2^T, \\
\Gamma_{122}^i &= S_2^i + A^T S_1^i - (1-\bar{\varsigma}^i) C^T N^{i^T} + S_1^i A - (1-\bar{\varsigma}^i) N^i C, \\
\Gamma_{211}^i &= P_1^i A + (1-\bar{\beta}^i) B_2 M^i, \\
\Gamma_{212}^i &= -(1-\bar{\beta}^i) B_2 M^i, \\
\Gamma_{222}^i &= S_1^i A - (1-\bar{\varsigma}^i) N^i C, \\
\Gamma_{221}^i &= diag\{\frac{1}{T_s} P_1^{i^{-1}}, \frac{1}{T_s} S_1^{i^{-1}}, \frac{1}{T_s} \alpha_1^{i^{-2}} P_1^{i^{-1}}, \frac{1}{T_s} \alpha_2^{i^{-2}} S_1^{i^{-1}}\},
\end{aligned}$$

where

$$P_1^i := U_1^{i^T} P_{11}^i U_1^i + U_2^{i^T} P_{22}^i U_2^i,$$

$U^i \in \mathbb{R}^{n\times n}$ *and* $V^i \in \mathbb{R}^{m\times m}$ *are orthogonal matrices satisfying*

$$U^i B_2 V^i = \begin{bmatrix} U_1^i \\ U_2^i \end{bmatrix} B_2 V^i = \begin{bmatrix} \Sigma^i \\ 0 \end{bmatrix}, \tag{13.64}$$

$\Sigma = diag\{\sigma_1, \dots, \sigma_m\}$, *and* σ_i $(i = 1, \dots, m)$ *are nonzero singular values of* B_2. *Then, the global system is exponentially mean-square stable and satisfies* H_∞ *constraint (13.14). Moreover, the controllers and observers are given by*

$$K^i = V^i \Sigma^{i^{-1}} P_{11}^{i^{-1}} \Sigma^i V^{i^T} M, \; L^i = S_1^{i^{-1}} N^i. \tag{13.65}$$

Proof. The proof is similar to that of Theorem 13.15 in [156] and omitted.

All the aforementioned LMI conditions are in the delta-domain. If the sampling interval $\mathtt{T}_s = 1$ and $A_z = A - I$, we have the LMI conditions in the discrete domain which are the same as Theorem 13.15 in [156]. If the sampling interval $\mathtt{T}_s$ tends to zero, we have the LMI conditions in the continuous domain, which are shown in the following corollary.

Corollary 13.18. *Given the scalar* $\{\gamma_i\}_{i=1}^s$, *if there exist symmetric matrices* $\{P_{11}^i\}_{i=1}^s$, $\{P_{22}^i\}_{i=1}^s$, $\{P_2^i\}_{i=1}^s$, $\{S_1^i\}_{i=1}^s$ *and* $\{S_2^i\}_{i=1}^s$, *and real matrices* $\{M^i\}_{i=1}^s$ *and* $\{N^i\}_{i=1}^s$ *such that the following LMIs are feasible*

$$\Omega^i < 0, \tag{13.66}$$

$$\Omega^i = \begin{bmatrix} \Omega^i_{11} & * & * & * & * \\ \Omega^i_{21} & \Omega^i_{22} & * & * & * \\ \bar{\beta}^i M^{i^T} B_2^T & 0 & -P_2^i & * & * \\ -\bar{\beta}^i M^{i^T} B_2^T & -\bar{\varsigma}^i C^T N^{i^T} & 0 & -S_2^i & * \\ B_1^T P_1^i & B_1^T S_1^i & 0 & 0 & -{\gamma_i}^2 I \end{bmatrix},$$

$$\Omega^i_{11} = P_2^i + A^T P_1^i + P_1^i A + (1-\bar{\beta}^i) M^{i^T} B_2^T \\ +(1-\bar{\beta}^i) B_2 M^i,$$

$$\Omega^i_{22} = S_2^i + A^T S_1^i - (1-\bar{\varsigma}^i) C^T N^{i^T} + S_1^i A - (1-\bar{\varsigma}^i) N^i C,$$

$$\Omega^i_{21} = -(1-\bar{\beta}^i) M^{i^T} B_2^T,$$

where

$$P_1^i := U_1^{i^T} P_{11}^i U_1^i + U_2^{i^T} P_{22}^i U_2^i,$$

U_1^i and U_2^i are defined in (13.64), then system (13.13) is globally exponentially mean-square stable with $\mathtt{T}_s \to 0$ and satisfies

$$\int_0^\infty \mathbb{E}_{\varsigma^i,\beta^i}\{z_k^T z_k\} dk < {\gamma_i}^2 \int_0^\infty \mathbb{E}_{\varsigma^i,\beta^i}\{\omega_k^T \omega_k\} dk. \tag{13.67}$$

Proof. Let the sampling time $\mathtt{T}_s \to 0$ in (13.63) and (13.66) can be obtained.

The LMI conditions in Theorem 13.17 actually suggest the convex optimization problem as follows:

$$J_I^i(F_p, a_q) = \inf_{\substack{P_{11}^i>0, P_{22}^i>0, P_2^i>0 \\ S_1^i>0, S_2^i>0, M^i, N^i}} \gamma_i \tag{13.68}$$

subject to (13.63).

Now that the optimal policies in both physical layer and cyber layer have been obtained, we are ready to present the coupled design methodology.

13.4.3 Joint Optimal Policy Design

In this subsection, a cross layer coupled design approach is provided using the previous results. It demonstrates the methodology to design a resilient controller from a holistic view. To be specific, we incorporate the control performance into the Markov game by letting

$$r(i, F_p, a_q; \mathbf{J}_I) = J_I^i(F_p, a_q). \tag{13.69}$$

We denote

$$\mathbf{H}^i = [H(i, F_p, a_q)]_{F_p \in \bar{\mathcal{L}}, a_q \in \mathcal{A}}, \tag{13.70}$$
$$\mathbf{W}^i = [W(i, F_p, a_q)]_{F_p \in \bar{\mathcal{L}}, a_q \in \mathcal{A}}. \tag{13.71}$$

Functions $\mathbf{H}^i$ and $\mathbf{W}^i$ can be extended to the set of all strategies. Let function $H_1^i : \mathbf{F} \times \mathbf{G} \to \mathbb{R}$ and $W_1^i : \mathbf{F} \times \mathbf{G} \to \mathbb{R}$ be the extension, then it is defined as:

$$H_1^i(\mathbf{f}(i), \mathbf{g}(i)) := \mathbf{f}(i)^T \mathbf{H}^i \ \mathbf{g}(i), \tag{13.72}$$
$$W_1^i(\mathbf{f}(i), \mathbf{g}(i)) := \mathbf{f}(i)^T \mathbf{W}^i \ \mathbf{g}(i). \tag{13.73}$$

Now we are ready to present the following theorem such that $\mathbf{J}_I^*$ and $\mathbf{J}_{II}^*$ are achieved simultaneously.

Before presenting the proof of Theorem 13.7, we give the following definition and lemma.

Lemma 13.19. *[14] Let A be an $m \times n$ matrix game. Introduce the following two LMIs*

$$\begin{aligned} \textit{Primal problem:} \quad & V_p = \max\ \tilde{y}^T l_m \\ \textit{subject to} \quad & A^T \tilde{y} \le l_n, \\ & \tilde{y} \ge 0, \\ \textit{where} \quad & l_n = \{1, \dots, 1\}^T \in \mathbb{R}^n. \end{aligned}$$

$$\begin{aligned} \textit{Dual problem:} \quad & V_d = \min\ \tilde{z}^T l_n \\ \textit{subject to} \quad & A\tilde{z} \ge l_m, \\ & \tilde{z} \ge 0, \\ \textit{where} \quad & l_m = \{1, \dots, 1\}^T \in \mathbb{R}^m. \end{aligned}$$

with their optimal values (if existed) denoted by V_p and V_d, respectively. Then, we have the following results:

*i Both LMIs admit a solution, and $V_p = V_d = 1/\boldsymbol{val}(A)$. **val** is a function that yields the game value of a zero-sum matrix game.*
ii If $\tilde{y}^*/\boldsymbol{val}(A)$ solves the primal problem, and $\tilde{z}^*/\boldsymbol{val}(A)$ solves the dual problem, $(\tilde{y}^*, \tilde{z}^*)$ is the optimal mixed strategy of the matrix A, in which $\tilde{y}^*$ (resp. $\tilde{z}^*$) is the optimal strategy of the minimizer (resp. maximizer). The tuple $(\tilde{y}^*, \tilde{z}^*)$ can be further expressed as $(\tilde{y}^*, \tilde{z}^*) = \boldsymbol{arg\ val}\{A\}$.

Definition 13.20. Let $\mathbf{Q}_1 \in \mathcal{Q}$ (resp. $\mathbf{Q}_2 \in \mathcal{Q}$) and $\mathbf{Q}_1^i$ (resp. $\mathbf{Q}_2^i$) be the ith element of $\mathbf{Q}_1$ (resp. $\mathbf{Q}_2$). We define

$$\mathbf{Q}_1^i = [Q_1(i, F_p, a_q)]_{F_p \in \bar{\mathcal{L}}, a_q \in \mathcal{A}},$$
$$\mathbf{Q}_2^i = [Q_2(i, F_p, a_q)]_{F_p \in \bar{\mathcal{L}}, a_q \in \mathcal{A}}.$$

The norm $\|\cdot\|$ is defined as:

$$\begin{aligned}&\left\|\mathbf{Q}_1^i - \mathbf{Q}_2^i\right\| \\ &= \max_{F_p \in \bar{\mathcal{L}}, a_q \in \mathcal{A}} \left|Q_1(i, F_p, a_q) - Q_2(i, F_p, a_q)\right|. \end{aligned} \tag{13.74}$$

Note that $|\cdot|$ here represents the absolute value of $'\cdot'$. Then, we can further define the norm $\|\cdot\|_\infty : \mathcal{Q} \to \mathbb{R}$ as:

$$\|\mathbf{Q}_1 - \mathbf{Q}_2\|_\infty = \max_{i \in \Theta} \|\mathbf{Q}_1^i - \mathbf{Q}_2^i\|. \tag{13.75}$$

Lemma 13.21. *[125] Let the mapping $P_N : \mathcal{Q} \to \mathcal{Q}$ satisfy*

$\boldsymbol{D}_1 : \quad \boldsymbol{Q}_* = \mathbb{E}\{P_N \boldsymbol{Q}_*\},$
$\boldsymbol{D}_2 : \quad \|P_N \boldsymbol{Q} - P_N \boldsymbol{Q}_*\|_\infty$
$\quad \leq \varpi \|\boldsymbol{Q} - \boldsymbol{Q}_*\|_\infty + \lambda_N,$
where $\lambda_N > 0$, $Pr\{\lim_{N \to \infty} \lambda_N = 0\} = 1$, $\varpi \in (0,1)$. Then, the following iteration converges to $\boldsymbol{Q}_*$ with probability 1.

$$\begin{aligned}\boldsymbol{Q}^{N+1} &= (1 - \rho_N)\boldsymbol{Q}^N \\ &\quad + \rho_N(P_N \boldsymbol{Q}^N). \end{aligned} \tag{13.76}$$

Theorem 13.22. *Given $\boldsymbol{H}^i$ and $\boldsymbol{W}^i$, if there exist $\boldsymbol{F}_s^*$ and $\boldsymbol{G}_s^*$ such that (13.18) is satisfied, and $\{K^i\}_{i=1}^s$ and $\{L^i\}_{i=1}^s$ such that (13.68) is solved with*

$$\bar{\varsigma}^i = \bar{\varsigma}_*^i := H_1^i(\boldsymbol{f}^*(i), \boldsymbol{g}^*(i)), \tag{13.77}$$
$$\bar{\beta}^i = \bar{\beta}_*^i := W_1^i(\boldsymbol{f}^*(i), \boldsymbol{g}^*(i)), \tag{13.78}$$

then the joint optimality $\boldsymbol{J}_I^*$ and $\boldsymbol{J}_{II}^*$ are achieved simultaneously.

Proof. Since $\mathbf{F}_s^*$ and $\mathbf{G}_s^*$ satisfy (13.18), they are saddle point equilibrium. According to [14], we have

$$J_{II}^{i*} = \inf_{\mathbf{F}_s} \sup_{\mathbf{G}_s} J_{II}^i = \sup_{\mathbf{G}_s} \inf_{\mathbf{F}_s} J_{II}^i \tag{13.79}$$

Thus, we arrive at $\mathbf{J}_{II}^*$ using $\mathbf{F}_s^*$ and $\mathbf{G}_s^*$. On the other hand, if the defender and attacker are reasonable and use their respective optimal policies $\mathbf{F}_s^*$ and $\mathbf{G}_s^*$, we predict the S-C delay and C-A delay to be

$$\bar{\varsigma}_*^i = \mathbb{E}_{\mathbf{f}^*(i),\, \mathbf{g}^*(i)}\{\mathbf{H}_1^i\} = \mathbf{f}^*(i)^T \mathbf{H}^i\, \mathbf{g}^*(i), \ \forall i \in \Theta, \tag{13.80}$$
$$\bar{\beta}_*^i = \mathbb{E}_{\mathbf{f}^*(i),\, \mathbf{g}^*(i)}\{\mathbf{W}_1^i\} = \mathbf{f}^*(i)^T \mathbf{W}^i\, \mathbf{g}^i(i), \ \forall i \in \Theta. \tag{13.81}$$

Substituting the estimated delays into (13.63) and solving (13.68), we can get $\mathbf{J}_I^*$. The proof is completed.

The following corollary follows directly from Theorem 13.7 and Theorem 13.22.

Corollary 13.23. *In the coupled design, we propose the following algorithms according to the availability of the transition probability $\mathbb{P}(j|i, F_p, a_q)$.*

C1: If the transition probability $\mathbb{P}(j|i, F_p, a_q)$ is known, Algorithm 11 is proposed:

Algorithm 11 Algorithm for Coupled Design with Known Transition Probability

---

Given: $\mathbf{H}^i$, $\mathbf{W}^i$ and $\mathbb{P}(j|i, F_p, a_q)$, $\forall i, j \in \Theta, \forall F_p \in \bar{\mathcal{L}}, \forall a_q \in \mathcal{A}$,
Output: $\{K^i\}_{i=1}^s$ and $\{L^i\}_{i=1}^s$; $\mathbf{F}_s^*$ and $\mathbf{G}_s^*$.

1. **Initialization**:
2. Initialize $\mathbf{J}_{II}^0$ and $\beta = 0.5$.
3. **Iterative update**:
4. **while** $(\mathbf{J}_{II}^{N+1} - \mathbf{J}_{II}^N > [\varepsilon, \varepsilon, \cdots, \varepsilon]')$ **do**
5. Solve the convex optimization problem (13.68) using $H(i, F_p, a_q)$ and $W(i, F_p, a_q)$, and obtain $J_I^i(F_p, a_q)$ to establish $r(i, F_p, a_q; \mathbf{J}_I) = J_I^i(F_p, a_q)$.
6. Calculate the Q-function $Q(i, F_p, a_q; \mathbf{J}_I)$ using (13.34).
7. Update $\mathbf{J}_{II}$ using (13.31).
8. **end while**
9. Obtain $\mathbf{F}_s^*$ and $\mathbf{G}_s^*$ using (13.36).
10. Use (13.77) and (13.78) to estimate the DoS inducing delays, and apply Theorem 13.22 to obtain $\{K^i\}_{i=1}^s$ and $\{L^i\}_{i=1}^s$.

C2: If the transition probability $\mathbb{P}(j|i, F_p, a_q)$ is completely unknown, Algorithm 12 is proposed:

- ***Case 1:*** *Resilient control design with full cyber layer information*
- ***Case 2:*** *Resilient control design with partial cyber layer information*

13.5 Numerical Simulation

In this section, the resilient control problem is investigated associated with the uninterrupted power system (UPS). UPS is used to provide uninterrupted, high quality and reliable power for safety critical systems, such as data storage systems, life supporting systems, and emergency systems. Thus, resilience and robustness is extremely vital for UPS. We perform a coupled design of the optimal defense policy for IDSs and optimal control strategy such that the output voltage maintains desired setting under the influence of DoS attacks.

Algorithm 12 Algorithm for Coupled Design with Unknown Transition Probability

Given: $\mathbf{H}^i$ and $\mathbf{W}^i$, $\forall i \in \Theta$,
Output: $\{K^i\}_{i=1}^s$ and $\{L^i\}_{i=1}^s$; $\mathbf{F}_s^*$ and $\mathbf{G}_s^*$.

1. **Initialization**:
2. Initialize $Q^0(i, F_p, a_q; \mathbf{J}_I) = 0$, $\rho_N = 0.1$ and $\beta = 0.5$.
3. **Loop**:
4. Choose actions F_p^N and a_q^N.
5. Solve the convex optimization problem (13.68) using $H(i, F_p, a_q)$ and $W(i, F_p, a_q)$, and then observe the outcome of $r(i, F_p, a_q; \mathbf{J}_I)$.
6. Update $Q(i, F_p, a_q; \mathbf{J}_I)$ using (13.35).
7. Let N:=N+1
8. Obtain $\mathbf{F}_s^*$ and $\mathbf{G}_s^*$ using (13.36).
9. Use (13.77) and (13.78) to estimate the DoS inducing delays, and apply Theorem 13.22 to obtain $\{K^i\}_{i=1}^s$ and $\{L^i\}_{i=1}^s$.

By using the delta operator and letting the sampling period be $\mathtt{T}_s = 0.01s$, the delta-domain model is obtained with matrices to be:

$$A = \begin{bmatrix} -7.7400 & -63.3000 & 0 \\ 100.0000 & -100.0000 & 0 \\ 0 & 100.0000 & -100.0000 \end{bmatrix},$$

$$B_1 = \begin{bmatrix} 50 \\ 0 \\ 20 \end{bmatrix}, \quad B_2 = \begin{bmatrix} 100 \\ 0 \\ 0 \end{bmatrix},$$

$$D = \begin{bmatrix} 0.1 \; 0 \; 0 \end{bmatrix}, \; C = \begin{bmatrix} 23.738 \; 20.287 \; 0 \end{bmatrix}. \tag{13.82}$$

For the cyber layer, two states are considered: a normal state 1 and a compromised state 2. We use library l_1 to detect a_1 and use l_2 to detect a_2. Suppose that the system can only load one library at a time. We provide the following tables with elements to be the pairs $(H(i, F_p, a_q), W(i, F_p, a_q))$, $\forall i, p, q \in \{1, 2\}$. At state 1, we have

	a_1	a_2
F_1	(0.02,0.02)	(0.06,0.06)
F_2	(0.04,0.04)	(0.02,0.02)

and the transition probabilities are

	a_1	a_2
F_1	(0.8,0.2)	(0.2,0.8)
F_2	(0.2,0.8)	(0.8,0.2)

At state 2, we have

	a_1	a_2
F_1	(0.05,0.05)	(0.1,0.1)
F_2	(0.08,0.08)	(0.05,0.05)

The transition probabilities are the same as those of state 1. Then, solving the convex optimization problem 13.68, we present the following cost/reward table with element to be $r(i, F_p, a_q; \mathbf{J}_I)$. At state 1, we have:

	a_1	a_2
F_1	0.0110	0.0196
F_2	0.0141	0.0110

At state 2, we have

	a_1	a_2
F_1	0.0164	0.0808
F_2	0.0325	0.0164

Using Algorithm 11, we obtain $\mathbf{J}_{II}^* = [0.0354 \quad 0.0520]^T$. The iterative process is shown in Figure13.3. It can be see from Figure 13.4 that $\mathbf{J}_{II}^* =$

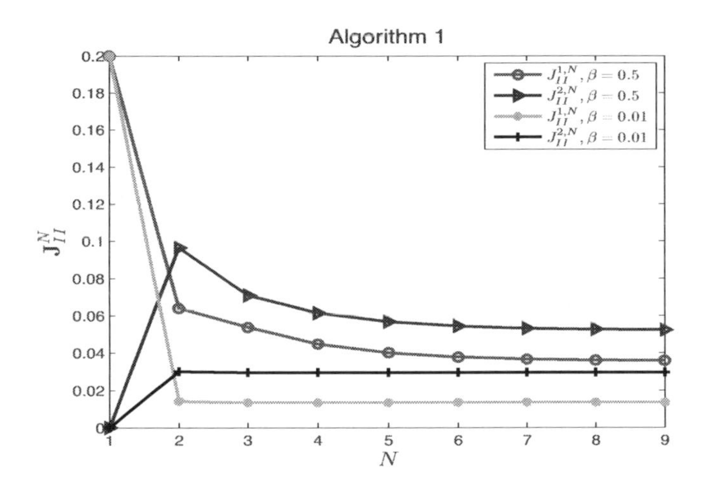

Figure 13.3 The trajectories of $\mathbf{J}_{II}^*$ using value iterative method.

$[0.0357 \quad 0.0523]^T$ is obtained by using Algorithm 12. This result is very close to the result of Algorithm 11, and the inconsistency is due to the randomness of the experiment. Comparing Figure 13.4 with Figure 13.3, we can see that Figure 13.4 needs more iterations to find the optimal value since Algorithm 12 requires less expert information. The transition probability in Algorithm 12 has to be learnt through a large number of experiments. Thus, Algorithm 11 and Algorithm 12 each has its own advantage. The optimal IDSs configuration policies using Algorithm 11 and Algorithm 12 are shown in Table 13.1.

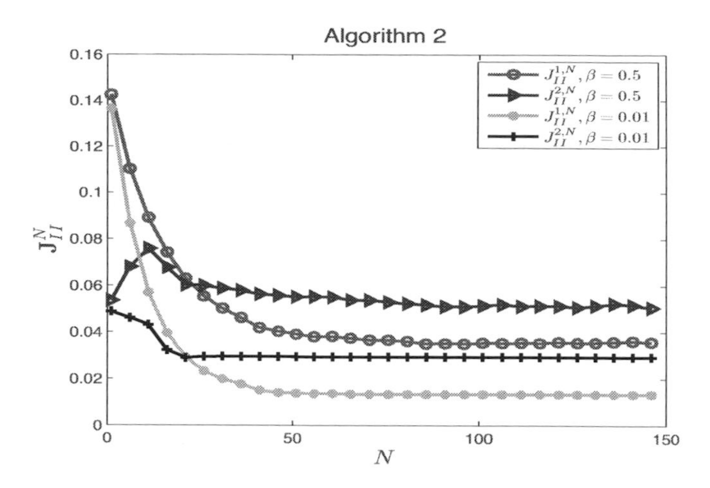

Figure 13.4 The trajectories of $\mathbf{J}_{II}^*$ using Q-learning method.

Table 13.1 The comparison of optimal IDSs configuration policies using Algorithm 11 and Algorithm 12

	Algorithm 11	Algorithm 12
$\mathbf{f}^*(1)$	$[0.3729\ 0.6271]^T$	$[0.3616\ 0.6384]^T$
$\mathbf{f}^*(2)$	$[0.2304\ 0.7696]^T$	$[0.2319\ 0.7681]^T$
$\mathbf{g}^*(1)$	$[0.6271\ 0.3729]^T$	$[0.6116\ 0.3884]^T$
$\mathbf{g}^*(2)$	$[0.7696\ 0.2304]^T$	$[0.7745\ 0.2255]^T$

Without loss of generality, we will adopt the optimal strategies obtained using Algorithm 11 henceforth. An Adversary Simulation Unit (ASU) is designed as

$$\begin{cases} o(n) = G_s^*, & \forall n, \\ o(k) = \sin(0.2k), & \forall k. \end{cases}$$

where $o(n)$ is the output of the ASU in the cyber layer, and $o(k)$ is the output of the ASU in the physical layer. Let the defender choose $\mathbf{F}_s^*$ as the defense policy, and the expectation of time delays can be calculated using 13.77 and 13.78.

$$\bar{\varsigma}_*^1 = \bar{\beta}_*^1 = 0.1286, \;\; \bar{\varsigma}_*^2 = \bar{\beta}_*^2 = 0.2468.$$

Then, we can use Theorem 13.17 to find the coupled H_∞ optimal controller and observer. The joint optimality and optimal policies in the physical layer are summarized in Table 13.2. The experimental results are shown in Figure 13.5-Figure 13.7. Figure 13.5 shows the H_∞ optimal control result on different cyber states. It can be seen that the H_∞ performance of the normal state is better than that of the compromised state. In Figure 13.6, the evolution

Table 13.2 Joint optimality and joint optimal policies in the physical layer

	State 1	State 2
$\mathbf{J}_I^*$	0.1286	0.2468
$\mathbf{J}_{II}^*$	0.0354	0.0520
K	$[-0.9412\ 0.6465\ \ -0.0000]$	$[-0.7198\ 0.7653\ \ -0.0002]$
L	$[2.7329\ 2.9661\ 1.2409]^T$	$[1.4930\ 2.4224\ 1.1834]^T$

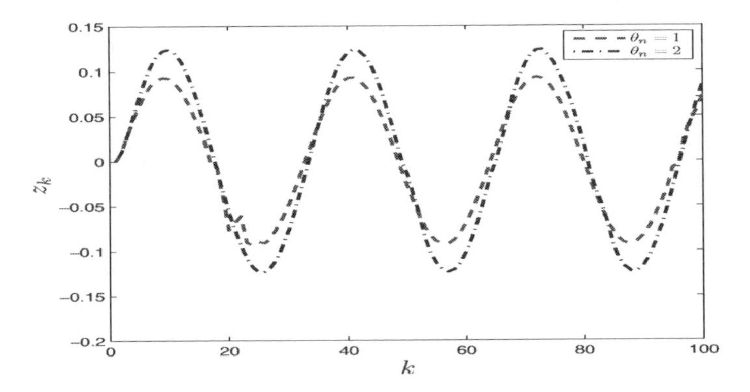

Figure 13.5 H_∞ optimal control result under different cyber states.

of the cyber state under the SPE configuration policy is shown, which can also be seen as the switching signal. Figure 13.7 is actually a combination of Figure 13.5 and Figure 13.6, and demonstrates the steady-state performance of system 13.82 under the co-designed H_∞ optimal control strategy and SPE configuration policy.

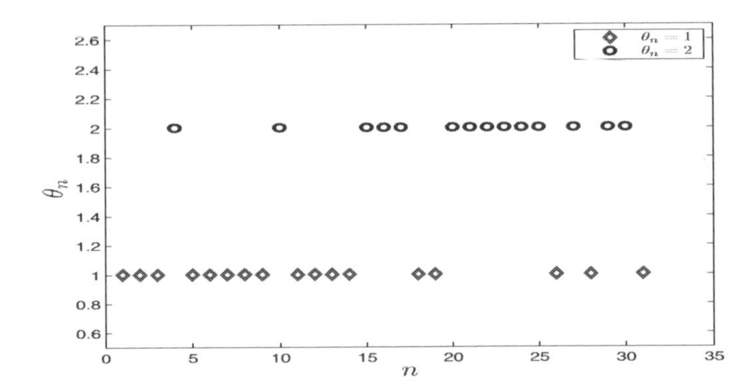

Figure 13.6 The evolution of the cyber state under the SPE configuration.

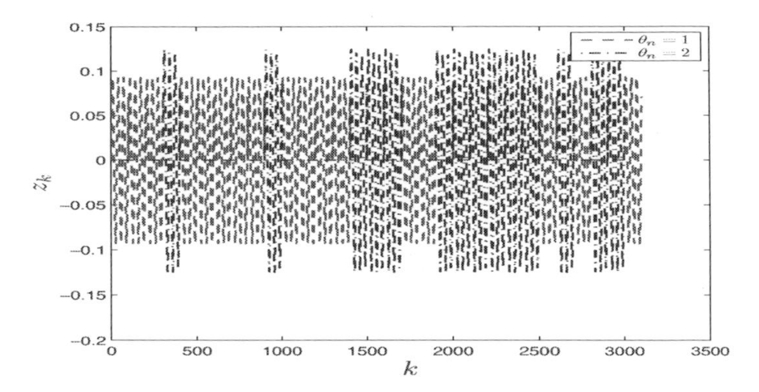

Figure 13.7 Steady-state performance of the dynamical system under co-designed controller when it switches between two cyber states.

13.6 Conclusion

CPSs in many critical infrastructures are exposed to the public and could be targeted by adversaries to disrupt normal operation. The goal of resilient control is to protect the system against cyber attacks and maintain or provide graceful degradation operational goals. In this chapter, a methodology to co-design the optimal IDSs configuration policy and the H_∞ optimal strategy has been proposed. The algorithms to find the optimal control and security policies have been unified into the framework of ILMIs. The proposed approach has been applied to the control of the UPS and we have shown that our method outperforms the controller, which has no resilience. This chapter has focused on the DoS attacks and their impact on the cyber security policies and performance of the dynamical system. As future work, we can consider different cyber attack models and study more sophisticated defense strategies.

References

1. M. Abouheaf, F. Lewis, K. Vamvoudakis, S. Haesaert, and R. Babuska. Multi-agent discrete-time graphical games and reinforcement learning solutions. *Automatica*, 50(12):3083–3053, 2014.
2. T. Alpcan and T. Başar. A game theoretic approach to decision and analysis in network intrusion detection. In *Proceedings of the 42nd IEEE Conference on Decision and Control*, pages 2595–2600, Maui, Hawaii, USA, 2003.
3. T. Alpcan and T. Başar. A game theoretic analysis of intrusion detection in access control systems. In *Proceedings of the 43rd IEEE Conference on Decision and Control*, pages 1568–1573, 2004.
4. T. Alpcan and T. Başar. *Network Security: A Decision and Game Theoretic Approach*. Cambridge, U.K.: Cambridge Univ. Press, 2010.
5. T. Alpcan, T. Basar, and R. Srikant. Cdma uplink power control as a noncooperative game. *Wireless Networks*, 8(6):659–670, 2002.
6. E. Altman, K. Avrachenkov, and A. Garnaev. Jamming in wireless networks under uncertainty. *Mobile Networks and Applications*, 16(2):246–254, 2011.
7. S. Amin, A. A. Cardenas, and S. S. Sastry. *Hybrid Systems: Computation and Control*. Springer-Verlag, Berlin Heidelberg, 2009.
8. S. Amin, S. A. Galina, and S. S. Sastry. Security of interdependent and identical networked control systems. *Automatica*, 49(1):186–192, 2013.
9. S. Amin, X. Litrico, S. Sastry, and A. M. Bayen. Cyber security of water scada systems-part i: Analysis and experimentation of stealthy deception attacks. *IEEE Transactions on Control Systems Technology*, 21(5):1963–1970, 2013.
10. S. Amin, X. Litrico, S. S. Sastry, and A. M. Bayen. Stealthy deception attacks on water scada systems. In *Proceedings of the 13th ACM International Conference on Hybrid Systems: Computation and Contro*, pages 161–170, New York, USA, 2010.
11. S. Amin, G. A. Schwartz, and A. Hussain. In quest of benchmarking security risks to cyber-physical systems. *IEEE Network*, 27(1):19–24, 2013.
12. B. D. O. Anderson and S. Vongpanitlerd. *Network analysis and synthesis: A modern systems theory approach*. Prentice-Hall, Upper Saddle River, NJ, 1973.
13. T. Başar. A dynamic games approach to controller design: Disturbance rejection in discrete-time. *IEEE Transactions on Automatic Control*, 36(8):936–952, 1991.
14. T. Başar and P. Bernhar. *H^∞ Optimal Control and Related Minmax Design Problems*. Berlin: Birkhauser-Verlag, 1995.
15. T. Basar. A dynamic games approach to controller design: Disturbance rejection in discrete-time. In *Proceedings of the 28th Conference on Decision and Control*, volume 36, pages 936–952, Tampa, Florida, 1989.
16. T. Basar and G. J. Olsder. *Dynamic noncooperative game theory*. London: Academic Press, 1995.

17. T. Basar and Q. Zhu. Prices of anarchy, information, and cooperation in differential games. *Dynamic Games and Applications*, 1(1):50–73, 2011.
18. G. K. Befekadu, V. Gupta, and P. J. Antsaklis. Risk-sensitive control under Markov modulated denial-of-service (dos) attack strategies. *IEEE Transactions on Automatic Control*, 60(12):3299–3304, 2015.
19. H. Bourles and A. Heniche. The inverse lqr problem and its application to analysis of a robust coordinated avr/pss. In *Proceedings of the 12th Power Systems Computation Conference*, pages 63–69, 1996.
20. R. Caballero, A. Hermoso, and J. Linares. Optimal state estimation for networked systems with random parameter matrices, correlated noises and delayed measurements. *Automatica*, 44(2):142–154, 2015.
21. G. Carl, G. Kesidis, R. R. Brooks, and S. Rai. Denial-of-service attack-detection techniques. *IEEE Internet Computing Magazine*, 10(1):82–89, 2006.
22. C. Chen, M. Song, C. Xin, and J. Backens. A game-theoretical anti-jamming scheme for cognitive radio networks. *IEEE Network*, 27(3):22–27, 2013.
23. H. Chen, J. Gao, T. Shi, and R. Lu. H_∞ control for networked control systems with time delay, data packet dropout and disorder. *Neurocomputing*, 179:211–218, 2016.
24. P. Chen and Q. Han. On designing a novel self-triggered sampling scheme for networked control systems with data losses and communication delays. *IEEE Transactions on Industrial Electronics*, 63(2):1239–1248, 2016.
25. W. Chen, J. Yang, L. Guo, and S. Li. Disturbance-observer-based control and related methods-an overview. *IEEE Transactions on Industrial Electronics*, 63(2):1083–1095, 2016.
26. Z. Chen, K. Reg, and L. Jiang. Jensen's inequality for g-expectation:i part i. *Comptes Rendus Mathematique*, 337:725–730, 2003.
27. D. J. Chmielewski and A. M. Manthanwar. On the tuning of predictive controllers: Inverse optimality and the minimum variance covariance constrained control problem. *Industrial & Engineering Chemistry Research*, 43(24):7807–7814, 2004.
28. A. K. Chorppath, T. Alpcan, and H. Boche. Bayesian mechanisms and detection methods for wireless network with malicious users. *IEEE Transactions on Mobile Computing*, 15(10):2452–2465, 2016.
29. S. Coogan, L. Ratliff, D. Calderone, C. Tomlin, and S. Sastry. Energy management via pricing in lq dynamic games. In *Proceedings of American Control Conference*, pages 443–448, 2013.
30. A. Crdenas, S. Amin, and S. Sastry. Research challenges for the security of control systems. In *Usenix Workshop on Hot Topics in Security*, pages 629–633, San Jose, CA, USA, 2008.
31. G. Dan and H. Sandberg. Stealth attacks and protection schemes for state estimators in power systems. In *Proceedings of IEEE International Conference on Smart Grid Communications*, pages 214–219, 2010.
32. D. Ding, Z. Wang, H. Dong, Y. Liu, and B. Ahmad. Performance analysis with network-enhanced complexities: On fading measurements, event-triggered mechanisms, and cyber attacks. *Abstract and Applied Analysis*, Article ID 461261:1–10, 2014.
33. G. Ding, Q. Wu, and J. Wang. Sensing confidence level-based joint spectrum and power allocation in cognitive radio networks. *Wireless Personal Communications*, 72(1):283–298, 2013.
34. H. Dong, Z. Wang, and H. Gao. Distributed H_∞ filtering for a class of Markovian jump nonlinear time-delay systems over lossy sensor networks. *IEEE Transactions on Industrial Electronics*, 60(10):4665–4672, 2016.
35. S. Du, X. Sun, and W. Wang. Guaranteed cost control for uncertain networked control systems with predictive scheme. *IEEE Transactions on Automation and Engineering*, 11(3):740–748, 2014.

36. E. Eyisi, X. Koutsoukos, and N. Kottenstette. Passivity-based trajectory tracking control with adaptive sampling over a wireless network. In *Proceedings of the 5th International Symposium on Resilient Control Systems*, pages 130–136, Salt Lake, Utah, 2012.
37. D. Geer. Security of critical control systems sparks concern. *Computer*, 39(1):20–23, 2006.
38. D. Ginoya, P. D. Shendge, and S. B. Phadke. Disturbance observer based sliding mode control of nonlinear mismatched uncertain systems. *Communications in Nonlinear Science and Numerical Simulation*, 26(1):98–107, 2015.
39. S. Giorgi, F. Saleheen, F. Ferrese, and C. H. Won. Adaptive neural replication and resilient control despite malicious attacks. In *Proceedings of the 5th International Symposium on Resilient Control Systems*, pages 112–117, Salt Lake, Utah, 2012.
40. S. Gorman. Electricity grid in U.S. penetrated by spies. *Wall Street Journal, April 8, 2009,*, pages http://online.wsj.com/article/SB123914805204099085.html,, 2013.
41. S. Greengard. *The new face of war.* Commun. ACM, 2010.
42. G. Guo. Linear system with medium-access constraint and Markov actuator assignment. *IEEE Transactions on Circuits and Systems-I: Regular Papers*, 57(11):2999–3010, 2010.
43. G. Guo and H. Jin. A switching system approach to actuator assignment with limited channels. *International Journal of Robust and Nonlinear Control*, 20(12):1407–1426, 2010.
44. G. Guo, Z. Lu, and Q. Han. Control with Markov sensor/actuator assignment. *IEEE Transactions on Automatic Control*, 57(7):1799–1804, 2012.
45. L. Guo and S. Cao. Anti-disturbance control theory for systems with multiple disturbances: A survey. *ISA Transactions*, 53(4):846–849, 2014.
46. L. Guo and W. Chen. Disturbance attenuation and rejection for systems with nonlinearity via dobc approach. *International Journal of Robust and Nonlinear Control*, 15:109–125, 2005.
47. Z. Guo, D. Shi, K. H. Johansson, and L. Shi. Optimal linear cyber-attack on remote state estimation. *IEEE Transactions on Control of Network Systems*, pages 1–1, 2016.
48. R. Gupta and M. Chow. Networked control system: overview and research trends. *IEEE Transactions on Industrial Electronics*, 57(7):2527–2535, 2010.
49. H. Yang H. Li, F. Sun and Y. Xia. Gain scheduling control of delta operator system using network-based measurements. *IEEE Control Systems Magazine.*, 63(3):538–547, 2014.
50. G. Hablinger and O. Hohlfeld. The Gilbert-Elliott model for packet loss in real time services on the internet. In *Proceedings of Measuring, Modelling and Evaluation of Computer and Communication Systems*, pages 1–15, 2008.
51. S. Han, M. Xie, H. Chen, and Y. Ling. Intrusion detection in cyber-physical systems: Techniques and challenges. *IEEE Systems Journal*, 8(4):1049–1059, 2014.
52. Z. Han, D. Niyato, W. Saad, T. Başar, and A. Hjørungnes. *Game Theory in Wireless and Communication Networks.* Cambridge University Press, 2012.
53. X. He, Z. Wang, X. Wang, and D. Zhou. Networked strong tracking filtering with multiple packet dropouts: algorithms and applications. *IEEE Transactions on Industrial Electronics*, 61(3):1454–1463, 2014.
54. J. P. Hespanha, P. Naghshtabrizi, and Y. Xu. A survey of recent results in networked control systems. In *International Symposium on Resilient Control Systems*, pages 138–162, 2007.
55. H. Hirano, M. Mukai, T. Azuma, and M. Fujiata. Optimal control of discrete-time linear systems with network-induced varying delay. In *Proceedings of American Control Conference*, pages 1419–1424, Portland, USA, 2005.
56. G. Hu, T. Wee, and Y. Wen. Cloud robotics: Architecture, challenges and applications. *IEEE Network*, 26(3):21–28, 2012.

57. J. Hu, Z. Wang, S. Liu, and H. Gao. A variance-constrained approach to recursive state estimation for time-varying complex networks with missing measurements. *Automatica*, 64:155–162, 2016.
58. J. Hu and M. P. Wellman. Nash q-learning for general-sum stochastic games. *Journal of Machine Learning Research*, 4:1039–1069, 2003.
59. S. Hu and Q. Zhu. Stochastic optimal control and analysis of stability of networked control systems with long delay. *Automatica*, 39(11):1877–1884, 2003.
60. T. Hu and Z. Lin. *Control Systems with Actuator Saturation: Analysis and Design*. Boston: Birkhäuser, 2001.
61. H. Huang, N. Ahmed, and P. Karthik. On a new type of denial of service attack in wireless networks: The distributed jammer network. *IEEE Transactions on Wireless Communications*, 10(7):2316–2324, 2011.
62. Y. L. Huang, A. A Cardenas, S. Amin, Z. S. Lin, H. Y. Tsai, and S. Sastry. Denial-of-service attack-detection techniques. *IEEE Internet Computing Magazine*, 2(3):73–83, 2009.
63. V. M. Igure, S. A. Laughter, and R. D. Williams. Security issues in scada networks. *Computers & Security*, 25(7):498–506, 2006.
64. C. Imer, S. Yuksel, and T. Basar. Optimal control of lti systems over communication networks. *Automatica*, 42:1429–1439, 2006.
65. O. Jackson, R. Tomas, and T. Xu. Epsilon-equilibria of perturbed games. *Games and Economic Behavior*, 75:198–216, 2012.
66. L. Jia, F. Yao, Y. Sun, Y. Niu, and Y. Zhu. Bayesian Stackelberg game for anti-jamming transmission with incomplete information. *IEEE Communication Letters*, (10):1991–1994, 2016.
67. Z. Jiang and Y. Wang. Input-to-state stabilization for discrete-time nonlinear systems. *Automatica*, 37(6):857–869, 2001.
68. M. Jimenez and A. Poznyak. ϵ-equilibrium in lq differential games with bounded uncertain disturbances: robustness of standard strategies and new strategies with adaptation. *International Journal of Control*, 79(7):786–797, 2006.
69. D. Kilinc, M. Ozger, and O. B. Akan. On the maximum coverage area of wireless networked control systems with maximum cost-efficiency under convergence constraint. *IEEE Transactions on Automatic Control*, 60(7):1910–1914, 2015.
70. R. A. Kisner, W. W. Manges, L. P. MacIntyre, J. J. Nutaro, J. K. Munro, P. D. Ewing, M. Howlader, P. T. Kuruganti, R. M. Wallace, and M. M. Olama. *Cybersecurity through Real-Time Distributed Control Systems*. Office of Scientific & Technical Information Technical Reports, 1995.
71. S. Kluge, K. Reif, and M. Brokate. Stochastic stability of the extended Kalman filter with intermittent observations. *IEEE Transactions on Automatic Control*, 55(2):514–518, 2010.
72. R. Langner. Stuxnet: dissecting a cyberwarfare weapon. *IEEE Security & Privacy*, 9(3):49–51, 2011.
73. Y. W. Law, T. Alpcan, and M. Palaniswami. Security games for risk minimization in automatic generation control. *IEEE Transactions on Power Systems*, 30(1):223–232, 2015.
74. P. Lee, A. Clark, L. Bushnell, and R. Poovendran. A passivity framework for modeling and mitigating wormhole attacks on networked control systems. *IEEE Transactions on Automatic Control*, 59(12):3224–3237, 2014.
75. H. Li, M. Y. Chow, and Z. Sun. Optimal stabilizing gain selection for networked control systems with time delays and packet losses. *IEEE Transactions on Control Systems Technology*, 17(5):1154–1162, 2009.
76. H. Li, Z. Li, H. Yang, Y. Zhang, and F. Sun. Stabilisation of networked delta operator systems with uncertainty. *IET Control Theory and Applications*, 8:2289–2296, 2015.
77. H. Li, H. Yang, F. Sun, and Y. Xia. A network-bound-dependent stabilization method of networked control systems. *Automatica*, 49(8):2561–2566, 2013.

78. H. Li, H. Yang, F. Sun, and Y. Xia. Sliding-mode predictive control of networked control systems under a multiple-packet transmission policy. *IEEE Transactions on Industrial Electronics*, 61(11):6234–6243, 2014.
79. L. Li, B. Hu, and M. D. Lemmon. Resilient event triggered systems with limited communication. Maui, Hawaii, USA.
80. S. Li and J. Yang. Robust autopilot design for bank-to-turn missiles using disturbance observers. *IEEE Transactions on Aerospace and Electronic Systems*, 49:558–579, 2013.
81. Y. Li, D. E. Quevedo, S. Dey, and L. Shi. SINR-based dos attack in remote state estimation: A game-theoretic approach. *IEEE Transactions on Control of Network Systems*, pages 1–1, 2016.
82. Y. Li, D. E. Quevedo, V. Lau, and L. Shi. Optimal periodic transmission power schedules for remote estimation of arma processes. *IEEE Transactions on Signal Processing*, 61(24):6164–6174, 2013.
83. Y. Li, L. Shi, P. Cheng, J. Chen, and D. E. Quevedo. Jamming attack on cyber-physical systems: A game-theoretic approach. In *Proceedings of the IEEE International Conference on Cyber Technology in Automation, Control and Intelligent Systems*, pages 252–257, Nanjing, China, 2013.
84. Y. Li, L. Shi, P. Cheng, J. Chen, and D. E. Quevedo. Jamming attacks on remote state estimation in cyber-physical systems: A game-theoretic approach. *IEEE Transactions on Automatic Control*, 60(10):2831–2836, 2015.
85. D. Liberzon and A. S. Morse. Basic problems in stability and design of switched systems. *IEEE Control Systems Magazine*, 19(4):59–70, 1999.
86. H. Lin, H. Su, P. Shi, R. Lu, and Z. Wu. LQG control for networked control systems over packet drop links without packet acknowledgment. *Journal of the Franklin Institute*, 352(11):5042–5060, 2015.
87. M. L. Littman. Markov games as a framework for multi-agent reiforcement learning. In *Proceedings of the Eleventh International Conference*, pages 10–13, Rutgers University, New Brunswick, NJ, 1994.
88. A. Liu, Q. Zhang, L. Yu, S. Liu, and M. Chen. New results on stabilization of networked control systems with packet disordering. *Automatica*, 52:255–259, 2015.
89. S. Liu, P. X. Liu, and A. E. Saddik. A stochastic game approach to the security issue of networked control systems under jamming attacks. *Journal of the Franklin Institute*, 351(9):4570–4583, 2014.
90. M. Long, C. H. Wu, and J. Y. Hung. Denial-of-service attack-detection techniques. *IEEE Internet Computing Magazine*, 1(2):85–96, 2005.
91. K. Ma, G. Hu, and C. Spanos. Distributed energy consumption control via real-time pricing feedback in smart grid. *IEEE Transactions on Control Systems Technology*, 22(5):2904–2914, 2014.
92. A. Maass, F. Vargas, and E. Silva. Optimal control over multiple erasure channels using a data dropout compensation scheme. *Automatica*, 68:155–161, 2016.
93. M. Mahmoud and A. Memon. Aperiodic triggering mechanisms for networked control systems. *Information Sciences*, 296(1):282–306, 2015.
94. J. Matusitza and E. Mineib. In quest of benchmarking security risks to cyber-physical systems. *Journal of Digital Forensic Practice*, 2(4):161–171, 2009.
95. A. Melin, R. Kisner, D. Fugate, and T. McIntyre. Minimum state awareness for resilient control systems under cyber-attack. In *Proceedings of Future of Instrumentation International Workshop*, pages 978–982, Gatlinburg, Tennessee, 2012.
96. Y. Mi, Y. Fu, C. Wang, and P. Wang. Decentralized sliding mode load frequency control for multi area power systems. *IEEE Transactions on Power Systems*, 28(4):4301–4309, 2013.
97. R. Middleton and G.C. Goodwin. *Digital control and estimation: A unified approach*. New Jersey, US: Englewood Cliffs, 1990.
98. Y. Mo, R. Chabukswar, and B. Sinopoli. Detecting integrity attacks on scada systems. *IEEE Transactions on Control Systems Technology*, 61(11):1396–1407, 2014.

99. Y. Mo, G. Emanele, and S. Bruno. LQG control with Markovian packet loss. In *Proceedings of European Control Conference*, pages 252–257, Zrich, Switzerland, 2013.
100. J. Moon and T. Başar. Control over TCP-like lossy networks: A dynamic game approach. In *American Control Conference*, pages 1578–1583, Washington, USA, 2013.
101. Y. Niu and D. W. C. Ho. Control strategy with adaptive quantizer's parameters under digital communication channels. *Automatica*, 50(10):2665–2671, 2014.
102. Z. Pang, G. Liu, D. Zhou, F. Hou, and D. Sun. Two-channel false data injection attacks against output tracking control of networked control systems. *IEEE Transactions on Industrial Electronics*, 63(5):3242–3251, 2016.
103. Z. Pang and G. P. Liu. Design and implementation of secure networked predictive control systems under deception attacks. *IEEE Transactions on Control Systems Technology*, 20(5):1334–1342, 2012.
104. Z. Pang, G. P. Liu, and D. Zhou. Design and performance analysis of incremental networked predictive control systems. *IEEE Transactions on Cybernetics*, 46(6):1400–1410, 2016.
105. K. Park, J. Kim, H. Lim, and Y. Eun. Robust path diversity for network quality of service in cyber-physical systems. *IEEE Transactions on Industrial Informatics*, 10(4):2204–2215, 2014.
106. F. Pasqualetti and F. Dorfler. Control-theoretic methods for cyberphysical security: geometric principles for optimal cross-layer resilient control systems. *IEEE Control Systems*, 35(1):110–127, 2015.
107. Y. Peng, C. Jiang, F. Xie, Z. Dai, Q. Xiong, and Y. Gao. Industrial control system cybersecurity research. *Journal of Tsinghua University*, 52(10):1396–1408, 2012.
108. E. Peters, D. Quevedo, and J. Ostergaard. Shaped Gaussian dictionaries for quantized networked control systems with correlated dropouts. *IEEE Transactions on Signal Processing*, 64(1):203–213, 2016.
109. J. Proakis and M. Salehi. *Digital Communication*. McGraw-Hill, 2007.
110. L. Qiu, S. Li, B. Xu, and G. Xu. h_∞ control of networked control systems based on Markov jump unified model. *International Journal of Robust and Nonlinear Control*, 26:2770–2786, 2015.
111. X. Qiu, L. Yu, and D. Zhang. Stabilization of supply networks with transportation delay and switching topology. *Neurocomputing*, 155:247–252, 2015.
112. T. E. S. Raghavan and J. A. Filar. Algorithms for stochastic games - a survey. *Methods Models Operations Research*, 35(6):437–472, 1991.
113. T. Raghaven, T. Ferguson, T. Parthasarathy, and O. Vrieze. *Stochastic Games and Related Topics: In Honor of Professor L.S. Shapley*. Berlin: Springer-Verlag, 1990.
114. L. J. Ratliff, S. Coogan, D. Calderone, and S. S. Sastry. Pricing in linear-quadratic dynamic games. In *Proceedings of the 50th Annual Allerton Conference on Communication, Control, and Computing*, pages 1798–1805, Illinois, USA, 2012.
115. C. G. Rieger, D. I. Gertman, and M. A. McQueen. Resilient control systems: Next generation design research. In *Conference on Human System Interactions,*, pages 629–633, Catania, Italy, 2009.
116. W. Saad, Z. Han, HV. Poor, and T. Basar. Game-theoretic methods for the smart grid: An overview of microgrid systems, demand-side management, and smart grid communications. *IEEE Signal Processing Magazine*, 9(6):86–105, 2012.
117. Y. E. Sagduyu, R. A. Berry, and A. Ephremides. Jamming games in wireless networks with incomplete information. *IEEE Communications Magazine*, 49(8):112–118, 2011.
118. H. Sandberg, S. Amin, and K. Johansson. Cyberphysical security in networked control systems: An introduction to the issue. *IEEE Control Systems Magazine*, 31(1):20–23, 2015.
119. L. Schenato, B. Sinopoli, M. Franceschetti, K. Poolla, and S. Sastry. Foundations of control and estimation over lossy networks. *Proceedings of the IEEE*, 95(1):163–187, 2007.
120. L. S. Shapley. Stochastic games. In *Proceedings of the National Academy of Sciences of the United States of America*, pages 1095–1100, Nashville, 1953.

121. V. Singh. A novel LMI-based criterion for the stability of direct-form digital filters utilizing a single twos complement nonlinearity. *Nonlinear Analysis: Real World Applications*, 14(1):684–689, 2013.
122. R. S. Smith. Covert misappropriation of networked control systems: presenting a feedback structure. *IEEE Control Systems Magazine*, 35(1):82–92, 2015.
123. L. Sun, Y. Wang, and G. Feng. Control design for a class of affine nonlinear descriptor systems with actuator saturation. *IEEE Transactions on Automatic Control*, 60(8):2195–2200, 2015.
124. X. Sun, D. Wu, G. P. Liu, and W. Wang. Input-to-state stability for networked predictive control with random delays in both feedback and forward channels. *IEEE Transactions on Industrial Electronics*, 61(7):3519–3526, 2014.
125. C. Szepesvari and M. L. Littman. A unified analysis of value-function-based reinforcement-learning algorithms. *Neural Computation*, 11(8):2017–2059, 1999.
126. X. H. Tan and B. C. Jose. Adaptive noncooperative-person games with unknown general quadratic objectives. *IEEE Transactions on Control Systems Technology*, 18(5):1033–1043, 2010.
127. A. Teixeira, D. Pérez, H. Sandberg, and K. H. Johansson. Attack models and scenarios for networked control systems. In *Proceedings of the 1st International Conference on High Confidence Networked Systems*, pages 55–64, 2012.
128. A. Teixeira, I. Shames, H. Sandberg, and K. Johansson. A secure control framework for resource-limited adversaries. *Automatica*, 51:135–148, 2015.
129. E. M. Tieghi. Integrating electronic security into the control. *Wall Street Journal*,.
130. K. C. Toh, M. J. Todd, and R. H. Tutuncu. Sdpt3—a Matlab software package for semidefinite programming. *Optimization Methods and Software*, 11(1-4):545–581, 1999.
131. T. T. Tran and Y. Rasis. Obeservers for nonlinear stochastic systems. *IEEE Transactions on Automatic Control*, 21(4):441–447, 1976.
132. D. Q. Truong and K. K. Ahn. Robust variable sampling period control for networked control systems. *IEEE Transactions on Industrial Electronics*, 62(9):5630–5643, 2015.
133. K. Tsumura, H. Ishii, and H. Hoshina. Tradeoffs between quantization and packet loss in networked control of linear systems. *Automatica*, 45:2963–2970, 2009.
134. N. Virvilis and D. Gritzalis. The big four-what we did wrong in advanced persistent threat detection? In *International Conference on Availability, Reliability and Security*, pages 248–254, Greece, 2013.
135. B. Wang, Y. Wu, K. Liu, and T. Clancy. An anti-jamming stochastic game for cognitive radio networks. *IEEE Journal on Selected Areas in Communications*, 29(4):877–889, 2011.
136. D. Wang, D. Liu, Q. Wei, D. Zhao, and N. Jin. Optimal control of unknown nonaffine nonlinear discrete-time systems based on adaptive dynamic programming. *Automatica*, 48:1825–1832, 2012.
137. Z. Wang, X. Wang, and X. Liu. Stochastic optimal linear control of wireless networked control systems with delays and packet losses. *IET Control Theory & Applications*, 10(7):742–751, 2016.
138. Z. Wang, X. Wang, X. Liu, and M. Huang. Optimal state feedback control for wireless networked control systems with decentralised controllers. *IET Control Theory & Applications*, 9(6):852–862, 2014.
139. X. Wei, N. Chen, C. Deng, X. Liu, and M. Tang. Composite stratified anti-disturbance control for a class of mimo discrete-time system with nonlinearity. *International Journal of Robust and Nonlinear Control*, 22:453–472, 2012.
140. G. Willamann, D. F. Coutiho, L. F. A. Pereira, and F. B. Libano. Multiple-loop h_∞ control design for uninterruptible power supplies. *IEEE Transactions on Industrial Electronics*, 54(3):1591–1602, 2007.
141. J. Wu, K. Ota, M. Dong, and C. Li. A hierarchical security framework for defending against sophisticated attacks on wireless sensor networks in smart cities. *IEEE Access*, 4:416–424, 2016.

142. L. Wu, J. Lam, X. Yao, and J. Xiong. Robust guaranteed cost control of discrete-time networked control systems. *Optimal Control Application and Methods*, 32(1):95–112, 2011.
143. Y. Wu, X. He, J. Wu, and L. Xie. Consensus of discrete-time multi-agent systems with adversaries and time delays. *International Journal of General Systems*, 43(3):402–411, 2014.
144. Y. Xia, M. Fu, and G. Liu. *Analysis and Synthesis of Networked Control Systems*. Springer-Verlag, Berlin Heidelberg, 2011.
145. Y. Xia, W. Xie, B. Liu, and X. Wang. Data-driven predictive control for networked control systems. *Information Sciences*, 235(20):45–54, 2013.
146. B. Xiao, Q. Hu, and P. Shi. Attitude stabilization of spacecrafts under actuator saturation and partial loss of control effective. *IEEE Transactions on Control Systems Technology*, 21(6):2251–2263, 2013.
147. L. Xiao, T. Chen, J. Liu, and H. Dai. Anti-jamming transmission Stackelberg game with observation errors. *IEEE Communications Letters*, 19(6):949–952, 2015.
148. N. Xiao, L. Xie, and L. Qiu. Feedback stabilization of discrete-time networked systems over fading channels. *IEEE Transactions on Automatic Control*, 57:2176–2189, 2012.
149. X. Xie, D. Yue, and S. Hu. Fuzzy control design of nonlinear systems under unreliable communication links: A systematic homogenous polynomial approach. *Information Sciences*, 370-371(20):763–771, 2016.
150. B. Xu, F. Sun, C. Yang, D. Gao, and J. Ren. Adaptive Kriging controller design for hypersonic flight vehicle via back-stepping. *IET Control Theory & Applications*, 6(4):487–497, 2012.
151. H. Xu, S. Jagannathan, and F. Lewis. Stochastic optimal control of unknown linear networked control system in the presence of random delays and packet losses. *Automatica*, 48:1017–1030, 2012.
152. W. Xu, W. Trappe, Y. Zhang, and T. Wood. The feasibility of launching and detecting jamming attacks in wireless networks. In *Proceedings of the 6th ACM international symposium on Mobile ad hoc networking and computing*, pages 46–57, 2005.
153. J. Yan, Y. Xia, and L. Li. Stabilization of fuzzy systems with quantization and packet dropout. *International Journal of Robust and Nonlinear Control*, 24(10):1563–1583, 2014.
154. C. Yang, X. Ren, W. Yang, H. Shi, and L. Shi. Jamming attack in centralized state estimation. In *Proceedings of the 34th Chinese Control Conference*, pages 28–30, Hangzhou, China, 2015.
155. D. Yang, G. Xue, J. Zhang, A. Richa, and X. Fang. Coping with a smart jamming in wireless networks: A Stackelberg game approach. *IEEE Transactions on Wireless Communications*, 12(8):4038–4047, 2013.
156. F. W. Yang, Z. D. Wang, Y. S. Hung, and M. Gani. h_∞ control for networked systems with random communication delays. *IEEE Transactions on Automatic Control*, 51(3):511–518, 2006.
157. H. Yang and Y. Xia. Low frequency positive real control for delta operator systems. *Automatica*, 48(8):1791–1795, 2012.
158. H. Yang, Y. Xia, and P. Shi. Observer-based sliding mode control for a class of discrete systems via delta operator approach. *Journal of the Franklin Institute*, 347(7):1199–1213, 2010.
159. H. Yang, Y. Xia, and P. Shi. Stabilization of networked control systems with nonuniform random sampling periods. *International Journal of Robust and Nonlinear Control*, 21(5):501–526, 2011.
160. H. Yang, Y. Xia, P. Shi, and M. Fu. Stability analysis for high frequency networked control systems. *IEEE Transactions on Automatic Control*, 57(10):2694–2700, 2012.
161. H. Yang, Y. Xia, P. Shi, and B. Liu. Guaranteed cost control of networked control systems based on delta operator Kalman filter. *International Journal of Adaptive Control and Signal Processing*, 27(8):707–717, 2013.

162. H. Yang, Y. Xia, P. Shi, and L. Zhao. *Analysis and Synthesis of Delta Operator Systems.* Springer-Verlag, Berlin Heidelberg, 2012.
163. H. Yang, C. Yan, Y. Xia, and J. Zhang. Stabilization on null controllable region of delta operator systems subject to actuator saturation. *International Journal of Robust and Nonlinear Control*, 26(7):3481–3506, 2016.
164. J. Yang, S. Li, C. Sun, and L. Guo. Nonlinear-disturbance-observer-based robust flight control for air breathing hypersonic vehicles. *IEEE Trans. Aerosp. Electron. Syst.*, 49:1263–1275, 2013.
165. R. Yang, G. P. Liu, P. Shi, C. Thomas, and M. V. Basin. Predictive output feedback control for networked control systems. *IEEE Transactions on Industrial Electronics*, 61(1):512–520, 2014.
166. R. Yang, P. Shi, G. P. Liu, and H. Gao. Network-based feedback control for systems with mixed delays based on quantization and dropout compensation. *Automatica*, 47(12):2805–2809, 2011.
167. Y. Yang, X. Fan, and T. Zhang. Anti-disturbance tracking control for systems with nonlinear disturbances using t-s fuzzy modeling. *Neurocomputing*, 171:1027–1037, 2015.
168. X. Yao and L. Guo. Composite anti-disturbance control for Markovian jump nonlinear systems via disturbance observer. *Automatica*, 49(8):2538–2545, 2013.
169. X. Yin, D. Yue, S. Hu, C. Peng, and Y. Xue. Model-based event-triggered predictive control for networked systems with data dropout. *SIAM Journal on Control and Optimization*, 54(2):567–586, 2016.
170. Y. Yuan and F. Sun. Data fusion-based resilient control system underi dos attacks: A game theoretic approach. *International Journal of Control Automation and Systems*, 13(3):513–520, 2015.
171. Y. Yuan, F. Sun, and H. Liu. Resilient control of cyber-physical systems against intelligent attacker: A hierarchal Stackelberg game approach. *International Journal of Systems Science*, 47(9):1–11, 2015.
172. Y. Yuan, S. Sun, and Q. Zhu. Resilient control in the presence of dos attack: Switched system approach. *International Journal of Control, Automation, and Systems*, 13(6):1423–1436, 2015.
173. Y. Yuan, H. Yuan, L. Guo, H. Yang, and S. Sun. Resilient control of networked control system under dos attacks: A unified game approach. *IEEE Transactions on Industrial Informatics*, 12(5):1786–1794, 2016.
174. Y. Yuan, H. Yuan, G. Lei, H. Yang, and S. Sun. Resilient control of cyber-physical systems against intelligent attacker: A hierarchal Stackelberg game approach. *IEEE Transactions on Industrial Informatics*, 47(9):2067–2077, 2016.
175. Y. Yuan, Q. Zhu, F. Sun, Q. Wang, and T. Başar. Resilient control of cyber-ohysical systems against denial-of-service attacks. In *Proceedings of the 6th International Symposium on Resilient Control Systems*, pages 54–59, Beijing, China, 2013.
176. M. Zakai. On the ultimate boundness of moments associated with solutions of stochastic differential equations. *Siam Journal on Control*, 5(4):588–593, 1967.
177. H. Zhang, P. Cheng, L. Shi, and J. Chen. Optimal dos attack scheduling in wireless networked control system. *IEEE Transactions on Control Systems Technology*, 24(3):843–852, 2016.
178. H. Zhang, P. Cheng, L. Shi, and J. M. Chen. Price of anarchy and price of information in n-person linear-quadratic differential games. In *Proceedings of Decision and Control*, pages 5444–5449, 2013.
179. H. Zhang, G. Feng, H. Yan, and Q. Chen. Sampled-data control of nonlinear networked systems with time-delay and quantization. *International Journal of Robust and Nonlinear Control*, 26:919–933, 2016.
180. H. Zhang, Y. Qi, J. Wu, L. Fu, and L. He. Dos attack energy management against remote state estimation. *IEEE Transactions on Control of Network Systems*, (99):1–1, 2016.

181. J. Zhang, J. Lam, and Y. Xia. Output feedback delay compensation control for networked control systems with random delays. *Information Sciences*, 265(1):154–166, 2014.
182. J. Zhang, Y. Xia, and P. Shi. Design and stability analysis of networked predictive control systems. *IEEE Transactions on Control Systems Technology*, 21(4):1495–1501, 2013.
183. L. Zhang, H. Gao, and O. Kaynak. Network-induced constraints in networked control systemsa survey. *IEEE Transactions on Industrial Informatics*, 9(1):403–416, 2013.
184. W. Zhang, L. Yu, and H. Song. h_∞ filtering of networked discrete-time systems with random packet losses. *Information Sciences*, 179(22):3944–3955, 2009.
185. X. Zhang, C. Lu, X. Xie, and Z. Dong. Stability analysis and controller design of a wide-area time-delay system based on the expectation model method. *IEEE Transactions on Smart Grid*, 7(1):520–529, 2016.
186. L. Zhao, Y. Yang, Y. Xia, and Z. Liu. Active disturbance rejection position control for a magnetic rodless pneumatic cylinder. *IEEE Transactions on Industrial Electronics*, 62:5838–5846, 2015.
187. Y. Zhao and H. Gao. Fuzzy-model-based control of an overhead crane with input delay and actuator saturtion. *IEEE Transactions on Fuzzy Systems*, 20(1):181–186, 2012.
188. M. Zhu and S. Martnez. On the performance analysis of resilient networked control systems under replay attacks. *IEEE Transactions on Automatic Control*, 59(3):804–808, 2014.
189. M. H. Zhu and M. Sonia. Stackelberg-game analysis of correlated attacks in cyber-physical systems. In *American Control Conference*, pages 4063–4068, San Francisco, CA, USA, 2011.
190. Q. Zhu and T. Başar. Dynamic policy-based ids configurationi. In *IEEE Conference on Decision and Control*, pages 8600–8605, Atlanta, Georgia, USA, 2009.
191. Q. Zhu and T. Başar. Robust and resilient control design for cyber-physical systems with an application to power systems. In *IEEE Conference on Decision and Control*, pages 4066–4071, Nashville, 2011.
192. Q. Zhu and T. Başar. A dynamic game-theoretic approach to resilient control system design for cascading failures. In *Proceedings of the 1st International conference on High Confidence Networked Systems*, pages 41–46, Beijing, China, 2012.
193. Q. Zhu and T. Başar. Game-theoretic methods for robustness, security, and resilience of cyberphysical control systems: Games-in-games principle for optimal cross-layer resilient control systems. *IEEE Control Systems Magazine*, 35(1):46–65, 2015.
194. Q. Zhu, C. Rieger, and T. Başar. Stealthy deception attacks on water scada systems. In *Proceedings of International Symposium on Resilient Control Systems*, pages 161–170, Stockholm, Sweden, 2011.

Index